Physical Organic Chemistry

The Fundamental Concepts
Second Edition, Revised and Expanded

Calvin D. Ritchie
Department of Chemistry
State University of New York at Buffalo
Buffalo, New York

MARCEL DEKKER, INC.   New York and Basel

Library of Congress Cataloging-In-Publication Data

Ritchie, Calvin D.
Physical organic chemistry : the fundamental concepts /
Calvin D. Ritchie. -- 2nd ed., rev. and expanded.
p. cm.
Includes bibliographical references.
ISBN 0-8247-8307-7 (alk. paper)
1. Chemistry, Physical organic. I. Title.
QD476.R57 1990
547.1'3--dc20 89-23312
CIP

This book is printed on acid-free paper.

MARCEL DEKKER, INC.
270 Madison Avenue, New York, New York 10016

Current printing (last digit)
10 9 8 7 6 5 4 3 2 1

PRINTED IN THE UNITED STATES OF AMERICA

PREFACE

The first edition of this book evolved from a set of notes for a one-semester course offered to senior undergraduate and first-year graduate students in the Department of Chemistry, State University of New York at Buffalo. The first edition has been used for more than ten years in that course with only minor changes. In that time, the background education of students, the **hot** areas of physical organic chemistry, and the author's interests and perceptions have all undergone some changes. For these reasons, it was felt that a thorough revision of the entire book might be useful. The result of this revision is substantially different from the first edition, but the rigor of the material has not changed; it is not intended for bedtime reading. Even the very best students should read with pencil and paper at hand.

The book strongly reflects my own philosophy of science. Much of the difficulty in learning science arises from the fact that the concepts and theories have developed over a long period of time and that approximations and shortcuts which were introduced during that development, and which frequently resulted from much trial and error rather than straightforward logic, lead to a terrifying amount of detailed esoterica. For example, molecular orbital theory at the *ab initio* Hartree-Fock level is very easily developed from first principles. If, however, one looks at some of the semi-empirical methods, such as CNDO, MINDO, etc., the amount of detail and complexity due to the approximations which are used is beyond anyone but the true practitioners of the methods. To the best of my ability, within the confines of reasonable time and space, I have tried to present all concepts at the most fundamental level. Not only is the material easier to learn when presented in this way, but it stays longer.

Several friends at other universities have commented that their students found the problems given at the ends of the chapters of the first edition to be impossibly difficult. After using the text for some time, I must agree that average students seem unable to complete other than the simplest ones. At least some of the trouble is that students frequently have not learned how to approach complex problems in a logical, stepwise fashion. It is probably also true that a lack of confidence prevents serious attempts to **think** about such problems. In order to build confidence and provide simple reinforcement of text material, I have added a number of easy problems at the end of each chapter. To help students learn how to approach problems, hints are now given with most problems to get the student started on the correct path. I still believe that the only way to learn the material in the text is to spend the time necessary to work the difficult problems. They do require considerable time.

It is assumed that students using this book have a working knowledge of calculus through simple differential equations and have completed the first three years of a normal chemistry curriculum including a full year of study of physical chemistry. Courses in linear algebra, thermodynamics, and quantum mechanics would certainly be useful, but the essential background in these subjects is presented in the text.

The purpose of the book is to introduce the most important of the fundamental concepts of physical organic chemistry. In my opinion, this requires a heavy emphasis on the physical side of the subject rather than on reaction mechanisms. I recognize, however, that students at the level taking this course are more comfortable with reaction mechanisms than with, for example, the operator formalism applied to a harmonic oscillator. For this reason, I have attempted to give the book a "story line" involving thorough studies of mechanisms. There is no intent to present a survey of mechanisms of organic reactions.

The material is divided roughly into two major mechanistic themes. Chapters 1 through 5 develop the topics of kinetics, salt, solvent, and structural effects on reactions in the loose context of SN1 and SN2 type reaction mechanisms. In Chapter 5, the context begins to move to other types of reactions, and Chapter 6 deals with acid-base reactions; the concepts of catalysis, reaction rate theories, and isotope effects are presented primarily in that context in Chapters 7 through 9.

Matrix algebra is introduced early in Chapter 1 in the context of kinetics and is then used throughout the text, particularly in the discussions of quantum mechanics and normal mode analysis. Sufficient introduction to the basics of matrix algebra is given so that diligent students with no prior knowledge of the subject should be able to understand the material presented. It is hoped that the examples of applications of matrix algebra will provide sufficient motivation for the students to spend the time necessary to develop a facility with this valuable tool.

My debts to my own teachers are enormous. Among these teachers, in addition to Profs. W. F. Sager and E. S. Lewis with whom I have been privileged to have had extended personal contact, I include Profs. L. P. Hammett, A. Frost and R. G. Pearson, and S. Glasstone, whose texts *Physical Organic Chemistry, Kinetics and Mechanisms of Reactions*, and *Theoretical Chemistry*, respectively, have greatly influenced my thinking.

I have been fortunate in having had, for many years, two exceptionally gifted departmental colleagues. Profs. Harry King and Jim McIver have been responsible for my education in mathematics and modern quantum mechanics, and have read and criticized sections of this book; their knowledge, patience, enthusiasm, and general intellectual abilities are greatly appreciated. Chapters 5, 8, and 9 could not have been written without many sessions spent with Harry King; the original material on tunneling, in Section 9.6, was developed in close collaboration with him.

Calvin D. Ritchie

CONTENTS

CHAPTER 1
Kinetics: Integration of Rate Equations

1.1 Introduction

Among other things, physical organic chemistry is concerned with the study of all factors, structural or environmental, which affect the rates and equilibria of organic reactions. This sub-area of physical organic chemistry includes a tremenduous range of interests, but the first step in the study of a reaction, regardless of final interest, nearly always involves some consideration of *mechanism* of the reaction. At the most elementary level, the *mechanism* of a reaction will specify any intermediates involved in the overall reaction. This specification of intermediates usually involves the determination of the empirical rate law for the reaction.

An empirical rate law states how the rates of formation of products, and rates of disappearance of reactants, depend on the concentrations of reactants, products, and catalysts, while all other variables, such as temperature, pressure, and ionic strength, are held constant. The empirical rate law allows us to begin testing postulated mechanisms.

The testing of a postulated mechanism using knowledge of the empirical rate law depends on the concept of an *elementary step* of a reaction. The essential feature of an *elementary step* of a reaction is that the rate law for that step **must** be directly related to the stoichiometry of that step. That is, if the reaction:

$$wA + xB \underset{k_r}{\overset{k_f}{\rightleftharpoons}} yC + zD \tag{1.1}$$

is an elementary step, then the rate law for this step **must** be:

$$\begin{aligned} \frac{1}{z}\frac{d[D]}{dt} = \frac{1}{y}\frac{d[C]}{dt} &= -\frac{1}{w}\frac{d[A]}{dt} \\ &= -\frac{1}{x}\frac{d[B]}{dt} = k_f[A]^w[B]^x - k_r[C]^y[D]^z . \end{aligned} \tag{1.2}$$

More practically, if the empirical rate law for some reaction does not correspond to the stoichiometry for the reaction, the reaction **must** involve more than one elementary step, and *must involve at least one intermediate.*

This characteristic of an elementary step of a reaction can be deduced from any of the theories of reaction rates. At the simplest level, the frequencies of encounters of particles are directly proportional to the concentrations of the particles, and if any type of particle is required for some reaction, then it must encounter the other types required for the reaction.

We write some postulated set of elementary steps for a reaction of interest, and see if the resulting implied rate law is consistent with the empirical rate law. In some cases, we can integrate the rate laws, such as Eq. (1.2), to obtain expressions for concentrations of the various species as functions of time. These expressions can then be compared with actually measured concentrations as functions of time as the reaction proceeds. In the general case where a reaction may consist of several elementary steps, the rate laws for the elementary steps form a set of simultaneous differential equations which must be integrated to obtain expressions for concentrations as functions of time.

Let's start with some cases of reactions involving a single elementary step.

1.2 Simple First-Order Reactions

The simplest case is that of a reversible first-order reaction, such as an isomerization reaction:

$$A \underset{k_{-1}}{\overset{k_{+1}}{\rightleftharpoons}} B \tag{1.3}$$

with the rate law given by:

$$\frac{d[B]}{dt} = -\frac{d[A]}{dt} = k_{+1}[A] - k_{-1}[B]. \tag{1.4}$$

The integration of the last equation can be done quite easily. Suppose the concentrations of A and B at time equal zero are $[A]_0$ and $[B]_0$, respectively. Since reaction of one mole of A produces one mole of B, the concentrations of A and B, at any time, are related by:

$$[A] + [B] = [A]_0 + [B]_0 \quad \text{or} \quad [A] = [A]_0 + [B]_0 - [B] \tag{1.5}$$

Substitution of this result into the rate law gives:

$$\begin{aligned} \frac{d[B]}{dt} &= k_{+1}\Big([A]_0 + [B]_0 - [B]\Big) - k_{-1}[B], \\ &= -\{k_{+1} + k_{-1}\} \cdot [B] + k_{+1}\Big([A]_0 + [B]_0\Big) \end{aligned} \tag{1.6}$$

which can be put into standard form by defining:

$$k_\psi \equiv \{k_{+1} + k_{-1}\} \quad \text{and} \quad X \equiv k_\psi[B] - k_{+1}\Big([A]_0 + [B]_0\Big)$$

so that:

$$\frac{d[B]}{dt} = \frac{1}{k_\psi}\frac{dX}{dt} \quad \text{and} \quad \frac{dX}{X} = -k_\psi dt$$

which integrates immediately to:

$$\ln(X) = -k_\psi t + \ln(X_0) \quad \text{or} \quad \frac{X}{X_0} = \exp(-k_\psi t)$$

where $\ln(X_0)$ is the constant of integration.

After substituting for X in terms of the original parameters, and rearranging, we obtain:

$$\{k_{+1} + k_{-1}\}[B] - k_{+1}\Big([A]_0 + [B]_0\Big) = X_0 \exp(-k_\psi t) \tag{1.7}$$

The appearance of the left-hand side of this equation may be simplified by noting:

$$K_{eq} = \frac{k_{+1}}{k_{-1}} = \frac{[B]_\infty}{[A]_\infty} \tag{1.8}$$

This equation may be solved for k_{-1} in terms of the concentrations at infinite time and the result substituted in Eq. (1.7). After a little algebra involving the fact that $[A]_\infty + [B]_\infty = [A]_0 + [B]_0$:

$$[B] - [B]_\infty = \frac{X_0[B]_\infty}{k_{+1}\Big([A]_\infty + [B]_\infty\Big)} \exp(-k_\psi t) \tag{1.9}$$

Noting that at $t = 0$, $[B] = [B]_0$ we can deduce that the entire term multiplying the exponential and involving X_0 must be equal to $[B]_0 - [B]_\infty$, so that we obtain the final form:

$$\frac{[B] - [B]_\infty}{[B]_0 - [B]_\infty} = \exp(-k_\psi t) \tag{1.10}$$

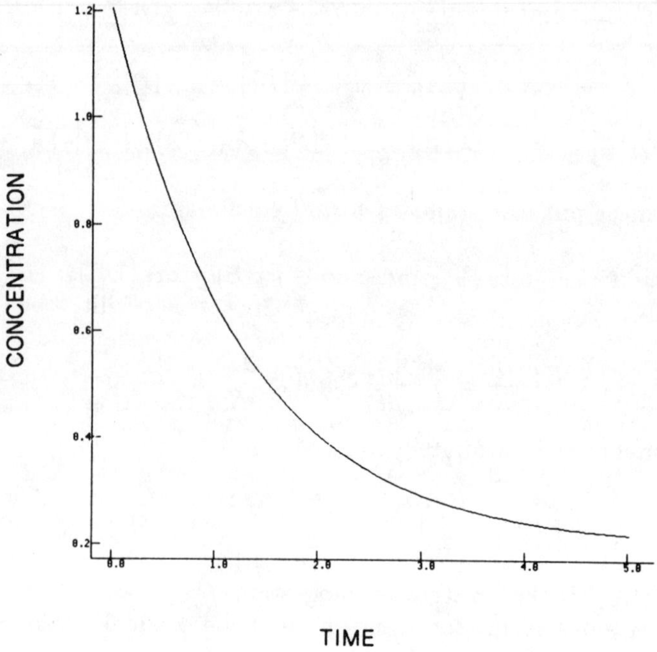

Figure 1.1: *First-order Decay. Plot of Eq.(1.11) with* $\beta_0 = 1.2$; $\beta_\infty = 0.2$; $k_\psi = 0.8$.

It is particularly convenient that any variable which is linearly related to the concentration of B can be substituted for [B] in this last equation; that is, for any constants a and c:

$$\text{If} \quad \beta = a[B] + c, \quad \text{then} \quad \frac{\beta - \beta_\infty}{\beta_0 - \beta_\infty} = exp(-k_\psi t) \qquad (1.11)$$

Figure 1 shows a typical plot of concentration as a function of time for a first-order process.

In the next few pages, it will become apparent that Eq. (1.11) is far more generally useful than one would expect if it applied only to the relatively few true first-order reactions. Because of the general utility, it is worth considering how one might examine experimental data to see whether or not the equation applies.

In the days before computers, Eq. (1.11) was usually written in the logarithmic form:

$$\ln(\beta - \beta_\infty) = -k_\psi t + \ln(\beta_0 - \beta_\infty).$$

A value for β_∞ was obtained from a final experimental determination when the reaction had essentially reached equilibrium. The values of $\ln(\beta - \beta_\infty)$ were then plotted vs time. A straight line plot indicated adherance of the data to first-order behavior, and the slope of the plot gave a value for k_ψ. This is an extremely easy treatment of data, and would be perfectly correct if the data were exact. Major practical problems arise because the logarithmic term "blows up" as β approaches β_∞ since $\ln(0) = -\infty$. As a result, small errors in β or β_∞ can cause extreme curvature in the plots. The situation is particularly bad in cases where β_∞ is difficult to obtain accurately because of instabilities of products. In all cases, it is improper to use a linear-least-squares fitting of the semi-logarithmic plot to obtain a value for k_ψ since errors in $ln(\beta - \beta_\infty)$ are not random.

With the availability of computers, it is now quite easy to be more correct in the data treatment. If a reaction is followed, for example, by observation of absorbance of the reaction solution as a function of time, Eq.(1.11) with β equal to the absorbance will apply for any case where the absorbances of reactants and products differ at the wavelength of observation. For reasonable absorbances from ca. 0 to 1.0, the expected errors is absorbance measurements will usually be nearly independent of the actual absorbance. The treatment of data can then involve a direct application of non-linear-least-squares fitting. We would have n values of β_i obtained at times t_i, and would like to choose values of k_ψ, β_0, and β_∞ such that the sum of the squares of the differences between observed β_i and those calculated from Eq.(1.11) using the values of t_i would be as small as possible.

Defining $\Re$ as the sum of the squares of the differences between observed and calculated absorbances:

$$\Re \equiv \sum_{i=1}^{n} [\beta_i - \Delta\beta_0 \exp(-k_\psi t_i) - \beta_\infty]^2 , \qquad (1.12)$$

$$\text{where } \Delta\beta_0 \equiv \beta_0 - \beta_\infty ,$$

the *best* values for k_ψ, $\Delta\beta_0$, and β_∞ will be such that the three partial derivatives, $\partial\Re/\partial k_\psi$, $\partial\Re/\partial(\Delta\beta_0)$, and $\partial\Re/\partial\beta_\infty$ are equal to zero. Defining the sums:

$$\overline{\mathcal{B}} = \sum_{i=1}^{n} [\beta_i] ; \quad \overline{\mathcal{E}} = \sum_{i=1}^{n} [exp(-k_\psi t_i)] ; \quad \overline{\mathcal{BE}} = \sum_{i=1}^{n} [\beta_i \exp(-k_\psi t_i)] ;$$

$$\overline{\mathcal{TE}} = \sum_{i=1}^{n} [t_i \exp(-k_\psi t_i)] ; \quad \overline{\mathcal{EE}} = \sum_{i=1}^{n} [\exp(-2k_\psi t_i)] ;$$

$$\overline{\mathcal{BTE}} = \sum_{i=1}^{n} [\beta_i t_i \exp(-k_\psi t_i)]\,; \quad \overline{\mathcal{TEE}} = \sum_{i=1}^{n} [t_i \exp(-2k_\psi t_i)]$$

and taking partial derivatives of $\Re$, we obtain:

$$\frac{\partial \Re}{\partial k_\psi} = 2\Delta\beta_0 \left\{\overline{\mathcal{BTE}} - \Delta\beta_0 \overline{\mathcal{TEE}} - \beta_\infty \overline{\mathcal{TE}}\right\} = 0$$

$$\frac{\partial \Re}{\partial(\Delta\beta_0)} = -2\left\{\overline{\mathcal{BE}} - \Delta\beta_0 \overline{\mathcal{EE}} - \beta_\infty \overline{\mathcal{E}}\right\} = 0$$

$$\frac{\partial \Re}{\partial \beta_\infty} = -2\left\{\overline{\mathcal{B}} - \Delta\beta_0 \overline{\mathcal{E}} - n\beta_\infty\right\} = 0$$

The last two equations may be solved for $\Delta\beta_0$ and β_∞:

$$\Delta\beta_0 = \frac{n\overline{\mathcal{BE}} - \overline{\mathcal{B}} \cdot \overline{\mathcal{E}}}{n\overline{\mathcal{EE}} - \overline{\mathcal{E}} \cdot \overline{\mathcal{E}}} \tag{1.13}$$

$$\beta_\infty = \frac{\overline{\mathcal{E}} \cdot \overline{\mathcal{BE}} - \overline{\mathcal{B}} \cdot \overline{\mathcal{EE}}}{\overline{\mathcal{E}} \cdot \overline{\mathcal{E}} - n\overline{\mathcal{EE}}} \tag{1.14}$$

Substitution of these results into the first of the three equations results in a single equation which must be satisfied by the *best* value for k_ψ:

$$\begin{aligned}\overline{\mathcal{BTE}}\left\{n\overline{\mathcal{EE}} - \overline{\mathcal{E}} \cdot \overline{\mathcal{E}}\right\} - \overline{\mathcal{TEE}}\left\{n\overline{\mathcal{BE}} - \overline{\mathcal{B}} \cdot \overline{\mathcal{E}}\right\} \\ + \overline{\mathcal{TE}}\left\{\overline{\mathcal{E}} \cdot \overline{\mathcal{BE}} - \overline{\mathcal{B}} \cdot \overline{\mathcal{EE}}\right\} = 0.\end{aligned} \tag{1.15}$$

A reasonable algorithm for finding k_ψ would involve the following steps:

- Choose an initial value for k_ψ equal to $0.69/t_{1/2}$, where $t_{1/2}$ is the approximate time at which the concentration is half-way between its initial and final values. Calculate the value of the left-hand-side of Eq.(1.15).
- Halve the value of k_ψ and recalculate the left-hand-side of the equation. If this value is larger magnitude and of the same sign as the first value, double the original value of k_ψ and again calculate the left-hand-side of the equation. Repeat the halving, or doubling, until the value calculated for the left-hand-side of the equation changes sign.
- Use repeated linear interpolations of k_ψ vs. the calculated value of the left-hand-side of the equation to whatever precision in k_ψ is desired.

In actual practice, Eq.(1.15) is a well-behaved function of k_ψ for reasonable data covering at least several half-lives of reaction, and convergence of the linear interpolations is quite rapid.

In order to judge whether or not the reaction is actually first-order, at least two tests should be applied: first, the root-mean-square error, $[\Re/n]^{1/2}$, should be no larger than the expected error in β; and, second, a plot of the experimental β_i against the values calculated using the selected parameters should show only random scatter from a straight line and no systematic curvature.

1.3 Second-Order Reversible Reactions

Now, let's consider the reaction:

$$A + B \underset{k_r}{\overset{k_f}{\rightleftharpoons}} C + D, \tag{1.16}$$

which, if it is an elementary step, gives the rate law:

$$\frac{d[D]}{dt} = \frac{d[C]}{dt} = -\frac{d[A]}{dt} = -\frac{d[B]}{dt} = k_f[A][B] - k_r[C][D]. \tag{1.17}$$

In order to integrate the rate law, we will follow essentially the same steps as we did above:

- Use the stoichiometry of the reaction to write the rate law in terms of a single variable concentration;
- write the differential equation in standard form with separated variables and integrate;
- eliminate the reverse rate constant by using $k_f/k_r = K_{eq}$, with K_{eq} expressed in terms of concentrations at equilibrium;
- evaluate the constant of integration from conditions at time equals zero.

For simplicity, we will assume that only A and B are present at $t = 0$, with concentrations $[A]_0$ and $[B]_0$ respectively. We then have the stoichiometric relationships:

$$[B] = [B]_0 - [A]_0 + [A]; \quad [C] = [D] = [A]_0 - [A]$$

which are substituted into the rate law to give:

$$\frac{d[A]}{dt} = -k_f[A]\{[B]_0 - [A]_0 + [A]\} + k_r\{[A]_0 - [A]\}^2,$$

and, sorting out powers of $[A]$:

$$\frac{d[A]}{dt} = -(k_f - k_r)[A]^2 - \{k_f([B]_0 - [A]_0) + 2k_r[A]_0\}[A] + k_r[A]_0^2.$$

At equilibrium, $d[A]/dt = 0$ and $[A] = [A]_\infty$, so, at $t = \infty$

$$0 = \frac{d[A]}{-(k_f - k_r)dt} = [A]_\infty^2 + \frac{k_f}{k_f - k_r}\{[B]_0 - [A]_0 + 2\frac{k_r}{k_f}[A]_0\}[A]_\infty - \frac{k_r}{k_f - k_r}[A]_0^2,$$

and, defining:

$$\mathcal{B} \equiv \frac{k_f}{k_f - k_r}\left\{[B]_0 - [A]_0 + 2\frac{k_r}{k_f}[A]_0\right\}$$

and

$$\mathcal{Q} \equiv \mathcal{B}^2 + 4\frac{k_r}{k_f - k_r}[A]_0^2 = \mathcal{B}^2 + 4\frac{[A]_0^2}{K_{eq} - 1},$$

we find:

$$[A]_\infty = \frac{\mathcal{Q}^{1/2} - \mathcal{B}}{2}$$

as the solution of the quadratic in $[A]_\infty$, and:

$$\frac{d[A]}{[A]^2 + \mathcal{B}[A] - \frac{1}{4}(\mathcal{Q} - \mathcal{B}^2)} = -(k_f - k_r)dt = -k_f\frac{K_{eq} - 1}{K_{eq}}dt.$$

The left-hand-side of this equation is in the standard form:

$$\int \frac{dx}{ax^2 + bx + c} = \frac{1}{\sqrt{b^2 - 4ac}} ln\left(\frac{2ax + b - \sqrt{b^2 - 4ac}}{2ax + b + \sqrt{b^2 - 4ac}}\right)$$

so that:

$$\frac{1}{\mathcal{Q}^{1/2}} \ln\left(\frac{2[A] + \mathcal{B} - \mathcal{Q}^{1/2}}{2[A] + \mathcal{B} + \mathcal{Q}^{1/2}}\right) = -k_f\frac{K_{eq} - 1}{K_{eq}}t + constant,$$

and, evaluating the constant of integration by setting $[A] = [A]_0$ at $t = 0$ and using $\mathcal{B} = \mathcal{Q}^{1/2} - 2[A]_\infty$:

$$\ln\left(\frac{\{[A] - [A]_\infty\}\{[A]_0 - [A]_\infty + \mathcal{Q}^{1/2}\}}{\{[A] - [A]_\infty + \mathcal{Q}^{1/2}\}\{[A]_0 - [A]_\infty\}}\right) = -k_f\mathcal{Q}^{1/2}\frac{K_{eq} - 1}{K_{eq}}t. \quad (1.18)$$

Several things of interest may not be immediately apparent in the development and final equation.

First, note that if $Q^{1/2}$ is much larger than $[A]_0 - [A]_\infty$, then Eq. (1.18) becomes the same as Eq. (1.10) (written in logarithmic form). This would occur if $[B]_0$ were much greater than $[A]_0$ and if K_{eq} were not too small. The point is more clearly seen in Eq. (1.17). If $[B]_0$ is much greater than $[A]_0$, and is large enough to force the reaction to proceed essentially completely to the right, then two things happen: (1) $[B]$ is essentially constant and equal to $[B]_0$ throughout the reaction; and (2) the term $k_f[A][B]$ is much larger than $k_r[C][D]$ throughout the measurable part of the reaction. Thus, Eq. (1.17) becomes a first-order differential equation with $k_\psi = k_f[B]_0$. By using a large excess of one reactant, we have forced the kinetics to be *pseudo-first-order*. We could determine the true order of the reaction with respect to $[B]$ by carrying out several kinetic measurements at various initial concentrations of B, all in large excess, and seeing how the measured values of k_ψ depend on $[B]_0$. This is an extremely useful general technique for studying complex reactions.

Second, note that if $Q = 0$, then the left-hand-side of Eq. (1.18) is always equal to zero. This would occur if $[B]_0 = [A]_0$ and if K_{eq} were very large. Looking again at Eq. (1.17), if $[B]_0 = [A]_0$, then $[A] = [B]$ throughout the reaction, and, if K_{eq} is very large, the term in $k_r[C][D]$ can be neglected. Then, Eq. (1.17) becomes $d[A]/dt = -k_f[A]^2$, which can be integrated immediately to obtain:

$$\frac{1}{[A]} - \frac{1}{[A]_0} = k_f t,$$

which is much simpler than Eq. (1.18).

These two things are examples of how one can save a lot of trouble by thinking about a reaction before carrying out a kinetic study. In most cases, there is no reason to work with an equation as complex as Eq. (1.18). Nearly always, the use of pseudo-first-order conditions will be the simplest and surest way to determine the empirical rate law for a reaction if it can be determined straightforwardly.

1.4 Consecutive Reactions with Unstable Intermediates

It should be obvious that most chemical reactions do not proceed in a single elementary step; that's what makes them interesting. Let's now examine what happens when several elementary steps are involved.

Consider a reaction proceeding by two elementary steps:

$$A \underset{k_{-1}}{\overset{k_1}{\rightleftharpoons}} B \underset{k_{-2}}{\overset{k_2}{\rightleftharpoons}} C \qquad (1.19)$$

As always for elementary steps, we can immediately write the differential rate laws:

$$\frac{d[A]}{dt} = -k_1[A] + k_{-1}[B],$$

$$\frac{d[B]}{dt} = k_1[A] - (k_{-1} + k_2)[B] + k_{-2}[C],$$

$$\text{and} \quad \frac{d[C]}{dt} = k_2[B] - k_{-2}[C].$$

There is no general way, however, in which we can use stoichiometry to write any one of these equations in terms of a single variable concentration. We can note that conservation of mass leads to:

$$[A] + [B] + [C] = [A]_0 + [B]_0 + [C]_0$$

$$\text{and} \quad \frac{d[A]}{dt} + \frac{d[B]}{dt} + \frac{d[C]}{dt} = 0.$$

This still leaves us with two independent simultaneous differential equations to solve [2]. We will see shortly that it is possible to use matrix algebra to put any set of simultaneous first-order differential equations into integrable form, but let's look at some special cases before proceeding to the general one.

Suppose the intermediate B in Eq. (1.19) is so 'unstable' that its concentration never builds up to be comparable to those of A or C. If we started the reaction with only A present at concentration $[A]_0$, and throughout the reaction $[A] + [C] = [A]_0$ within our experimental error, then we can set $d[B]/dt = 0$ for all practical purposes, and obtain:

$$\frac{d[B]}{dt} = 0 = k_1[A] - (k_{-1} + k_2)[B] + k_{-2}[C].$$

This equation can be solved for $[B]$, giving:

$$[B] = \frac{k_1[A] + k_{-2}[C]}{k_{-1} + k_2},$$

which may be substituted into the equation for $d[A]/dt$ to obtain:

$$\frac{d[A]}{dt} = -k_1[A] + k_{-1}\frac{k_1[A] + k_{-2}[C]}{k_{-1} + k_2},$$

and, using $[C] = [A]_0 - [A]$, we obtain:

$$\frac{d[A]}{dt} = -\frac{k_1k_2 + k_{-1}k_{-2}}{k_{-1} + k_2}[A] + \frac{k_{-1}k_{-2}}{k_{-1} + k_2}[A]_0,$$

which can be integrated just as was Eq. (1.6) to give the same form as Eq. (1.10) with

$$k_\psi = \frac{k_1k_2 + k_{-1}k_{-2}}{k_{-1} + k_2}. \tag{1.20}$$

The set of approximations used here are called the *steady-state approximation* or the *Bodenstein approximation.* The approximation allows us to remove one of the independent concentrations from the set of simultaneous equations.

In any set of reactions corresponding to Eq. (1.19), if $[A] + [C] = [A]_0 + [C]_0$ within experimental error, the Bodenstein approximation will be valid, and the overall reaction will show the same kinetic behavior that a simple first-order conversion of A to C *with no intermediate* would show. *The kinetics alone will give no indication of the existence of the intermediate.* It is possible that such an intermediate could be detected, for example, by trapping it, or could be indicated by other techniques, such as isotope labelling. We will discuss some of these techniques in later chapters. It is important to recognize here that the observation of simple first-order kinetics for a reaction does not allow the exclusion of reactive intermediates in the reaction mechanism.

Any time that a reaction consists of more than one step, with unstable intermediates allowing the application of the Bodenstein approximation, we encounter the concept of *rate-limiting step.* In Eq. (1.20), k_ψ is the sum of the pseudo-first-order forward rate constant $k_1k_2/(k_{-1} + k_2)$ and the pseudo-first-order reverse rate constant $k_{-1}k_{-2}/(k_{-1} + k_2)$. If $k_2 >> k_{-1}$, then the pseudo-first-order forward rate constant becomes essentially equal to k_1, and the reverse rate constant becomes essentially equal to $(1/K_2)k_{-1}$. We speak of the first step as being the *rate-limiting,* or *rate-determining* step. At the other extreme, if $k_{-1} >> k_2$, the forward rate constant becomes K_1k_2, and the reverse rate constant becomes k_{-2}. In this case, the second step is *rate-limiting,* or *rate-determining.* If k_{-1} and k_2 are of comparable magnitudes, then neither step is solely rate-limiting. In the mechanism under discussion, Eq. (1.19), kinetics alone can give no information about which step, if either, is rate-determining.

In other cases involving the Bodenstein approximation, kinetics alone may give indications of the existence of intermediates and the identities of rate-determining steps. Suppose we have the set of reactions:

$$\begin{aligned} A + B &\underset{k_{-1}}{\overset{k_1}{\rightleftharpoons}} C \\ C + D &\underset{k_{-2}}{\overset{k_2}{\rightleftharpoons}} E \end{aligned} \tag{1.21}$$

in which C is an unstable intermediate so that we may set $d[C]/dt = 0$ to obtain:

$$\frac{d[C]}{dt} = 0 = k_1[A][B] - (k_{-1} + k_2[D])[C] + k_{-2}[E]$$

$$[C] = \frac{k_1[A][B] + k_{-2}[E]}{k_{-1} + k_2[D]}.$$

This result may be substituted into

$$\frac{d[A]}{dt} = -k_1[A][B] + k_{-1}[C]$$

to obtain:

$$\frac{d[A]}{dt} = -k_1[A][B] + k_{-1}\frac{k_1[A][B] + k_{-2}[E]}{k_{-1} + k_2[D]},$$

which may be put into a form containing a single concentration variable $[A]$ from the stoichiometry of the reactions:

$$[A]_0 - [A] = [B]_0 - [B] = [D]_0 - [D] = [E] - [E]_0.$$

This final differential equation is quite messy (particularly since it contains a variable concentration term in the denominator of the right-hand-side), but simplifies under some conditions. If, for example, both $[D]_0$ and $[B]_0$ are much larger than $[A]_0$ and k_{-2} is small enough that we can ignore the reverse reaction, we obtain:

$$\begin{aligned} \frac{d[A]}{dt} &= -k_1[B]_0[A] + k_{-1}\frac{k_1[B]_0[A]}{k_{-1} + k_2[D]_0} \\ &= -\frac{k_1k_2[B]_0[D]_0}{k_{-1} + k_2[D]_0}[A], \end{aligned}$$

which is just our old friend again ... a first-order differential equation with

$$k_\psi = \frac{k_1k_2[B]_0[D]_0}{k_{-1} + k_2[D]_0}. \tag{1.22}$$

Unlike the k_ψ in Eq. (1.20), the one in Eq. (1.22) furnishes us with an experimental 'handle' in the quantity $[D]_0$. It may be possible to change from a limiting condition of $k_{-1} >> k_2[D]_0$ to that of $k_{-1} << k_2[D]_0$ by changing $[D]_0$. Whether or not this can be done in any particular situation will depend on the relative magnitudes of k_{-1} and k_2, and upon the practical range of concentrations of D that can be attained while still keeping it in large excess. If it can be done, then a series of measurements at various concentrations of D would show that k_ψ is proportional to $[D]_0$ at low concentrations, but becomes independent of $[D]_0$ at high concentrations. This would correspond to changing from step 2 being rate-determining at low $[D]_0$ to step 1 being rate-determining at high $[D]_0$. Note also that if $k_2[D]_0 >> k_{-1}$ were true under all attainable conditions, the reaction would be zero-order in $[D]$ and the rate law would not correspond to the stoichiometry of the overall reaction, clearly implicating the existence of an intermediate.

Another interesting case is given by the set of reactions:

$$\begin{aligned} A \quad &\underset{k_{-1}}{\overset{k_1}{\rightleftharpoons}} \quad B + C \\ B + D \quad &\underset{k_{-2}}{\overset{k_2}{\rightleftharpoons}} \quad E \end{aligned} \tag{1.23}$$

in which B is an unstable intermediate. Let's simplify by assuming that k_{-2} is so small that it can be neglected, and that C and D are present in large excess. We can then write:

$$\frac{d[B]}{dt} = 0 = k_1[A] - \{k_{-1}[C]_0 + k_2[D]_0\}[B], \quad \text{and}$$

$$[B] = \frac{k_1}{k_{-1}[C]_0 + k_2[D]_0}[A]$$

and then

$$\frac{d[A]}{dt} = -k_1[A] + k_{-1}[C]_0[B] = -\frac{k_1 k_2[D]_0}{k_{-1}[C]_0 + k_2[D]_0}[A],$$

which is, again, a first-order differential equation now with

$$k_\psi = \frac{k_1 k_2[D]_0}{k_{-1}[C]_0 + k_2[D]_0}. \tag{1.24}$$

Note particularly that $[C]_0$ appear only in the denominator in this equation. If $k_{-1}[C]_0 >> k_2[D]_0$, k_ψ will be inversely proportional to $[C]_0$.

The resulting effect of decreasing rate with increasing concentration of C is called a *mass law effect.* Under the assumption of $k_{-1}[C]_0 >> k_2[D]_0$, the first step of Eq. (1.23) is a rapid equilibrium followed by the rate-determining second step. The inverse dependence of k_ψ on $[C]_0$ arises from the fact that the equilibrium concentration of B must decrease with increasing concentration of C . . . thus the name *mass law effect.*

The mechanism shown by Eq. (1.23) is just the mechanism of SN1 reactions [1] when A is an alkyl halide or alkyl tosylate, D is a nucleophilic reagent such as azide ion, and B is a carbocation. The most convincing evidence for the existence of carbocation intermediates in some of these reactions is the observation of mass law effects. In those cases where mass law effects are not observed, but the kinetics are zero-order with respect to concentration of nucleophile, an intermediate is clearly indicated, but could be something other than a carbocation.

1.5 Basics of Matrix Algebra

In order to consider the general case of consecutive or simultaneous first-order reactions, of which Eq. (1.19) is the simplest example, we will need to use some matrix algebra. Let's start from the beginning.

A matrix is a rectangular array of numbers or variables arranged into rows and columns. For a general matrix,$\mathcal{M}$, with elements m_{ij}:

$$\mathcal{M} = \begin{pmatrix} m_{11} & m_{12} & \dots & m_{1c} \\ m_{21} & m_{22} & \dots & m_{2c} \\ \vdots & \vdots & \ddots & \vdots \\ m_{r1} & m_{r2} & \dots & m_{rc} \end{pmatrix}$$

The matrix $\mathcal{M}$ shown here has r rows and c columns. The element m_{ij} is found in the i'th row and the j'th column. The first index is always the row index, and the second one is the column index. A mnemonic device is to remember the **R**oman **C**atholic system of indexing. The *dimensions* of a matrix are the number of rows by the number of columns. We speak of M as being an $r \times c$ dimensioned matrix, or simply as an 'r by c' matrix. A matrix having the same number of rows as it has columns, an $r \times r$ matrix, is called a *square* matrix. In a square matrix, the elements m_{ii} form the *principal diagonal* of the matrix.

Two matrices, $\mathcal{M}$ and $\mathcal{N}$, are equal if and only if they have the same dimensions and $m_{ij} = n_{ij}$ for all i and j.

The addition of two matrices, $\mathcal{M}$ and $\mathcal{N}$, produces a matrix $\mathcal{S}$ in which $s_{ij} = m_{ij} + n_{ij}$. The addition is defined only when the two matrices $\mathcal{M}$ and $\mathcal{N}$ have the same dimensions, and the matrix $\mathcal{S}$ then also has the same dimensions. A few examples should make the operation clear:

$$\begin{pmatrix} 1 & 2 \\ 3 & 4 \end{pmatrix} + \begin{pmatrix} 3 & 1 \\ 2 & 5 \end{pmatrix} = \begin{pmatrix} 4 & 3 \\ 5 & 9 \end{pmatrix} \qquad \begin{pmatrix} 1 \\ 2 \\ 3 \end{pmatrix} + \begin{pmatrix} 4 \\ 5 \\ 6 \end{pmatrix} = \begin{pmatrix} 5 \\ 7 \\ 9 \end{pmatrix}$$

$$\begin{pmatrix} 1 & 2 & 3 & 4 \\ 4 & 3 & 2 & 1 \end{pmatrix} + \begin{pmatrix} 1 & 2 & 1 & 2 \\ 2 & 1 & 2 & 1 \end{pmatrix} = \begin{pmatrix} 2 & 4 & 4 & 6 \\ 6 & 4 & 4 & 2 \end{pmatrix}$$

$$\begin{pmatrix} m_{11} & m_{12} & \dots & m_{1c} \\ m_{21} & m_{22} & \dots & m_{2c} \\ \vdots & \vdots & \ddots & \vdots \\ m_{r1} & m_{r2} & \dots & m_{rc} \end{pmatrix} + \begin{pmatrix} n_{11} & n_{12} & \dots & n_{1c} \\ n_{21} & n_{22} & \dots & n_{2c} \\ \vdots & \vdots & \ddots & \vdots \\ n_{r1} & n_{r2} & \dots & n_{rc} \end{pmatrix}$$

$$= \begin{pmatrix} m_{11}+n_{11} & m_{12}+n_{12} & \dots & m_{1c}+n_{1c} \\ m_{21}+n_{21} & m_{22}+n_{22} & \dots & m_{2c}+n_{2c} \\ \vdots & \vdots & \ddots & \vdots \\ m_{r1}+n_{r1} & m_{r2}+n_{r2} & \dots & m_{rc}+n_{rc} \end{pmatrix}$$

Similarly, the subtraction of two matrices is defined only for matrices having the same dimensions, and $\mathcal{M}-\mathcal{N} = \mathcal{D}$ implies that $d_{ij} = m_{ij}-n_{ij}$ for all i and j:

$$\begin{pmatrix} 1 & 2 \\ 3 & 4 \end{pmatrix} - \begin{pmatrix} 3 & 1 \\ 2 & 5 \end{pmatrix} = \begin{pmatrix} -2 & 1 \\ 1 & -1 \end{pmatrix} \qquad \begin{pmatrix} 1 \\ 2 \\ 3 \end{pmatrix} - \begin{pmatrix} 4 \\ 5 \\ 6 \end{pmatrix} = \begin{pmatrix} -3 \\ -3 \\ -3 \end{pmatrix}$$

$$\begin{pmatrix} 1 & 2 & 3 & 4 \\ 4 & 3 & 2 & 1 \end{pmatrix} - \begin{pmatrix} 1 & 2 & 1 & 2 \\ 2 & 1 & 2 & 1 \end{pmatrix} = \begin{pmatrix} 0 & 0 & 2 & 2 \\ 2 & 2 & 0 & 0 \end{pmatrix}$$

$$\begin{pmatrix} m_{11} & m_{12} & \dots & m_{1c} \\ m_{21} & m_{22} & \dots & m_{2c} \\ \vdots & \vdots & \ddots & \vdots \\ m_{r1} & m_{r2} & \dots & m_{rc} \end{pmatrix} - \begin{pmatrix} n_{11} & n_{12} & \dots & n_{1c} \\ n_{21} & n_{22} & \dots & n_{2c} \\ \vdots & \vdots & \ddots & \vdots \\ n_{r1} & n_{r2} & \dots & n_{rc} \end{pmatrix}$$

$$= \begin{pmatrix} m_{11}-n_{11} & m_{12}-n_{12} & \dots & m_{1c}-n_{1c} \\ m_{21}-n_{21} & m_{22}-n_{22} & \dots & m_{2c}-n_{2c} \\ \vdots & \vdots & \ddots & \vdots \\ m_{r1}-n_{r1} & m_{r2}-n_{r2} & \dots & m_{rc}-n_{rc} \end{pmatrix}$$

Matrix multiplication becomes just a little more complicated. $\mathcal{MN} = \mathcal{P}$ implies

$$p_{ij} = \sum_{k=1}^{c} m_{ik} n_{kj}$$

for all i and j. The matrix $\mathcal{N}$ must have the same number, c, of rows that $\mathcal{M}$ has of columns. It is important to note that $\mathcal{MN} \neq \mathcal{NM}$ and that the product $\mathcal{NM}$ may not even be defined if both matrices are not square. The above equation for the product elements says that to obtain the ij'th element of $\mathcal{P}$, we take the i'th row of the matrix on the left and the j'th column of the matrix on the right, multiply corresponding elements and add them together. A few examples should make the process clear:

$$(1 \quad 2 \quad 3)\begin{pmatrix} 1 \\ 2 \\ 3 \end{pmatrix} = 14; \quad \begin{pmatrix} 1 \\ 2 \\ 3 \end{pmatrix}(1 \quad 2 \quad 3) = \begin{pmatrix} 1 & 2 & 3 \\ 2 & 4 & 6 \\ 3 & 6 & 9 \end{pmatrix}$$

$$\begin{pmatrix} 2 & 1 \\ 3 & 2 \\ 1 & 4 \end{pmatrix}\begin{pmatrix} 3 & 2 & 1 \\ 1 & 4 & 2 \end{pmatrix} = \begin{pmatrix} 7 & 8 & 4 \\ 11 & 14 & 7 \\ 7 & 18 & 9 \end{pmatrix};$$

$$\begin{pmatrix} 1 & 2 & 1 \\ 3 & 1 & 2 \\ 1 & 2 & 3 \end{pmatrix}\begin{pmatrix} 1 & 2 & 3 \\ 3 & 2 & 1 \\ 1 & 2 & 2 \end{pmatrix} = \begin{pmatrix} 8 & 8 & 7 \\ 8 & 12 & 14 \\ 10 & 12 & 11 \end{pmatrix}.$$

The distributive and associative laws follow directly from the above definitions:

$$\mathcal{A}(\mathcal{B} + \mathcal{C}) = \mathcal{AB} + \mathcal{AC}; \quad (\mathcal{B} + \mathcal{C})\mathcal{A} = \mathcal{BA} + \mathcal{CA}$$

$$(\mathcal{AB})\mathcal{C} = \mathcal{A}(\mathcal{BC}) = \mathcal{ABC}$$

The number one in ordinary algebra has an analog in matrix algebra which is called the *identity matrix*, symbolized by $\mathcal{I}$. Multiplication of any matrix by the appropriately dimensioned identity matrix leaves the matrix unchanged: $\mathcal{IM} = \mathcal{MI} = \mathcal{M}$; $\mathcal{I}^n = \mathcal{I}$ for all n. An identity matrix is a square matrix with ones along the principal diagonal and zeros in all other positions. The 3×3 identity matrix, for example, is

$$\begin{pmatrix} 1 & 0 & 0 \\ 0 & 1 & 0 \\ 0 & 0 & 1 \end{pmatrix}, \quad \text{and obviously}$$

$$\begin{pmatrix} 1 & 0 & 0 \\ 0 & 1 & 0 \\ 0 & 0 & 1 \end{pmatrix} \begin{pmatrix} 1 & 2 & 3 \\ 3 & 2 & 1 \\ 1 & 2 & 2 \end{pmatrix} = \begin{pmatrix} 1 & 2 & 3 \\ 3 & 2 & 1 \\ 1 & 2 & 2 \end{pmatrix}.$$

In the notation usually used in matrix algebra, any multiple of the identity matrix ... that is, any matrix with the same number, c at all positions along the principal diagonal and zeros at all other positions ... is simply written as a constant, c:

$$\begin{pmatrix} c & 0 & 0 & \dots & 0 \\ 0 & c & 0 & \dots & 0 \\ 0 & 0 & c & \dots & 0 \\ \vdots & \vdots & \vdots & \ddots & \vdots \\ 0 & 0 & 0 & \dots & c \end{pmatrix} = c.$$

The effect of multiplying a matrix by a constant is to multiply every element of the matrix by the constant. You should now be able to prove that a constant commutes with any matrix: $c\mathcal{M} = \mathcal{M}c$.

In matrix algebra, as, in fact, in ordinary algebra, division is not defined except in the sense of an inverse of multiplication. Given a matrix $\mathcal{M}$, if one can find a matrix $\mathcal{N}$ such that $\mathcal{MN} = \mathcal{I}$, then $\mathcal{N}$ is called a *right inverse* of $\mathcal{M}$, $\mathcal{M}^{-1}$. For a square matrix, a right inverse is also a left inverse; if $\mathcal{M}^{-1}\mathcal{M} = \mathcal{I}$, then $\mathcal{M}\mathcal{M}^{-1} = \mathcal{I}$. Not all matrices have inverses.

The *transpose* of a real matrix $\mathcal{M}$, written $\mathcal{M}^\dagger$, is the matrix whose rows are the columns of $\mathcal{M}$... that is, $m^\dagger_{ij} = m_{ji}$. [1] For example, if

$$\mathcal{M} = \begin{pmatrix} 1 & 2 & 3 \\ 4 & 5 & 6 \\ 7 & 8 & 9 \end{pmatrix}, \text{ then } \quad \mathcal{M}^\dagger = \begin{pmatrix} 1 & 4 & 7 \\ 2 & 5 & 8 \\ 3 & 6 & 9 \end{pmatrix}.$$

From this definition, it is easy to see that

$$(\mathcal{MN})^\dagger = \mathcal{N}^\dagger\mathcal{M}^\dagger, \quad (\mathcal{M}^\dagger)^\dagger = \mathcal{M}, \quad \text{and} \quad \mathcal{I}^\dagger = \mathcal{I}.$$

If $\mathcal{M}^\dagger = \mathcal{M}$, the matrix $\mathcal{M}$ is said to be *Hermitian.* Many, but not all, matrices which occur in physical problems are Hermitian.

---

1 For a matrix having imaginary or complex elements, $m^\dagger_{ij} = m^*_{ji}$ where m^*_{ji} is the complex conjugate of m_{ji}.

A very important type of matrix is a *unitary* matrix. A matrix $\mathcal{U}$ is unitary if $\mathcal{U}^\dagger\mathcal{U} = \mathcal{I}$ or $\mathcal{U}\mathcal{U}^\dagger = \mathcal{I}$. If $\mathcal{U}$ is a square unitary matrix, then $\mathcal{U}\mathcal{U}^\dagger = \mathcal{U}^\dagger\mathcal{U} = \mathcal{I}$. This definition requires that

$$\sum_{k=1}^{c} u_{ik}u^\dagger_{kj} = \sum_{k=1}^{c} u_{ik}u_{jk} = \delta_{ij},$$

where δ_{ij} is the *Kronecker delta*, $\delta_{ij} = 1$ if $i = j$, and $\delta_{ij} = 0$ if $i \neq j$.

A very large number of physical problems involve finding *eigenvectors* and *eigenvalues* of matrices. In order to define these things, we need a few more bits of language and notation. Vectors are simply matrices having a single row or a single column, and are designated row vectors and column vectors, resepctively. It is common practice to designate a column vector as $\vec{v}$, and a row vector as $\vec{v}^\dagger$. In one of the matrix multiplication examples, we saw that a row vector times a column vector produces a simple number [2], and that a column vector times a row vector produces a matrix. We now have a set of notation so that a fancy capital letter, such as $\mathcal{M}$, represents a matrix; a small letter with an arrow over it, such as $\vec{r}$, represents a vector; and, a simple lower-case letter, such as b, or other unadorned characters, such as λ, represent numbers.

If for a matrix $\mathcal{M}$ a vector $\vec{v}$ is found such that $\mathcal{M}\vec{v} = \lambda\vec{v}$, then $\vec{v}$ is said to be an eigenvector of $\mathcal{M}$ with eigenvalue λ. For row vectors, the equation would be written $\vec{v}^\dagger\mathcal{M} = \vec{v}^\dagger\lambda$, but we would still use the same language. Some examples of eigenvectors are:

$$\begin{pmatrix} 2 & 1 & 1 \\ 1 & 1 & 2 \\ 1 & 2 & 1 \end{pmatrix}\begin{pmatrix} 1 \\ 1 \\ 1 \end{pmatrix} = 4\begin{pmatrix} 1 \\ 1 \\ 1 \end{pmatrix}$$

$$(1 \quad 1 \quad 1)\begin{pmatrix} 2 & 1 & 1 \\ 1 & 1 & 2 \\ 1 & 2 & 1 \end{pmatrix} = 4(1 \quad 1 \quad 1)$$

$$\begin{pmatrix} 1 & 0 & 1 & 0 \\ 1 & 1 & 1 & -1 \\ 0 & 3 & -6 & 6 \\ 2 & 2 & 2 & 1 \end{pmatrix}\begin{pmatrix} 1 \\ 0 \\ 2 \\ 3 \end{pmatrix} = 3\begin{pmatrix} 1 \\ 0 \\ 2 \\ 3 \end{pmatrix}$$

Any matrix has at least one eigenvector (not necessarily real). The proof of this statement is given in Wilkenson's book, "The Algebraic

---

[2] This corresponds to the dot product of two vectors in vector algebra

Eigenvalue Problem" [3], and will not be repeated here. This reference is an excellent source of information on matrices.

Finding eigenvalues and eigenvectors of a matrix is not generally an easy task, and for anything larger than a 3×3 matrix, the computer is a virtual necessity. Nearly all modern computer centers and mathematical software packages have programs for the eigenvector problem. In the following, we shall use a brute force method and limit our examples to 3×3 matrices.

Let's take a specific matrix $\mathcal{M}$:

$$\mathcal{M} = \begin{pmatrix} 1 & -2 & 0 \\ -1 & 2 & 0 \\ 1 & 0 & 1 \end{pmatrix} \quad \text{with} \quad \vec{v} = \begin{pmatrix} v_1 \\ v_2 \\ v_3 \end{pmatrix}$$

so that

$$\mathcal{M}\vec{v} = \lambda\vec{v}, \quad \text{or} \quad [\mathcal{M} - \lambda\mathcal{I}]\vec{v} = 0,$$

$$\begin{pmatrix} (1-\lambda) & -2 & 0 \\ -1 & (2-\lambda) & 0 \\ 1 & 0 & (1-\lambda) \end{pmatrix} \begin{pmatrix} v_1 \\ v_2 \\ v_3 \end{pmatrix} = 0$$

which is the three simultaneous equations:

$$\begin{aligned} (1-\lambda)v_1 - 2v_2 &= 0, \\ -v_1 + (2-\lambda)v_2 &= 0, \\ v_1 + (1-\lambda)v_3 &= 0. \end{aligned}$$

These equations have a solution, λ, if, and only if, the determinant:

$$\begin{bmatrix} (1-\lambda) & -2 & 0 \\ -1 & (2-\lambda) & 0 \\ 1 & 0 & (1-\lambda) \end{bmatrix} = 0.$$

Expansion of the determinant gives the polynomial

$$(1-\lambda)(2-\lambda)(1-\lambda) - 2(1-\lambda) = \lambda^3 - 4\lambda^2 + 3\lambda = 0,$$

which has the solutions $\lambda = 0$, 1, or 3. The solution $\lambda = 0$ substituted into the three simultaneous equations above gives

$$v_1 - 2v_2 = 0, \quad -v_1 + 2v_2 = 0, \quad \text{and} \quad v_1 + v_3 = 0,$$

which are satisfied if

$$2v_2 = v_1 = -v_3.$$

We are free to choose any value we wish for one on the vector components as long as this relationship is satisfied by values for the other two. It is easy to see that this is a general result by the following application of commutative and associative relationships.

$$\text{If} \quad \mathcal{N}\vec{x} = \lambda\vec{x}, \text{ then} \quad a\mathcal{N}\vec{x} = a\lambda\vec{x}, \quad \text{and} \quad \mathcal{N}a\vec{x} = \lambda a\vec{x},$$
$$\text{and} \quad \mathcal{N}[a\vec{x}] = \lambda[a\vec{x}].$$

In words: *If a vector, $\vec{x}$, is an eigenvector of a matrix with eigenvalue λ, then any constant times the vector, $a\vec{x}$, is also an eigenvector of the matrix, with the same eigenvalue, λ.* We may, therefore, choose the magnitudes for convenience.

In the present case, let's choose $v_1 = 2$. The eigenvector associated with the eigenvalue of zero is then

$$\vec{v} = \begin{pmatrix} 2 \\ 1 \\ -2 \end{pmatrix} \quad \text{for} \quad \lambda = 0.$$

For the $\lambda = 1$ solution of the polynomial substituted into the three equations for the vector components, we have

$$-2v_2 = 0, \quad -v_1 + v_2 = 0, \quad \text{and} \quad v_1 = 0,$$

and, arbitrarily choosing $v_3 = 1$,

$$\vec{v} = \begin{pmatrix} 0 \\ 0 \\ 1 \end{pmatrix} \quad \text{for} \quad \lambda = 1.$$

Finally, the solution $\lambda = 3$ gives

$$-2v_1 - 2v_2 = 0, \quad -v_1 - v_2 = 0, \quad \text{and} \quad v_1 - 2v_3 = 0,$$

and choosing $v_1 = 2$

$$\vec{v} = \begin{pmatrix} 2 \\ -2 \\ 1 \end{pmatrix} \quad \text{for} \quad \lambda = 3.$$

There are several different conventions used in different situations for the choice of magnitudes of eigenvector components. In the above example, we chose the components so that small whole numbers resulted.

This, of course, will not always be possible. A slightly more common convention is to choose the leading non-zero component to be unity. For the above example, this convention gives

$$\vec{v} = \begin{pmatrix} 1 \\ 1/2 \\ -1 \end{pmatrix}, \quad \begin{pmatrix} 0 \\ 0 \\ 1 \end{pmatrix}, \quad \begin{pmatrix} 1 \\ -1 \\ 1/2 \end{pmatrix}.$$

The most common convention is to *normalize the vectors to unit length.* That is, we choose the components such that

$$\sum_{k=1}^{n} v_k^2 = 1.$$

For the above example, we would obtain

$$\vec{v} = \begin{pmatrix} 2/3 \\ 1/3 \\ -2/3 \end{pmatrix}, \quad \begin{pmatrix} 0 \\ 0 \\ 1 \end{pmatrix}, \quad \begin{pmatrix} 2/3 \\ -2/3 \\ 1/3 \end{pmatrix}.$$

An important type of operation in matrix algebra which has no counterpart in ordinary algebra is *a similarity transformation of a matrix.* This is an operation on a matrix, $\mathcal{M}$, according to the equation

$$\mathcal{R}^{-1}\mathcal{M}\mathcal{R} = \mathcal{M}'.$$

The matrices $\mathcal{M}$ and $\mathcal{M}'$ are *equivalent* matrices. A particularly important type of similarity transformation is the *unitary transformation*

$$\mathcal{U}^{\dagger}\mathcal{M}\mathcal{U} = \Lambda$$

where Λ is a *diagonal matrix* ... a matrix having zeros for all elements except those along the principal diagonal:

$$\Lambda = \begin{pmatrix} \lambda_1 & 0 & \dots & 0 \\ 0 & \lambda_2 & \dots & 0 \\ \vdots & \vdots & \ddots & \vdots \\ 0 & 0 & \dots & \lambda_n \end{pmatrix}$$

and $\mathcal{U}$ is a unitary matrix. This is called *matrix diagonalization.* An important theorem of matrix algebra is that *any Hermitian matrix may be diagonalized by a unitary transformation.* [3]

---

[3] See Wilkenson, ref. 3, at the end of the Chapter.

Note that left multiplication of $\mathcal{U}^{\dagger}\mathcal{M}\mathcal{U} = \Lambda$ by $\mathcal{U}$ produces

$$\mathcal{M}\mathcal{U} = \mathcal{U}\Lambda,$$

which shows that the i'th column of $\mathcal{U}$ is an eigenvector of $\mathcal{M}$, with eigenvalue of λ_i. As a specific example, consider the matrix

$$\mathcal{M} = \begin{pmatrix} 11/6 & -4/6 & -1/6 \\ -4/6 & 14/6 & -4/6 \\ -1/6 & -4/6 & 11/6 \end{pmatrix}.$$

You may verify, by actually carrying out the operations, that this matrix has the following eigenvectors and eigenvalues:

$$\vec{v}_1 = \begin{pmatrix} 1/\sqrt{3} \\ 1/\sqrt{3} \\ 1/\sqrt{3} \end{pmatrix}, \quad \lambda_1 = 1; \qquad \vec{v}_2 = \begin{pmatrix} 1/\sqrt{2} \\ 0 \\ -1/\sqrt{2} \end{pmatrix}, \quad \lambda_2 = 2;$$

$$\vec{v}_3 = \begin{pmatrix} 1/\sqrt{6} \\ -2/\sqrt{6} \\ 1/\sqrt{6} \end{pmatrix}, \quad \lambda_3 = 3.$$

You may also verify, by carrying out $\mathcal{U}^{\dagger}\mathcal{U}$ that the matrix having these three vectors as columns is indeed unitary. Now, you can see that

$$\begin{pmatrix} 1/\sqrt{3} & 1/\sqrt{3} & 1/\sqrt{3} \\ 1/\sqrt{2} & 0 & -1/\sqrt{2} \\ 1/\sqrt{6} & -2/\sqrt{6} & 1/\sqrt{6} \end{pmatrix} \begin{pmatrix} 11/6 & -4/6 & -1/6 \\ -4/6 & 14/6 & -4/6 \\ -1/6 & -4/6 & 11/6 \end{pmatrix}$$
$$\times \begin{pmatrix} 1/\sqrt{3} & 1/\sqrt{2} & 1/\sqrt{6} \\ 1/\sqrt{3} & 0 & -2/\sqrt{6} \\ 1/\sqrt{3} & -1/\sqrt{2} & 1/\sqrt{6} \end{pmatrix} = \begin{pmatrix} 1 & 0 & 0 \\ 0 & 2 & 0 \\ 0 & 0 & 3 \end{pmatrix}.$$

As an example of how matrix diagonalization can be used in some problems of simultaneous differential equations, suppose we have

$$\frac{dx_1}{dt} \equiv \dot{x}_1 = m_{11}x_1 + m_{12}x_2 + m_{13}x_3$$
$$\frac{dx_2}{dt} \equiv \dot{x}_2 = m_{21}x_1 + m_{22}x_2 + m_{23}x_3$$
$$\frac{dx_3}{dt} \equiv \dot{x}_3 = m_{31}x_1 + m_{32}x_3 + m_{33}x_3.$$

These equations can be written in matrix form as

$$\begin{pmatrix} m_{11} & m_{12} & m_{13} \\ m_{21} & m_{22} & m_{23} \\ m_{31} & m_{32} & m_{33} \end{pmatrix} \begin{pmatrix} x_1 \\ x_2 \\ x_3 \end{pmatrix} = \begin{pmatrix} \dot{x}_1 \\ \dot{x}_2 \\ \dot{x}_3 \end{pmatrix};$$

$$\text{or} \quad \mathcal{M}\vec{x} = \dot{\vec{x}}.$$

If $\mathcal{M}$ is a Hermitian matrix, ($m_{31} = m_{13}$, $m_{32} = m_{23}$, and $m_{21} = m_{12}$.) then it can be diagonalized. Assume that we have, by hook or crook, found the matrix $\mathcal{U}$ such that $\mathcal{U}^\dagger\mathcal{M}\mathcal{U} = \Lambda$. We first left-multiply our matrix equation by $\mathcal{U}^\dagger$ to obtain $\mathcal{U}^\dagger\mathcal{M}\vec{x} = \mathcal{U}^\dagger\dot{\vec{x}}$. Now, since $\mathcal{U}\mathcal{U}^\dagger = \mathcal{I}$, and since $\mathcal{I}$ can be inserted anywhere in an equation without changing it, we have

$$\mathcal{U}^\dagger\mathcal{M}[\mathcal{U}\mathcal{U}^\dagger]\vec{x} = [\mathcal{U}^\dagger\mathcal{M}\mathcal{U}]\mathcal{U}^\dagger\vec{x} = \mathcal{U}^\dagger\dot{\vec{x}},$$

and

$$\Lambda\mathcal{U}^\dagger\vec{x} = \mathcal{U}^\dagger\dot{\vec{x}}.$$

Now, note carefully that the time derivatives of the elements of the vector $\mathcal{U}^\dagger\vec{x}$ are just the elements of $\mathcal{U}^\dagger\dot{\vec{x}}$ since $\mathcal{U}$ is time independent. We define

$$\vec{y} \equiv \mathcal{U}^\dagger\vec{x}, \quad \text{and have} \quad \dot{\vec{y}} = \mathcal{U}^\dagger\dot{\vec{x}}$$

so that

$$\Lambda\vec{y} = \dot{\vec{y}}.$$

This equation is equivalent to the three simple differential equations

$$\lambda_1 y_1 = \dot{y}_1, \quad \lambda_2 y_2 = \dot{y}_2, \quad \text{and} \quad \lambda_3 y_3 = \dot{y}_3,$$

each of which can be integrated independently to give

$$y_i = y_i^0 e^{\lambda_i t} \quad i = 1, 2, 3$$

where y_i^0 is the value of y_i at $t = 0$. Each y variable is a linear combination of the original x variables:

$$y_i = \sum_{k=1}^{3} u_{ik}^\dagger x_k \quad \text{from} \quad \vec{y} = \mathcal{U}^\dagger\vec{x}.$$

If we wish to write the integrated equations in terms of the original x variables, we left multiply this last equation defining $\vec{y}$ by $\mathcal{U}$ to obtain

$$\mathcal{U}\vec{y} = \mathcal{U}\mathcal{U}^\dagger\vec{x} = \mathcal{I}\vec{x} = \vec{x}, \quad \text{and} \quad x_i = \sum_{k=1}^{3} u_{ik} y_k \quad \text{for} \quad i = 1, 2, 3.$$

Another useful example of an application of matrix diagonalization occurs in parabolic functions of the form

$$f(x_1, x_2, \dots x_n) = a_{11}x_1^2 + a_{12}x_1x_2 + \dots a_{1n}x_1x_n \dots a_{22}x_2^2 + \dots a_{nn}x_n^2$$
$$= \sum_{k=1}^{n}\sum_{l=1}^{n} a_{kl}x_kx_l.$$

This equation can be written in matrix form as

$$\vec{x}^\dagger \mathcal{A}\vec{x} = f(x_1, x_2, \dots x_n)$$

where the diagonal elements of $\mathcal{A}$ are just equal to the coefficients of the squared terms [4], and the off-diagonal elements are 1/2 the coefficients of the cross terms. In this case, $\mathcal{A}$ is Hermitian, by definition, and can be diagonalized by a unitary transformation. Again, suppose we have found $\mathcal{U}$ such that $\mathcal{U}^\dagger \mathcal{A}\mathcal{U} = \Lambda$, where Λ is a diagonal matrix. Step-by-step, we develop

$$\begin{aligned} f(x_1, x_2, \dots x_n) &= \vec{x}^\dagger \mathcal{A}\vec{x} \\ &= \vec{x}^\dagger [\mathcal{U}\mathcal{U}^\dagger]\mathcal{A}\vec{x} \\ &= \vec{x}^\dagger [\mathcal{U}\mathcal{U}^\dagger]\mathcal{A}[\mathcal{U}\mathcal{U}^\dagger]\vec{x} \\ &= [\vec{x}^\dagger \mathcal{U}]\,[\mathcal{U}^\dagger \mathcal{A}\mathcal{U}]\,[\mathcal{U}^\dagger \vec{x}] \\ &= \vec{y}^\dagger \Lambda\vec{y} = f(y_1, y_2, \dots y_n) \\ &= \sum_{k=1}^{n} \lambda_k y_k^2 \end{aligned}$$

We have *rotated* the original x coordinates into a new set of y coordinates so that the function, f, contains no cross terms, y_iy_j. The function f is *elliptically parabolic*, and we have found a set of coordinates y which are parallel to the major elliptical axes of the function.

The rotation of axes in two dimensions is shown in Figure 2. The point, P, having coordinates x and y in the original coordinate system has coordinates x' and y' in the new coordinate system.

---

[4] It is interesting, and pertinent, to note that the elements of $\mathcal{A}$ may be written as second derivatives of f with respect to the x's:

$$a_{ij} = \frac{1}{2}\frac{d^2 f}{dx_i dx_j}$$

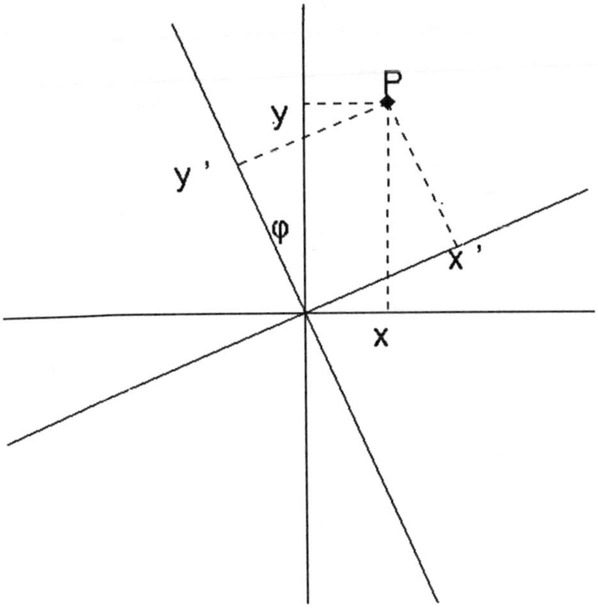

Figure 1.2: *Rotation of Cartesian Coordinates. The original axes are rotated counterclockwise by the angle ϕ.*

The coordinates are related by

$$x' = x\cos\phi + y\sin\phi, \text{ and } \quad y' = -x\sin\phi + y\cos\phi,$$

or, in matrix notation

$$\begin{pmatrix} x' \\ y' \end{pmatrix} = \begin{pmatrix} \cos\phi & \sin\phi \\ -sin\phi & cos\phi \end{pmatrix} \begin{pmatrix} x \\ y \end{pmatrix}$$

Note that the 2×2 matrix here is a unitary matrix. In fact, any 2×2 unitary matrix may be written in this form where the elements are in terms of $\cos\theta$ and $\sin\theta$ for some θ. It is for this reason that unitary matrices are frequently called *rotation* matrices.

The important theorem that we have been using is that any Hermitian matrix may be diagonalized by a unitary transformation. Unfortunately, not all physical problems involve Hermitian matrices. The problem of consecutive first-order reaction kinetics, Eq. (1.19), for example, can be written quite nicely in matrix notation as

$$\mathcal{K}\vec{c} = \dot{\vec{c}}, \qquad \begin{pmatrix} -k_1 & k_{-1} & 0 \\ k_1 & -(k_{-1}+k_2) & k_{-2} \\ 0 & k_2 & -k_{-2} \end{pmatrix} \begin{pmatrix} [A] \\ [B] \\ [C] \end{pmatrix} = \begin{pmatrix} d[A]/dt \\ d[B]/dt \\ d[C]/dt \end{pmatrix},$$

but the matrix $\mathcal{K}$ is not Hermitian (however, see Prob. 1.16).

1.6 Triangularization of Matrices

Fortunately, there is something for square non-Hermitian matrices almost as good as diagonalization ...they can be *triangularized* by a similarity transformation. Given any square matrix $\mathcal{M}$, it is always possible to find a matrix $\mathcal{R}$, with inverse $\mathcal{R}^{-1}$, such that

$$\mathcal{R}^{-1}\mathcal{M}\mathcal{R} = \mathcal{T},$$

where $\mathcal{T}$ is of the form

$$\mathcal{T} = \begin{pmatrix} t_{11} & 0 & \dots & 0 \\ t_{21} & t_{22} & \dots & 0 \\ \vdots & \vdots & \ddots & \vdots \\ t_{n1} & t_{n2} & \dots & t_{nn} \end{pmatrix}, \quad \text{or} \quad \begin{pmatrix} t_{11} & t_{12} & \dots & t_{1n} \\ 0 & t_{22} & \dots & t_{2n} \\ \vdots & \vdots & \ddots & \vdots \\ 0 & 0 & \dots & t_{nn} \end{pmatrix},$$

called *lower-triangular* or *upper-triangular* forms, respectively.

Since matrix triangularization is not such a common operation as matrix diagonalization, we shall go into enough detail that, if you really wanted to, you could write a general algorithm. Basically, triangularizing a matrix involves finding eigenvectors, and it is assumed that you have some way to find an eigenvector of any matrix. Thus, given a matrix $\mathcal{M}$, find $\vec{v}$ and λ_1 such that

$$\begin{pmatrix} m_{11} & m_{12} & \dots & m_{1n} \\ m_{21} & m_{22} & \dots & m_{2n} \\ \vdots & \vdots & \ddots & \vdots \\ m_{n1} & m_{n2} & \dots & m_{nn} \end{pmatrix} \begin{pmatrix} 1 \\ v_2 \\ \vdots \\ v_n \end{pmatrix} = \lambda_1 \begin{pmatrix} 1 \\ v_2 \\ \vdots \\ v_n \end{pmatrix},$$

where the vector $\vec{v}$ has been normalized to have its first element equal to unity.

This is a rather unusual normalization, and, in fact, is not possible if the first element of the eigenvector must be zero. This can be taken care of fairly easily, however. If an eigenvector is found whose first element is zero, we simply transform the problem using a *permutation matrix*. A simple permutation matrix is a matrix which has all elements along the diagonal equal to unity, except for the ii'th and jj'th ones, which are equal to zero, and all off-diagonal elements equal to zero except for the ij'th and ji'th ones, which are equal to unity. Such a matrix multiplying a vector simply interchanges the i'th and j'th elements of the vector. A simple permutation matrix is its own inverse: $\mathcal{PP} = \mathcal{I}$, and the transformation of any matrix, $\mathcal{M}$, with a simple permutation matrix, $\mathcal{PMP}$, simply interchanges the i'th and j'th rows and columns of the matrix.

If we find, for some matrix $\mathcal{M}$, an eigenvector whose first element is zero, we will transform the problem, using a permutation matrix $\mathcal{P}$ which interchanges the first vector element with the first non-zero vector element, as follows:

$$\mathcal{M}\vec{v} = \lambda\vec{v}, \quad \mathcal{PM}[\mathcal{PP}]\vec{v} = \lambda\mathcal{P}\vec{v}, \quad [\mathcal{PMP}][\mathcal{P}\vec{v}] = \lambda[\mathcal{P}\vec{v}],$$

and defining

$$\mathcal{PMP} \equiv \mathcal{M}', \quad \text{and} \quad \mathcal{P}\vec{v} \equiv \vec{v}', \quad \text{we have} \quad \mathcal{M}'\vec{v}' = \lambda\vec{v}'.$$

This operation can be inserted wherever necessary in the triangularization procedure.

Returning to the main problem; having an eigenvector $\vec{v}$ of the matrix $\mathcal{M}$, we construct a matrix $\mathcal{A}$

$$\mathcal{A} = \begin{pmatrix} 1 & 0 & 0 & \dots & 0 \\ v_2 & 1 & 0 & \dots & 0 \\ v_3 & 0 & 1 & \dots & 0 \\ \vdots & \vdots & \vdots & \ddots & \vdots \\ v_n & 0 & 0 & \dots & 1 \end{pmatrix},$$

whose inverse is

$$\mathcal{A}^{-1} = \begin{pmatrix} 1 & 0 & 0 & \dots & 0 \\ -v_2 & 1 & 0 & \dots & 0 \\ -v_3 & 0 & 1 & \dots & 0 \\ \vdots & \vdots & \vdots & \ddots & \vdots \\ -v_n & 0 & 0 & \dots & 1 \end{pmatrix}.$$

It should then be easy for you to show that the matrix $\mathcal{N} = \mathcal{A}^{-1}\mathcal{MA}$ has $n_{11} = \lambda_1$ and all other elements in the first column equal to zero:

$$\mathcal{N} = \begin{pmatrix} \lambda_1 & n_{12} & \dots & n_{1n} \\ 0 & n_{22} & \dots & n_{2n} \\ \vdots & \vdots & \ddots & \vdots \\ 0 & n_{n2} & \dots & n_{nn} \end{pmatrix}$$

Now, define the matrix $\mathcal{N}_r$

$$\mathcal{N}_r = \begin{pmatrix} n_{22} & n_{23} & \dots & n_{2n} \\ n_{32} & n_{33} & \dots & n_{3n} \\ \vdots & \vdots & \ddots & \vdots \\ n_{n2} & n_{n3} & \dots & n_{nn} \end{pmatrix},$$

in which the first row and column of $\mathcal{N}$ are omitted. Following exactly the same procedure as above, find an eigenvector of $\mathcal{N}_r$,

$$\mathcal{N}_r \vec{w} = \lambda_2 \vec{w},$$

where, again, the first element of $\vec{w}$ is set equal to unity. Construct the matrices $\mathcal{B}$ and $\mathcal{B}^{-1}$ similar to, but a little different from, the matrices $\mathcal{A}$ and $\mathcal{A}^{-1}$ above,

$$\mathcal{B} = \begin{pmatrix} 1 & 0 & 0 & \dots & 0 \\ 0 & 1 & 0 & \dots & 0 \\ 0 & w_2 & 1 & \dots & 0 \\ \vdots & \vdots & \vdots & \ddots & \vdots \\ 0 & w_k & 0 & \dots & 1 \end{pmatrix}, \quad \mathcal{B}^{-1} = \begin{pmatrix} 1 & 0 & 0 & \dots & 0 \\ 0 & 1 & 0 & \dots & 0 \\ 0 & -w_2 & 1 & \dots & 0 \\ \vdots & \vdots & \vdots & \ddots & \vdots \\ 0 & -w_k & 0 & \dots & 1 \end{pmatrix},$$

$$\text{where} \quad k = n - 1.$$

Now, you should convince yourself that $\mathcal{B}^{-1}\mathcal{N}\mathcal{B} = \mathcal{N}'$ is a matrix of the form

$$\mathcal{N}' = \begin{pmatrix} \lambda_1 & n'_{12} & n'_{13} & \dots & n'_{1n} \\ 0 & \lambda_2 & n'_{23} & \dots & n'_{2n} \\ 0 & 0 & n'_{33} & \dots & n'_{3n} \\ \vdots & \vdots & \vdots & \ddots & \vdots \\ 0 & 0 & n'_{n3} & \dots & n'_{nn} \end{pmatrix}.$$

It is also fairly obvious that

$$\mathcal{N}' = \mathcal{B}^{-1}\mathcal{N}\mathcal{B} = \mathcal{B}^{-1}\mathcal{A}^{-1}\mathcal{M}\mathcal{A}\mathcal{B}, \quad \text{and} \quad [\mathcal{A}\mathcal{B}]^{-1} = \mathcal{B}^{-1}\mathcal{A}^{-1}$$

We can continue this process by constructing a new matrix which omits the first two rows and columns of $\mathcal{N}'$, finding its eigenvector, then constructing the matrices $\mathcal{C}$ and $\mathcal{C}^{-1}$ in which the eigenvector is substituted below the diagonal into the third column of an identity matrix, analogous to matrices $\mathcal{A}$ and $\mathcal{B}$, above to obtain $\mathcal{C}^{-1}\mathcal{B}^{-1}\mathcal{A}^{-1}\mathcal{M}\mathcal{A}\mathcal{B}\mathcal{C}$ having only zeros below the principal diagonal in the first three columns. After $n-1$ steps of the process, we will have an upper triangular matrix

$$\mathcal{T} = [\mathcal{L}^{-1} \dots \mathcal{P} \dots \mathcal{B}^{-1}\mathcal{A}^{-1}]\mathcal{M}[\mathcal{A}\mathcal{B} \dots \mathcal{P} \dots \mathcal{L}] = \mathcal{R}^{-1}\mathcal{M}\mathcal{R},$$

$$\text{where} \quad \mathcal{R} = [\mathcal{A}\mathcal{B} \dots \mathcal{P} \dots \mathcal{L}],$$

where the $\mathcal{P}$'s arise from any necessary permutations, as discussed above.

Now that we have developed the necessary tools, we are ready to tackle the problem of simultaneous and/or consecutive first-order reactions.

1.7 Simultaneous and/or Consecutive First-Order Reactions

With matrix notation, it is easy to consider a completely general reaction scheme in which we have n species X_i, which may be reactants, intermediates, or products, interconverting according to:

$$X_i \underset{k_{ij}}{\overset{k_{ji}}{\rightleftharpoons}} X_j, \quad \text{with } i \text{ and } j = 1, n.$$

If we make the special definition

$$k_{ii} = -\sum_{j \neq i} k_{ji}$$

for the diagonal elements of a matrix $\mathcal{K}$ whose off-diagonal elements are the rate constants in the scheme, we have

$$\mathcal{K}\vec{x} = \dot{\vec{x}},$$

where the elements of $\vec{x}$ are the concentrations of the species, and the elements of $\dot{\vec{x}}$ are the various dX_i/dt's, expressing the complete set of the n simultaneous differential rate equations.

We now find the matrix $\mathcal{R}$ such that $\mathcal{R}^{-1}\mathcal{K}\mathcal{R} = \mathcal{G}$, where $\mathcal{G}$ is a lower triangular matrix[5], and proceed as follows:

$$\begin{aligned}
\dot{\vec{x}} &= \mathcal{K}\vec{x} \\
\mathcal{R}^{-1}\dot{\vec{x}} &= \mathcal{R}^{-1}\mathcal{K}\vec{x} \\
&= \mathcal{R}^{-1}\mathcal{K}\left[\mathcal{R}\mathcal{R}^{-1}\right]\vec{x} \\
\left[\mathcal{R}^{-1}\dot{\vec{x}}\right] &= \left[\mathcal{R}^{-1}\mathcal{K}\mathcal{R}\right]\left[\mathcal{R}^{-1}\vec{x}\right] \\
\dot{\vec{y}} &= \mathcal{G}\vec{y}.
\end{aligned}$$

[5] You might wish to think about how a lower triangular matrix is obtained rather than the upper triangular one which was discussed above. It really doesn't make any difference whether we use the upper or lower triangular forms, so we'll choose one or the other for notational convenience in the discussions here.

The last line gives the set of differential equations:

$$\dot{\vec{y}}_i = \sum_{j=1}^{i} g_{ij} y_j\,, \quad \text{or, written out:}$$

$$\begin{aligned}
\dot{y}_1 &= g_{11}y_1 \\
\dot{y}_2 &= g_{21}y_1 + \quad g_{22}y_2 \\
&\vdots \qquad\qquad\qquad \ddots \\
\dot{y}_n &= g_{n1}y_1 + \quad g_{n2}y_2 \quad \cdots \quad g_{nn}y_n
\end{aligned}$$

where the first equation can be integrated immediately to give $y_1 = y_1^0 \exp(g_{11}t)$. This result can be substituted into the second equation to obtain

$$\dot{y}_2 = g_{21}y_1 + g_{22}y_2 = g_{21}y_1^0 e^{g_{11}t} + g_{22}y_2,$$

which now contains only the two variables, y_2 and t. The equation is a linear first-order differential equation which may be integrated quite readily [4], to obtain

$$y_2 = \frac{g_{21}y_1^0}{g_{11} - g_{22}} e^{g_{11}t} + \left[y_2^0 - \frac{g_{21}y_1^0}{g_{11} - g_{22}} \right] e^{g_{22}t}$$

This equation and the one for y_1 can, in turn, be substituted into the third differential equation to put it into the form having only the two variables y_3 and t, which, again, is an easily integrated linear first-order differential equation. The process can be repeated for each successive differential equation, allowing all of them to be integrated.

The process will probably be made a little clearer by a specific example. Let's consider the fairly simple case:

$$X_1 \underset{k_{12}}{\overset{k_{21}}{\rightleftarrows}} X_2 \xrightarrow{k_{32}} X_3. \tag{1.25}$$

For this scheme, we have

$$\mathcal{K} = \begin{pmatrix} -k_{21} & k_{12} & 0 \\ k_{21} & -(k_{12} + k_{32}) & 0 \\ 0 & k_{32} & 0 \end{pmatrix}$$

An eigenvector of $\mathcal{K}$ must satisfy the equations

$$\begin{aligned}
-(k_{21} + \lambda)v_1 + k_{12}v_2 &= 0 \\
k_{21}v_1 - (k_{12} + k_{32} + \lambda)v_2 &= 0 \\
k_{32}v_2 - \lambda v_3 &= 0,
\end{aligned}$$

where λ can be found from expansion of the determinant

$$\begin{bmatrix} -(k_{21}+\lambda) & k_{12} & 0 \\ k_{21} & -(k_{12}+k_{32}+\lambda) & 0 \\ 0 & k_{32} & -\lambda \end{bmatrix} = 0$$

$$-(k_{21}+\lambda)(k_{12}+k_{32}+\lambda)\lambda + k_{21}k_{12}\lambda = 0$$

giving

$$\lambda = 0, \quad -(1/2)\left\{k_{21}+k_{12}+k_{32}+\left[(k_{21}+k_{12}+k_{32})^2 - 4k_{21}k_{32}\right]^{1/2}\right\},$$

$$-(1/2)\left\{k_{21}+k_{12}+k_{32}-\left[(k_{21}+k_{12}+k_{32})^2 - 4k_{21}k_{32}\right]^{1/2}\right\}.$$

Choosing λ_1 to be the last solution on the right, and choosing the normalization $v_1 = 1$, a little algebra produces

$$\vec{v} = \begin{pmatrix} 1 \\ (k_{21}+\lambda_1)/k_{12} \\ k_{32}(k_{21}+\lambda_1)/\lambda_1 k_{12} \end{pmatrix}.$$

Constructing the matrix $\mathcal{A}$ as described in our discussion of triangularization

$$\mathcal{A} = \begin{pmatrix} 1 & 0 & 0 \\ v_2 & 1 & 0 \\ v_3 & 0 & 1 \end{pmatrix},$$

we obtain, with considerable algebra

$$\mathcal{A}^{-1}K\mathcal{A} = \mathcal{N} = \begin{pmatrix} \lambda_1 & k_{12} & 0 \\ 0 & -(k_{21}+k_{12}+k_{32}+\lambda_1) & 0 \\ 0 & -k_{21}k_{32}/\lambda_1 & 0 \end{pmatrix}.$$

You can now expand the determinant for the 2×2 matrix problem involving the last two rows and columns of $\mathcal{N}$ to obtain[6]

$$\vec{w} = \begin{pmatrix} 1 \\ -1 \end{pmatrix} \quad \text{with} \quad \lambda_2 = k_{21}k_{32}/\lambda_1.$$

---

6 Note that we have $-(k_{21}+k_{12}+k_{32}+\lambda_1) = k_{21}k_{32}/\lambda_1$ from the polynomial obtained in the expansion of the above 3×3 determinant. The solution of the 2×2 problem, in fact, gives the same eigenvalue as the second of the three found by expansion of the original determinant.

Now construct

$$\mathcal{B} = \begin{pmatrix} 1 & 0 & 0 \\ 0 & 1 & 0 \\ 0 & -1 & 1 \end{pmatrix},$$

to obtain

$$\mathcal{B}^{-1}\mathcal{A}^{-1}\mathcal{K}\mathcal{A}\mathcal{B} = \mathcal{G} = \begin{pmatrix} \lambda_1 & k_{12} & 0 \\ 0 & \lambda_2 & 0 \\ 0 & 0 & 0 \end{pmatrix}$$

and

$$\mathcal{R} = \mathcal{A}\mathcal{B} = \begin{pmatrix} 1 & 0 & 0 \\ (k_{21}+\lambda_1)/k_{12} & 1 & 0 \\ (k_{12}+\lambda_2)/k_{12} & -1 & 1 \end{pmatrix}$$

$$\mathcal{R}^{-1} = \begin{pmatrix} 1 & 0 & 0 \\ -(k_{21}+\lambda_1)/k_{12} & 1 & 0 \\ 1 & 1 & 1 \end{pmatrix}.$$

With $\vec{y} = \mathcal{R}^{-1}\vec{x}$, we have

$$y_2 = y_2^0 \exp(\lambda_2 t), \quad \text{and}$$

$$y_1 = k_{12}\frac{y_2^0}{\lambda_2 - \lambda_1}\exp(\lambda_2 t) + \left[y_1^0 - \frac{k_{12}y_2^0}{\lambda_2 - \lambda_1}\right]\exp(\lambda_1 t)$$

$$\text{and} \quad y_3 = y_3^0.$$

The equation for y_3 comes from $\lambda_3 = 0$, and arises from the fact that we really have only two independent rate equations since $\dot{x}_1 + \dot{x}_2 + \dot{x}_3 = 0$ for conservation of mass.

The equations for y's can be transformed into equations for the original x's by use of $\vec{x} = \mathcal{R}\vec{y}$. Carrying out the algebra with the simplifying assumption that $x_2^0 = x_3^0 = 0$, we obtain

$$x_1 = x_1^0\left\{-\frac{k_{21}+\lambda_2}{\lambda_1 - \lambda_2}e^{\lambda_1 t} + \frac{k_{21}+\lambda_1}{\lambda_1 - \lambda_2}e^{\lambda_2 t}\right\}$$

$$x_2 = x_1^0\left\{\frac{(k_{21}+\lambda_2)(k_{21}+\lambda_1)}{k_{12}(\lambda_1 - \lambda_2)}\right\}\left\{e^{\lambda_2 t} - e^{\lambda_1 t}\right\}$$

Now, after that big job, let's look at a few numerical examples. First, consider a case for which

$$k_0 \equiv k_{21} = k_{12} = k_{32}, \quad \mathcal{K} = k_0\begin{pmatrix} -1 & 1 & 0 \\ 1 & -2 & 0 \\ 0 & 1 & 0 \end{pmatrix}.$$

Obviously, k_0 is simply a time scale factor that does not need to be carried through the operations; we will work with $(1/k_0)\mathcal{K}$ which is just the numerical matrix. Simply plugging values into the equations for λ's above, we get

$$\lambda_1 = \frac{-3-\sqrt{5}}{2} \simeq -2.62 \quad \lambda_2 = \frac{-3+\sqrt{5}}{2} \simeq -0.38,$$

and plugging into the final integrated kinetic equations:

$$x_1 = x_1^0\{0.276e^{-2.62t} + 0.724e^{-0.38t}\}$$

$$x_2 = 0.446x_1^0\{e^{-0.38t} - e^{-2.62t}\}.$$

The concentration of the intermediate, X_2, will increase with time from the start of the reaction until $0.38\exp(-0.38t) = 2.62\exp(-2.62t)$, then start to decrease. Solving this equation for t, we find that X_2 reaches its maximum concentration of $0.276x_1^0$ at $t = 0.86$. At the same time, X_1 has reached a concentration of $0.541x_1^0$. A complete plot of concentrations for this specific case is shown in Figure 1.3.

Let's now look at a case where the forward rate constant for the first step is a factor of ten smaller than the other two rate constants:

$$10k_0 \equiv k_{12} = k_{32} = 10k_{21}$$

In units of k_0, we find $\lambda_1 = -20.5$ and $\lambda_2 = -0.49$. Substitution into the kinetic equations shows that

$$x_1 = x_1^0\{0.0255e^{-20.5t} + 0.975e^{-0.49t}\}, \quad \text{and}$$

$$x_2 = 0.0497x_1^0\{e^{-0.49t} - e^{-20.5t}\}.$$

In this case, X_2 reaches its maximum concentration of $0.045x_1^0$ at $t = 0.18$, at which time the concentration of X_1 is $0.87x_1^0$. Note particularly that the exponential term in λ_1 in the expression for x_1 is equal to only $0.0255x_1^0$ at $t = 0$ and decreases in magnitude relative to the term in λ_2 thereafter. Right at the start of the reaction, 97.5% of x_1 is given by the term in $\exp(-0.49t)$. By the time that X_2 reaches its maximum concentration, the term in λ_1 in the expression for x_1 is only $0.0013x_1^0$, while that in λ_2 is equal to $0.87x_1^0$. Thus, any time after approximately the first 10% reaction of X_1, kinetic behavior of X_1 is essentially completely governed by the term in $exp(\lambda_2 t)$, and we would

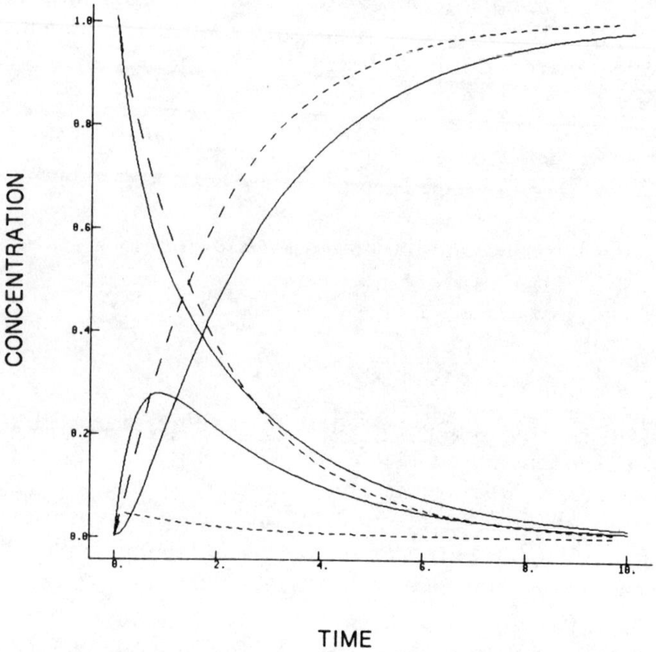

Figure 1.3: *Kinetic behavior of the reaction system shown in Eq. (1.25). Solid lines are for* $k_{21} = k_{12} = k_{32} = 1$. *Dashed lines are for* $k_{12} = k_{32} = 10$; $k_{21} = 1$.

see simple *first-order* decay. The behavior of concentrations is shown for this case also in Figure 1.3.

A little thought should now convince you that as either k_{12} or k_{32} increases relative to k_{21}:

- the time at which the intermediate reaches its maximum concentration decreases;
- the maximum concentration of the intermediate relative to x_1^0 decreases;
- the kinetic behavior of X_1 approaches more closely to simple first-order decay with apparent rate constant equal to the λ of smaller magnitude.

This is the true justification for the Bodenstein approximation.

1.8 More Complex Simultaneous and/or Consecutive Reactions

Since the advent of fast, high capacity computers, there has been nothing to hinder the treatment of any kinetic scheme that can be dreamed up. Once a reaction scheme is written in terms of supposed elementary steps, the differential equations governing all the concentrations can be written. These equations can always be integrated numerically to any degree of accuracy desired either by choosing extremely short time intervals, δt, for the simplest integrations:

$$x_{i(t+\delta t)} = x_{i(t)} + \left[\frac{dx_i}{dt}\right]_t \cdot \delta t,$$

or by using slightly greater time intervals in the longer equations

$$x_{i(t+\delta t)} = x_{i(t)} + \left\{\left[\frac{dx_i}{dt}\right]_t + \frac{1}{2}\left[\frac{d^2x_i}{dt^2}\right]_t \cdot \delta t\right\} \cdot \delta t,$$

or

$$x_{i(t+\delta t)} = x_{i(t)} + \left[\frac{dx_i}{dt}\right]_t \cdot \delta t + \frac{1}{2}\left[\frac{d^2x_i}{dt^2}\right]_t \cdot (\delta t)^2 + \ldots + \frac{1}{n!}\left[\frac{d^nx_i}{dt^n}\right]_t \cdot (\delta t)^n$$

where all of the higher order derivatives are available from the rate equations.

If we have a set of experimentally observed concentrations at various times, we can compare these with those calculated by the numerical integrations using some set of rate constants, which could be varied to produce a minimum in the sums of squares of differences between observed and calculated concentrations. Obviously, the more complex the reaction scheme and the more rate constants involved, the more tricky becomes the problem of deciding whether or not either the scheme or the set of rate constants uniquely describe the experimental data.

1.9 REFERENCES

All of the references to books in this and following chapters should be considered as strongly recommended supplementary reading for any course in physical organic chemistry.

[1]. J. Hine, **Physical Organic Chemistry**, McGraw-Hill Book Co., Inc., New York, New York, **1962**, Chapter 6. There is an unusually

complete listing of data and references concerning the kinetics of first- and second-order nucleophilic substitution reactions.

[2]. A. R. Fersht and W. P. Jencks, **J. Amer. Chem. Soc., 92**, 5432 (**1970**). This paper contains a discussion of a practical method for evaluation of the rate constants for a fairly simple two-step reaction.

[3]. J. H. Wilkinson, **The Algebraic Eigenvalue Problem**, Oxford University Press, Oxford, England, **1965**.

[4]. M. L. Boas, **Mathematical Methods in the Physical Sciences**, John Wiley & Sons, Inc., New York, New York, **1966**, pages 328ff.

[5]. A. A. Frost and R. G. Pearson, **Kinetics and Mechanisms**, 2nd Edition, John Wiley & Sons, Inc., New York, New York, **1961**. Chapters 2, 3, and 8 contain a thorough treatment of the integrations of reasonably simple rate laws.

[6]. S. W. Benson, **The Foundations of Chemical Kinetics**, McGraw-Hill Book Co., Inc., New York, New York, **1960**.

[7]. B. K. Carpenter, **Determination of Organic Reaction Mechanisms**, John Wiley & Sons, Inc., New York, New York, **1984**, Chapter 4.

1.10 PROBLEMS

Problem 1.1

For a first-order reaction, show that the time required for the concentration of reactant to decrease half-way from its value at some time, t, to its final value at $t = \infty$ is the same regardless of the time, t, that we choose.

Hint: Use the logarithmic form of Eq. (1.11). Let $\beta = \beta_i$ *at time* t_i, *and* $(\beta_j - \beta_\infty) = (1/2)(\beta_i - \beta_\infty)$ *at* t_j, *where* $t_j - t_i = t_{1/2}$.

Problem 1.2

Note that a pseudo-first-order rate constant, k_ψ, is equal to the forward rate constant **plus** the reverse rate constant. Yet, it is obvious that any reverse reaction must decrease the rate of conversion of reactants into products. Resolve the apparent paradox.

Hint: Sketch some concentration vs. time curves for a constant value of a forward rate constant and several increasing values of a reverse rate constant. Note particularly what happens to the concentration of product at equilibrium.

Problem 1.3

Derive the integrated rate law for a reaction where two reactants, A and B, reversibly produce C under the conditions where $[C]_0 = 0$, and $[A]_0 = [B]_0$.

Hint: The form of the differential equation is the same as that leading to Eq. (1.18).

Problem 1.4

The reaction of methyl perchlorate to produce methanol in aqueous solution at 0° C was followed by placing the reaction solution in a conductivity cell and measuring the resistance of the cell as a function of time. From the data given, calculate the first-order rate constant for the reaction.

Hint: First, think about how you will get something that is a linear function of some concentrations.

Time(sec)	Resistance(ohms)	Time(sec)	Resistance(ohms)
0	4.0×10^5	1250	2.521
60	39.63	1500	2.220
100	24.01	2000	1.853
200	12.31	2500	1.642
500	5.296	3000	1.508
1000	2.977	15000	1.171

Problem 1.5

The reaction of methyl iodide with bromide ion in dimethylformamide solution at 25° C was studied by measuring the concentration of iodide ion as a function of time. The initial concentrations of both bromide ion and of methyl iodide were $3.0 \times 10^{-3} M$. From the data given, calculate the forward and reverse rate constants.

Hint: This involves a very straight-forward use of Eq. (1.18).

Time(min)	$[I^-] \times 10^3 (M)$	Time(min)	$[I^-] \times 10^3 (M)$
0	0	4.00	1.104
0.50	0.209	6.00	1.373
1.00	0.391	8.00	1.554
1.50	0.549	10.00	1.679
2.00	0.688	20.00	1.932
3.00	0.929	200.00	2.000

Problem 1.6

The reactions of aryldiazonium ions with hydroxide ion in aqueous solution involve the steps

$$ArN_2^+ + OH^- \underset{k_{-1}}{\overset{k_1}{\rightleftharpoons}} syn\text{-}ArN_2OH$$

$$syn\text{-}ArN_2OH + OH^- \underset{k_{-2}}{\overset{k_2}{\rightleftharpoons}} syn\text{-}ArN_2O^- + H_2O.$$

Such reactions are studied in buffered solution so that the concentration of hydroxide ion remains constant throughout the reaction, even though it is not present in large excess. The second step of the reaction is a very rapid one, such that the ratio of concentrations of syn-ArN_2OH and syn-ArN_2O^- is the equilibrium ratio throughout the reaction. Syn-ArN_2OH is **not** a reactive intermediate; its concentration is appreciable except at very high concentrations of hydroxide ion. Derive an expression that shows how observed pseudo-first-order rate constants vary with varying hydroxide ion concentrations.

Hint: Note that the dependence of the ratio of concentrations of syn-ArN_2OH and syn-ArN_2^- on hydroxide ion concentration affects the observed rate constant only through its effect on the reverse rate, which depends on the concentration of syn-ArN_2OH only. Write the expression for $d[ArN_2^+]/dt$, then use the stoichiometry of the reaction along with the equilibrium ratio of the two products to put the expression in terms of a single variable concentration.

Problem 1.7

The hydrolysis reactions of alkyl halides in aqueous solution actually involve reaction of the alkyl halide with both water and hydroxide ion

$$RX + H_2O \xrightarrow{k_{water}} ROH + H^+ + X^-, \quad \text{and}$$

$$RX + OH^- \xrightarrow{k_{OH}} ROH + X^-.$$

Describe how you would determine both k_{water} and k_{OH}.

Hint: Consider how pseudo-first-order rate constants would depend on pH.

Problem 1.8

Most techniques for the study of very fast reactions involve perturbing a reaction initially at equilibrium by a rapid change in temperature, pressure, or electric field. The concentrations of reactants and products must then 'relax' to the new equilibrium concentrations. Usually, the changes in equilibrium concentrations caused by the perturbation are quite small. Show that even for a reaction such as

$$A + B + C \underset{k_r}{\overset{k_f}{\rightleftharpoons}} D + E + F$$

the kinetics of the process will be first-order.

Hint: We can let the concentrations before the perturbation be $[A]_0$, $[B]_0, \ldots, [F]_0$, *and those at equilibrium after the perturbation be* $[A]_\infty$, *etc. Set* $a = [A] - [A]_\infty$, *so that* $da = d[A]$, *and similarly for the other species, then note that all of these variables are quite small.*

How could you determine the order of the reaction with respect to each of the reactants?

Problem 1.9

Derive the integrated kinetic equation for the reaction scheme:

$$A + B \overset{K_1}{\rightleftharpoons} C$$

$$C \xrightarrow{k_f} \textit{Product} + B,$$

where the first step is a very rapid equilibrium. What would happen if B were present in extremely small concentration relative to detectable concentrations of A and K_1 were very large?

Hint: Watch out for what you mean by $[A]_0$. *It is usually defined as the number of moles of A added to the solution, divided by the volume of the solution.*

Problem 1.10

Consider the reaction scheme:

$$A \underset{k_{-1}}{\overset{k_1}{\rightleftharpoons}} B$$

$$B \underset{k_{-2}}{\overset{k_2}{\rightleftharpoons}} C$$

$$B \underset{k_1}{\overset{k_{-1}}{\rightleftharpoons}} A' \quad \text{an enantiomer of } A$$

$$C \underset{k_{-3}}{\overset{k_3}{\rightleftharpoons}} D + E$$

$$C \xrightarrow{k_4} P$$

$$D \xrightarrow{k_5} P$$

where B, C, and D are all reactive intermediates whose concentrations never become comparable to those of the other species. Derive the integrated kinetic expression for the scheme with the assumption that E is present in large excess.

Hint: Set $d[B]/dt$, $d[C]/dt$, *and* $d[D]/dt$ *all equal to zero, then solve the simultaneous algebraic equations for* $[B]$, $[C]$, *and* $[D]$ *in terms of* $[A]$, $[A']$, *and* $[P]$.

Problem 1.11

The *trace* of a matrix is the sum of the diagonal elements. That is,

$$Tr(\mathcal{M}) = \sum_{i=1}^{n} m_{ii}.$$

Show that the trace of a matrix is left unchanged by a unitary transformation; that is,

$$Tr(\mathcal{M}) = Tr(\mathcal{M}') \quad \text{where} \quad \mathcal{M}' = \mathcal{U}^{\dagger}\mathcal{M}\mathcal{U}.$$

Hint: Write out the equation for m'_{ii}.

Problem 1.12

Show that the largest element, λ_{max}, of a diagonal matrix Λ is greater than or equal to the largest diagonal element of the matrix M where $\mathcal{U}^\dagger \mathcal{M} \mathcal{U} = \Lambda$.

Hint: Consider the reverse transformation, $\mathcal{U}\Lambda\mathcal{U}^\dagger$, *and write the equation for* m_{ii}.

Problem 1.13

Prove the following statement: If a matrix $\mathcal{M}$ has two eigenvectors, $\vec{v}_1$ and $\vec{v}_2$, with the same eigenvalue (i. e. $\lambda_1 = \lambda_2$), then any linear combination of $\vec{v}_1$ and $\vec{v}_2$, $a\vec{v}_1 + b\vec{v}_2$, is also an eigenvector of $\mathcal{M}$ with the same eigenvalue.

Hint: Use the distributive law for matrix multiplication and the commutative law for constants and matrices.

Problem 1.14

Suppose that an optically active compound can racemize, rearrange with inversion, or rearrange with retention, through optically active intermediates as in the scheme:

$$A \underset{k_{-1}}{\overset{k_1}{\rightleftharpoons}} B$$

$$A' \underset{k_{-1}}{\overset{k_1}{\rightleftharpoons}} B'$$

$$B \underset{k_3}{\overset{k_3}{\rightleftharpoons}} B'$$

$$B \xrightarrow{k_2} C$$

$$B' \xrightarrow{k_2} C'$$

where the primed and unprimed species are enantiomers. Note the equivalences of the rate constants for reactions of enantiomers. Using matrix algebra, solve the kinetic equations to give integrated rate equations for each of the species involved.

Hint: Start with the concentration vector

$$\vec{x} = \begin{pmatrix} [A] \\ [B] \\ [C] \\ [C'] \\ [B'] \\ [A'] \end{pmatrix}$$

and carry out a unitary transformation on the appropriate $\mathcal{K}$ matrix with

$$\mathcal{U} = \begin{pmatrix} 1/\sqrt{2} & 0 & 0 & 0 & 0 & 1/\sqrt{2} \\ 0 & 1/\sqrt{2} & 0 & 0 & 1/\sqrt{2} & 0 \\ 0 & 0 & 1/\sqrt{2} & 1/\sqrt{2} & 0 & 0 \\ 0 & 0 & -1/\sqrt{2} & 1/\sqrt{2} & 0 & 0 \\ 0 & -1/\sqrt{2} & 0 & 0 & 1/\sqrt{2} & 0 \\ -1/\sqrt{2} & 0 & 0 & 0 & 0 & 1/\sqrt{2} \end{pmatrix}.$$

This will produce a new matrix, $\mathcal{K}'$, which is in what is called block form, having two 3×3 blocks.

$$\mathcal{K}' = \begin{pmatrix} \begin{pmatrix} x & x & x \\ x & x & x \\ x & x & x \end{pmatrix} & \begin{pmatrix} 0 & 0 & 0 \\ 0 & 0 & 0 \\ 0 & 0 & 0 \end{pmatrix} \\ \begin{pmatrix} 0 & 0 & 0 \\ 0 & 0 & 0 \\ 0 & 0 & 0 \end{pmatrix} & \begin{pmatrix} x & x & x \\ x & x & x \\ x & x & x \end{pmatrix} \end{pmatrix},$$

where the x's represent possibly non-zero elements. Most importantly, blocked matrices may be multiplied by similarly blocked matrices to produce a matrix having the same block form. This means that the above problem can be treated as two 3×3 matrix problems. Later on, you will see that the symmetry of some quantum mechanical problems allows this same kind of blocking-out to simplify the matrix problem.

Problem 1.15

A frequently encountered kinetic situation is one in which an initial reactant is in rapid equilibrium with another species which does not react directly to produce product:

$$A \underset{k_{-1}}{\overset{k_1}{\rightleftharpoons}} B, \qquad A \xrightarrow{k_2} Product$$

with both k_1, and $k_{-1} >> k_2$.

Under this condition, $[B]/[A] = K_1$ throughout the reaction and the stoichiometric relationship may be written:

$$[A]_0 - (1 + K_1)[A] = [Product],$$

where $[A]_0$ is the amount of A added to the solution, and $[A]_0/(1 + K_1)$ is the true concentration of $[A]$ at $t = 0$ (Actually, $[A]_0 = [A]_{t=0} + [B]_{t=0}$).

Now consider the two possible step by step developments:

(a)

$$\frac{d[A]}{dt} = -(k_1 + k_2)[A] + k_{-1}[B]$$
$$= -(k_1 + k_2)[A] + K_1 k_{-1}[A]$$
$$= -k_2[A] + (K_1 k_{-1} - k_1)[A],$$

but $K_1 = k_1/k_{-1}$, so

$$\frac{d[A]}{dt} = -k_2[A], \quad \text{and} \quad [A] = [A]_0(1 + K_1)e^{-k_2 t}$$

(b) Starting with

$$[A]_0 - (1 + K_1)[A] = [Product]$$
$$\frac{d[Product]}{dt} = -(1 + K_1)\frac{d[A]}{dt}$$
$$= k_2[A], \quad \text{therefore}$$
$$\frac{d[A]}{dt} = -\frac{k_2}{1 + K_1}[A]$$
$$\text{so} \quad [A] = [A]_0(1 + K_1)\exp\left(\frac{-k_2}{1 + K_1}t\right).$$

The second development, (b), is correct. Where is the error in the first one?

Hint: Consider what happens to $d[B]/dt$ when substitutions are made. Note, also, the careful definition of $[A]_0$.

Problem 1.16

Consider the general reaction scheme given at the beginning of Section 1.7 in which there are n species, X_i, all interconverting and reaching concentrations of $[X_i]_\infty$ at $t = \infty$.

a. How many **independent** rate constants are in the scheme?
Hint: Note that the equilibrium constant, K_{ji}, for conversion of X_i into X_j is equal to $k_{ji}/k_{ij} = [X_j]_\infty/[X_i]_\infty$. Then note that, for example, $K_{jk} = K_{ji}/K_{ki}$.

b. Construct the two diagonal matrices, $\mathcal{S}$ with $s_{ii} = \sqrt{[X_i]_\infty}$, and $\mathcal{S}^{-1}$ with $s_{ii}^{-1} = 1/\sqrt{[X_i]_\infty}$. Perform a similarity transformation on the rate constant matrix, $\mathcal{K}$, according to $\mathcal{H} = \mathcal{S}^{-1}\mathcal{K}\mathcal{S}$. Show that $\mathcal{H}$ is Hermitian. How might this be used to solve the kinetic problem?
Hint: Think about something similar to the development on pg. 23. If you get stuck, see ref. [7], pg. 52–56.

CHAPTER 2
Kinetics: Characterization of Transition States

2.1 Introduction

The determination of the empirical rate law for a reaction gives us some information concerning the existence of intermediates and the identities of elementary steps in an overall reaction. The next phase in our investigation of the mechanism of a reaction is usually an attempt to understand why the elementary steps proceed at whatever rates are observed, and to enquire how we might control the rates and/or directions of the reactions.

Not all elementary steps of reactions proceed at the same rates, and we might start this second phase of investigation by trying to understand exactly how an elementary step occurs. Obviously, in some manner the various reactant species come together to form some sort of aggregate that is somehow transformed into the products. We would like to understand the nature of this aggregate, how it is formed, and how it is transformed into products.

Before proceeding, it might be well to review a little of the symbolism and formalism of the thermodynamics of reactions.

2.2 Review of Some Thermodynamic Quantities

This section is intended primarily as a memory refresher; Hammett [1] presents a particularly pertinent and concise discussion of thermodynamics of solutions for those needing more detail than given here.

The standard free energies, enthalpies, and entropies which are most frequently of direct interest for reactions in solution are actually the partial molar free energies, enthalpies, and entropies. We will be particularly interested in the partial molar free energy which is often called the *chemical potential* of various solutes.

The partial molar free energy of a species is given by:

$$\mu_i \equiv G_i \equiv \left(\frac{\partial \mathcal{G}}{\partial m_i}\right)_{T,P,m_j},$$

where $\mathcal{G}$ is the Gibbs free energy of the total system, and m_i is the molarity of species i. The quantity μ_i is the chemical potential of species i. The symbol G_i will be used throughout this book to also mean chemical potential or partial molar free energy, and sometimes, when rigor is relaxed, we will simply call μ_i or G_i the free energy of i. The subscripts on the partial derivative emphasize that temperature, pressure, and concentrations of all species except the i'th one are held constant. We choose some value for all of these parameters, including the concentration of species i, to define what we call a *standard state*, and write:

$$\mu_i^\circ = \mu_i - RT \ln m_i$$

as the *standard chemical potential* of species i. For any state other than the standard state, we write

$$\mu_i = \mu_i^\circ + RT \ln \alpha_i = \mu_i^\circ + RT \ln m_i + RT \ln \gamma_i,$$

which defines the *activity*, α_i, and the *activity coefficient*, γ_i, of the species.

In elementary physical chemistry, it is customary to define the standard state of a solute species as '*one molar infinite dilution*', by setting

$$\mu_i^\circ = \lim_{m_i \to 0} (\mu_i - RT \ln m_i),$$

and the standard state of the solvent as the pure solvent. **We will usually ignore this convention.**

The criterion for thermodynamic equilibrium in any system is that the free energy of the system, $\mathcal{G}$, be at a minimum with respect to changes in the concentrations. For a reaction

$$aA + bB \overset{K}{\rightleftharpoons} cC + dD$$

this requires that

$$\frac{d\mathcal{G}}{dm_k} = 0 = \sum_{i=1}^{4} \frac{\partial \mathcal{G}}{\partial m_i} \frac{dm_i}{dm_k} = \sum_{i=1}^{4} \mu_i \frac{dm_i}{dm_k},$$

and, therefore

$$\begin{aligned} 0 &= d\mu_D + c\mu_C - b\mu_B - a\mu_A \\ &= d\mu_D^\circ + c\mu_C^\circ - b\mu_B^\circ - a\mu_A^\circ + dRT \ln \alpha_D \\ &\quad + cRT \ln \alpha_C - bRT \ln \alpha_B - aRT \ln \alpha_A, \\ &= \Delta\mu^\circ + RT \ln \left(\frac{\alpha_D^d \alpha_C^c}{\alpha_B^b \alpha_A^a} \right) \end{aligned}$$

where the last equivalence defines

$$\Delta\mu^\circ \equiv d\mu_D^\circ + c\mu_C^\circ - b\mu_B^\circ - a\mu_A^\circ.$$

Since $\Delta\mu^\circ$ is a constant, we may define a new constant, K, such that

$$\Delta\mu^\circ \equiv \Delta G^\circ = -RT\ln\left(\frac{\alpha_D^d\alpha_C^c}{\alpha_B^b\alpha_A^a}\right) \equiv -RT\ln K.$$

Frequently we will need to consider a reaction in several different solvents. This poses the problem of a consistent use of standard state definitions and an understanding of activity coefficients.

For any single species, A, in two different phases or states at equilibrium, $\mu_A(1) = \mu_A(2)$, or

$$\mu_A^\circ + RT\ln\alpha_A(1) = \mu_A^\circ + RT\ln\alpha_A(2),$$

therefore

$$\alpha_A(1) = \alpha_A(2) \quad \text{and} \quad m_A(1)\gamma_A^{(1)} = m_A(2)\gamma_A^{(2)}.$$

As an eye-opening example of the meaning of activity coefficients, suppose we have one mole of NaCl dissolved in one liter of water, which we will define as our standard state. We now place this solution in a separatory funnel with some heptane solvent and shake to establish equilibrium. Certainly, the concentration of NaCl in the heptane layer will be extremely small; suppose it is $10^{-8}M$. What is the activity coefficient of NaCl in the heptane solution? We have $\alpha_{NaCl}(water) = \alpha_{NaCl}(heptane)$, and since we have chosen the water solution to be our standard state, $m_{NaCl}(water) = m_{NaCl}(heptane)\gamma$, and putting in the values for concentrations at equilibrium, we see that $\gamma = 10^{+8}$ for the activity coefficient of NaCl in heptane.

If we have some solute species at the same concentration in two different states or phases, the change in free energy to transfer δm_A moles of the species from one phase or state to the other is

$$\delta\mathcal{G} = \{\mu_A(1) - \mu_A(2)\}\delta m_A = \left\{RT\ln\frac{\gamma_A^{(1)}}{\gamma_A^{(2)}}\right\}\delta m_A$$

or

$$\frac{\delta\mathcal{G}}{\delta m_A} \equiv \Delta G_{transfer}^A = -RT\ln\frac{\gamma_A^{(2)}}{\gamma_A^{(1)}},$$

where the last equation defines the free energy of transfer of the species between the two states. In the case where the state 1 is the standard state, we write

$$\Delta G_t^\circ = -RT\ln\gamma_A^{(2)},$$

where we have now shortened the subscript and added the "standard" superscript to G.

2.3 Analogies Between Kinetics and Thermodynamics

Since the very beginnings of the investigations of rates of reactions, analogies have been drawn between rates and equilibria.

It is always necessary to keep in mind that there are some important features of reactions, even those consisting of a single elementary step, which are completely different for rates and equilibria. It is not at all unusual, for example, for the less stable of two isomers to be formed at the greater rate as in reactions of some allylic cations [2]:

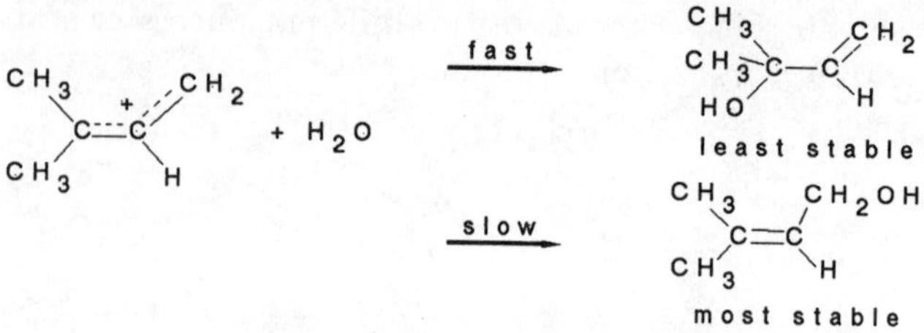

or for the less stable of two isomers to be an intermediate in the formation of the more stable one, as in the reactions of aryldiazonium ions with cyanide ion [3]:

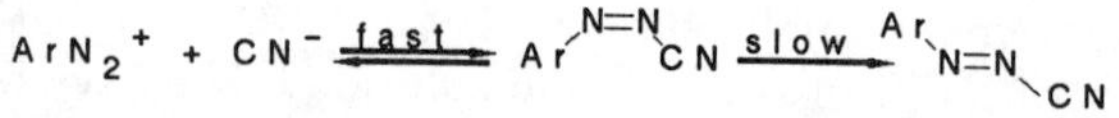

More frequently, however, there are close analogies between rate constants and equilibrium constants for elementary steps of reactions.

The kinetics of an overall reaction are not related to the stoichiometry, but we have seen that kinetics and stoichiometry *are* directly related for an elementary step. There is, then, a sort of analogy between the amount of product formed at equilibrium and the rate of the formation of products for such steps.

The effect of temperature on rates of reactions is found empirically to produce a linear relationship between the change in logarithm of the rate constant and the change in the reciprocal of the absolute temperature; $\Delta \ln k \propto \Delta(1/T)$. The relationship is of the same form as the differential of the thermodynamic equation for the effect of temperature on equilibrium constants:

$$\frac{d \ln K}{d(1/T)} = -T^2 \frac{d \ln K}{dT} = \frac{-\Delta H^\circ}{R}. \tag{2.1}$$

This similarity suggested writing

$$\frac{d \ln k}{d(1/T)} = \frac{-E_a}{R}, \tag{2.2}$$

where E_a is called the activation energy for the reaction. In integrated form, this can be written as

$$k = Ze^{-E_a/RT}, \tag{2.3}$$

which is known as the Arhennius equation. This relationship was rationalized by supposing that the reactants come together to form an *activated complex* having a definite structure and some minimum of energy. The pre-exponential term, Z, was supposed to be a measure of the probability of the complex having the proper orientations to lead to reaction, and E_a was supposed to be the minimum energy which the complex must posess in order to break bonds, etc., to form product.

In the case of ionic reactions

$$A^{+n} + B^{-m} \xrightarrow{k} P^{(n-m)+},$$

it is frequently found that the dependence of the rate of the reaction on ionic strength of the solution, or on change of solvent, is analogous to the effect on the equilibrium concentration of product, $[P]_{eq}$, as expressed in terms of activity coefficients [4]:

$$[P]_{eq} = K[A][B]\frac{\gamma_A \gamma_B}{\gamma_P}, \quad \text{and} \quad Rate = k[A][B]\frac{\gamma_A \gamma_B}{\gamma_P}$$

Again, the relationship suggests the formation of some type of activated complex having a total charge derived from the sum of the charges of the reactants.

There are also many cases where changes in structures of reactants in a series of similar reactions lead to similar changes in rate and equilibrium constants for the reactions. In fact, one frequently observes linear relationships between $\log k$'s and $\log K$'s in such series. In the reactions of *meta*- and *para*-substituted benzenediazonium ions with cyanide ion, for example, all of the substituents which increase the overall equilibrium constant of the reaction also increase the rate of the reaction and there is a reasonably precise linear relationship between the logarithms of the rate and equilibrium constants. Such features may be rationalized by postulating the formation of an activated complex having at least some of the features of the products of the reactions.

2.4 The Concept of Transition States

These early ideas of activated complexes culminated in *transition state theory* or the *theory of absolute reaction rates*. In a later chapter, we'll see that this theory can be developed from first principles, but for now, let's just look at the basic concept and operative conclusions. The central concepts of transition state theory are that

- some critical structure, called the transition state, must be attained in passing from reactants to products, or *vice-versa*, of each elementary step of a reaction,
- this critical state is in *thermodynamic* equilibrium with reactants, and,
- once formed, this critical structure decomposes to products with a *universal rate constant* equal to $\kappa T/h$, where κ is Boltzmann's constant, T is the absolute temperature, and h is Planck's constant.

For an elementary step of reaction, we write

$$aA + bB \rightleftharpoons TS^{\ddagger} \rightleftharpoons P,$$

where $TS^{\ddagger}$ is the transition state. The equilibrium postulate of transition state theory states that

$$K^{\ddagger} = \frac{\alpha_{\ddagger}}{\alpha_A^a \alpha_B^b} = \frac{[TS^{\ddagger}]}{[A]^a[B]^b} \frac{\gamma_{\ddagger}}{\gamma_A^a \gamma_B^b}, \tag{2.4}$$

where the α's are thermodynamic activities and the γ's are activity coefficients.

For a reversible reaction, transition states are being formed from both reactants and products, and are being transformed into both reactants and products. At equilibrium, the number of transformations from reactants to transition state to products must exactly equal those in the opposite direction. In general, not necessarily at equilibrium, the net fraction of conversions in the forward direction is given by (see Prob. 2.15)

$$1 - \frac{\alpha_P}{\alpha_A^a \alpha_B^b} \frac{1}{K_{eq}}$$

in which K_{eq} is the equilibrium constant for the conversion of A and B into P. The postulate concerning the rate of decomposition of transition states into products corrected by this net fraction of transition states moving from reactants to products means that:

$$\frac{d[P]}{dt} = \left(\frac{\kappa T}{h}\right)\left(1 - \frac{\alpha_P}{\alpha_A^a \alpha_B^b} \frac{1}{K_{eq}}\right)[TS^{\ddagger}]. \tag{2.5}$$

Solving Eq. (2.4) for $[TS^{\ddagger}]$ and substituting the result into Eq. (2.5) gives

$$\frac{d[P]}{dt} = \left(\frac{\kappa T}{h}\right) K^{\ddagger} \left\{[A]^a[B]^b \frac{\gamma_A^a \gamma_B^b}{\gamma_{\ddagger}} - \frac{1}{K_{eq}}[P]\frac{\gamma_P}{\gamma_{\ddagger}}\right\}. \tag{2.6}$$

Notice particularly that we have solved for *concentration* of the transition state, since the rate is always in terms of concentration, and that the activity coefficient of the transition state appears in the denominator of the right-hand-side of Eq. (2.6). If we compare Eq. (2.6) with the rate equation in terms of empirical rate constants k_f and k_r:

$$\frac{d[P]}{dt} = k_f[A]^a[B]^b - k_r[P]$$
$$= \left(\frac{\kappa T}{h}\right) K^{\ddagger}[A]^a[B]^b \frac{\gamma_A^a \gamma_B^b}{\gamma_{\ddagger}} - \left(\frac{\kappa T}{h}\right) \frac{K^{\ddagger}}{K_{eq}}[P]\frac{\gamma_P}{\gamma_{\ddagger}},$$

we see that

$$k_f = \left(\frac{\kappa T}{h}\right) K^{\ddagger} \frac{\gamma_A^a \gamma_B^b}{\gamma_{\ddagger}} \equiv k_f^{\circ} \frac{\gamma_A^a \gamma_B^b}{\gamma_{\ddagger}}, \tag{2.7}$$

and similarly,

$$k_r = \left(\frac{\kappa T}{h}\right) \frac{K^{\ddagger}}{K_{eq}} \frac{\gamma_P}{\gamma_{\ddagger}} \equiv k_r^{\circ} \frac{\gamma_P}{\gamma_{\ddagger}},$$

where the last equivalence in each of these two equations defines the k°'s as the rate constants measured for the standard state[1].

Finally, note that $k_f^{\circ}/k_r^{\circ} = K_{eq}$. Transition state theory requires that the transition state for the reverse reaction is the same as that for the forward reaction of every elementary step of a reaction. This trivial sounding statement is not trivial at all, and has some interesting consequences. The general statement is called the *principle of microscopic reversability* or the *principle of detailed balancing*. With transition state theory, it is quite easy to show that any reaction that proceeds in a forward direction through any number of particular steps must proceed in the reverse direction through the same set of steps [5]; it's simply a matter that the transition states which are of lowest energy for the forward direction are the same as those for the reverse direction.

Since we have postulated thermodynamic equilibrium between reactants and transition state, we can define any of the thermodynamic quantities, such as standard free energy, enthalpy, entropy, etc., for conversion of reactants into transition states by the usual relationships involving equilibrium constants. We solve Eq. (2.7) for $K^{\ddagger}$, and write

$$\Delta G^{\circ\ddagger} = -RT \ln K^{\ddagger} = -RT \ln k_f^{\circ} + RT \ln\left(\frac{\kappa T}{h}\right), \tag{2.8}$$

1 Although the o superscripts are not used on the equilibrium constants $K^{\ddagger}$ and K_{eq}, these are also for the standard state.

$$\Delta H^{\circ\ddagger} = -RT - \frac{Rd(\ln k_f^{\circ})}{d(1/T)}, \tag{2.9}$$

$$\Delta S^{\circ\ddagger} = \frac{\Delta H^{\circ\ddagger} - \Delta G^{\circ\ddagger}}{T}, \quad \text{and} \tag{2.10}$$

$$\Delta C_p^{\circ\ddagger} = \frac{d\Delta H^{\circ\ddagger}}{dT}. \tag{2.11}$$

We can also see from Eqs. (2.9) and (2.2) that the empirical activation energy E_a and *pre-exponential factor*, Z are related to the thermodynamic quantities by:

$$E_a = \Delta H^{\circ\ddagger} + RT, \quad \text{and}$$
$$\ln Z = \frac{\Delta S^{\circ\ddagger}}{R} + \ln\left(\frac{\kappa T}{h}\right) + 1. \tag{2.12}$$

Transition state theory transforms the question of what governs the rate of any reaction into the question of what governs the free energy change for conversion of reactants into transition state. The transition state for a reaction has all[2] the attributes of a normal species except that it decomposes into products with a rate essentially equal to a vibrational frequency[3]. Since we know a good deal about free energies and enthalpies of molecules, such as contributions of bond strengths, or about activity coefficients of solute species, such as contributions of ionic strength, we can infer what transition states *look like* from the determinations of the thermodynamic quantities.

2.5 Rapid Equilibria Among Species or States

For reactions proceeding by way of more than one elementary step, there will be a transition state for each step, and the free energies, enthalpies, and entropies of activation for each step provide some information about each of the transition states. In the special cases of such

[2] Well...nearly all. They do have one rather peculiar degree of freedom called the *reaction coordinate.*

[3] At 25°C, the quantity $\kappa T/h$ is approximately $6.3 \times 10^{12} sec^{-1}$. This, in fact, corresponds to an imaginary frequency for *vibration* along the *reaction coordinate.* All of these things will be discussed in a later chapter.

reactions in which all intermediates are very *unstable*[4], and in which one particular step is rate determining, the overall rate will depend only on the free energy difference between the transition state of the rate determining step and the original reactants.

Consider, for example, the two step process:

$$A \underset{k_{-1}}{\overset{k_1}{\rightleftharpoons}} B \xrightarrow{k_2} C,$$

where the reverse of the second step is eliminated for simplicity and we assume that all quantities are for the standard state. As we saw in Chapter 1, if B is a very reactive intermediate allowing the application of the Bodenstein approximation, the pseudo-first-order rate constant observed for the reaction will be given by

$$k_\psi = \frac{k_1 k_2}{k_{-1} + k_2}.$$

Suppose that $k_{-1} >> k_2$ so that the second step is rate determining. Then, $k_\psi \approx (k_1/k_{-1})k_2$, and, since the equilibrium constant for the first step, K_1, is equal to (k_1/k_{-1}), we have

$$k_\psi = K_1 k_2 = \left(\frac{\kappa T}{h}\right) K_1 K_2^\ddagger,$$

where $K_2^\ddagger$ is the equilibrium constant for conversion of B into the transition state for the second step. But $K_1 K_2^\ddagger$ is just the equilibrium constant for conversion of A into the transition state for the second step, and

$$-RT\ln(K_1 K_2^\ddagger) = -RT\ln K_1 - RT\ln K_2^\ddagger = \Delta G_1^\circ + \Delta G_2^{\circ\ddagger}.$$

The sum of the standard free energies is, of course, just the standard free energy difference between the initial reactant A and the transition state for the rate determining step, which, in this case, is the second step.

It is sometimes easy to fool yourself into such errors as stating that any change in reaction conditions or structure of reactants that stabilize an intermediate such as B relative to reactant A should *make*

---

4 It should be obvious from the discussion in Chapter 1 that an *unstable* intermediate is one whose concentration is negligible relative to those of the reactants and products.

the reaction easier. Nothing affects the rate of such a reaction, however, unless it affects the free energy difference between the initial reactant A and the transition state for the rate determining step.

As a more specific case, let's look at the mechanism of Scheme I for a substitution reaction at a carbonyl group

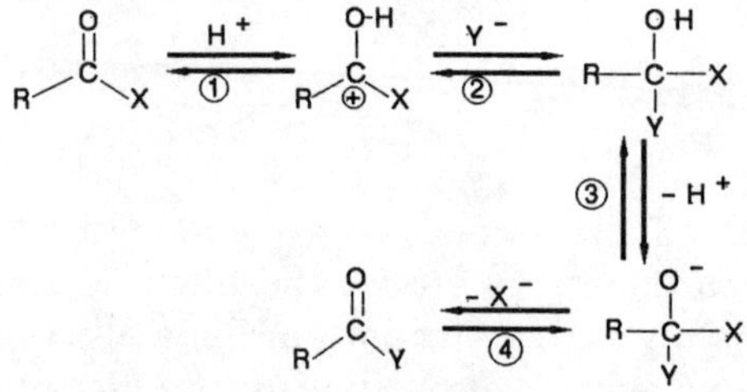

SCHEME I

where all of the intermediates are unstable, and step 4 is rate determining. Since the first step of the reaction involves protonation of the carbonyl oxygen, it might be thought that any change in the structure of R that increases the basicity of the carbonyl compound, or any increase in acidity of the solution would increase the rate of reaction. Such is not the case. Any change will affect the observed rate constant only as it affects the free energy of the transition state of step 4 relative to the free energy of the initial reactant $RCOX$. Since the transition state of step 4 has a negative charge, any change in structure of R which increases basicity by stabilizing the positive charge of the protonated carbonyl group is most likely to **decrease** the rate of reaction since the stabilization of positive charge usually implies destabilization of negative charge.

Scheme I is, in fact, somewhat contrived. Since the transition state for the rate determining step 4 does not contain the proton added in step 1, the rate of reaction will be independent of the concentration of H^+. The kinetics, or response of kinetics to structural and environmental changes, can give no indication of the existence of step 1 so long as step 4 remains rate determining. The mechanism of Scheme I is indistinguishable from one involving direct attack of Y^- on the carbonyl compound to form the anionic tetrahedral intermediate.

There are many practical cases in which an initial reactant is converted to some other species, or is transferred between states, rapidly with respect to the rate of formation of products. Let's consider some cases involving equilibria where the Bodenstein approximation is **not** valid.

As a first case, consider the reactions

$$\mathrm{A + B} \underset{fast}{\overset{K}{\rightleftharpoons}} \mathrm{C} \qquad \mathrm{A + B} \xrightarrow{k} \mathrm{P},$$

where C is not an intermediate, and may be formed in appreciable concentrations. Let's simplify by assuming that B is present in large excess over A. The initial amount of A added to the solution is $[A]_0$, so that we have

$$K[B]_0 = \frac{[C]}{[A]} \quad \text{and} \quad \frac{d[P]}{dt} = k[B]_0[A] = \frac{k[B]_0}{1 + K[B]_0}\left([A]_0 - [P]\right).$$

The final equation shows that the reaction kinetics will be pseudo-first-order with

$$k_\psi = \frac{k[B]_0}{1 + K[B]_0}.$$

At low concentrations of B, there is little C formed and the pseudo-first-order rate constant is simply $k[B]_0$; at high concentrations of B, we would reach a point where $k_\psi = k/K$. In terms of free energy, the chemical potential of A is dependent on the concentration of B.

As a second case, consider the reactions

$$\mathrm{A} + \mathrm{B} \underset{fast}{\overset{K}{\rightleftharpoons}} \mathrm{C}, \qquad \mathrm{C} \xrightarrow{k} \mathrm{P},$$

where C is an intermediate which may have an appreciable concentration. Again, let's assume that we have B present in large excess and that the initial amount of A added to the solution is $[A]_0$. We then find that

$$K[B]_0 = \frac{[C]}{[A]}, \quad \text{and} \quad \frac{d[P]}{dt} = k[C] = \frac{kK[B]_0}{1 + K[B]_0}\left([A]_0 - [P]\right).$$

Again, the reaction kinetics will be pseudo-first-order, and now

$$k_\psi = \frac{kK[B]_0}{1 + K[B]_0},$$

which differs from the first case only by the appearance of K in the numerator. Experimentally, the two cases are indistinguishable because both have the same dependence on $[B]_0$. So long as the equilibrium between the reactants and some other species, such as C, is fast with respect to the reaction being observed, we have similar situations. In the two cases just discussed, the fact that the pseudo-first-order rate constants have a complex dependence on $[B]_0$ would allow us to infer the existence of C even if we had not detected the equilibrium by other methods.

As a third, and final, case, let's consider the reactions

$$\mathrm{A} \underset{fast}{\overset{K}{\rightleftharpoons}} \mathrm{A}' \qquad \mathrm{A} \xrightarrow{k} \mathrm{P} \qquad \mathrm{A}' \xrightarrow{k'} \mathrm{P},$$

where A and A′ could be isomers, or could represent A in two different states, such as in two different solvents, at equilibrium with each other. Proceeding just as before, we find that

$$K = \frac{[A']}{[A]} \quad [A]_0 - [P] = [A](1+K) = [A'](1+1/K),$$

$$\frac{d[P]}{dt} = k[A] + k'[A'] = \left\{\frac{k}{1+K} + \frac{k'K}{1+K}\right\}([A]_0 - [P])$$

Again, the reaction shows pseudo-first-order kinetics, and

$$k_\psi = \frac{k + k'K}{1+K}$$

If the product formed from A′, call it P′, were different from that formed from A, the ratio of products formed would be

$$\frac{[P]}{[P']} = \frac{k}{k'K}$$

Taking $-RT$ times the logarithm of the right-hand side of this last equation, we find

$$\begin{aligned} -RT\ln k + RT\ln k' + RT\ln K &= \mu^\circ_{ts} - \mu^\circ_A - \mu^\circ_{ts'} + \mu^\circ_{A'} + \mu^\circ_A - \mu^\circ_{A'} \\ &= \mu^\circ_{ts} - \mu^\circ_{ts'}, \end{aligned}$$

where μ°_{ts} is the standard chemical potential of the transition state leading to P, and $\mu^\circ_{ts'}$ is that of the transition state leading to P′. Therefore,

$$\frac{[P]}{[P']} = \frac{k}{k'K} = \exp - \left\{\frac{\mu^\circ_{ts} - \mu^\circ_{ts'}}{RT}\right\}.$$

In words, the relative amounts of the two products formed is independent of the relative amounts of A and A′ present, and depends only on the difference in standard chemical potentials of the two transition states. This is sometimes called the Curtin-Hammett principle [6].

An example of the principle is shown in Scheme II.

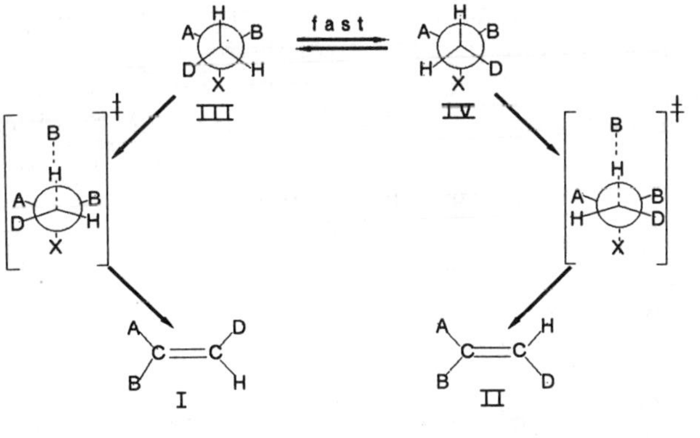

SCHEME II

Here, two conformations, *III* and *IV*, are in rapid equilibrium, and proceed through the two transition states shown to give the isomeric products, *I* and *II*. In this case, the principle states that the ratio of the rate constants for formation of the two different products depends only on the difference in free energies of the two different transition states, and is independent of the relative stabilities of the two conformations, *III* and *IV*, of the reactant.

Another application of the principle which occurs fairly commonly seems to be particularly deceptive. Suppose we have a reaction which can occur in two phases which are in contact with each other, such as a reaction in the presence of a micelle, or a reaction in two different solvents which are in contact with one another:

$$A(1) + B(1) \underset{k_r}{\overset{k_f}{\rightleftharpoons}} P(1) \quad \text{and} \quad A(2) + B(2) \underset{k'_r}{\overset{k'_f}{\rightleftharpoons}} P(2).$$

If the reactants are equilibrated between the two phases rapidly in comparison to the rate at which they are converted to product, then we can set

$$\alpha_A(1) = \alpha_A(2) \quad \text{and} \quad \alpha_B(1) = \alpha_B(2).$$

Now, note that this along with the basic postulate that transition states are in equilibrium with reactants requires that the activities of the transition states in the two different phases be equal:

$$\mu_{TS}(1) = \mu_{TS}(2), \quad \text{therefore} \quad \alpha_{TS}(1) = \alpha_{TS}(2)$$

and

$$[TS]^{(1)}\gamma_{TS}^{(1)} = [TS]^{(2)}\gamma_{TS}^{(2)},$$

Since the rate of reaction depends only on concentration of transition state, it is obvious that the rates in the two different phases differ **only** in that the activity coefficients of the transition state are different in the two states. Higher or lower concentrations, or differing activity coefficients, of the reactants in the two different phases have no effect on the relative rates in the two phases. It is also inconsequential whether or not, for example, a 'phase transfer catalyst' is used in one of the phases. So long as the reactants are at equilibrium between the phases, the **chemical potentials** of the transition states in the two phases must be the same. This must be true even if the transition state composition for the effective reaction path is different in the two phases.

Before proceeding on with our more general discussion, there is another point worth particular notice in regard to multistep processes with reactive intermediates. It is common to discuss these types of processes with the aid of reaction coordinate diagrams in which *energy* is shown vs. progress along a reaction coordinate. If one has only simple unimolecular steps in a process, such diagrams can be useful and almost never cause confusion. In other cases, however, one must be extremely careful in using them; in particular, if the vertical placement of intermediates and transition states is to indicate which step of the process is rate determining, the energy measure that is used must be specified in an unusual way because the relative placements of the intermediates and transition states, and thereby the identity of the rate determining step, are **concentration dependent** in the general case. Let's consider the reaction which we treated in Chapter 1, Eq. (1.21):

$$A + B \underset{k_{-1}}{\overset{k_1}{\rightleftharpoons}} C \qquad C + D \underset{k_{-2}}{\overset{k_2}{\rightleftharpoons}} E,$$

where pseudo-first-order conditions are used. We found that the pseudo-first-order rate constant is given by

$$k_\psi = \frac{k_1 k_2 [B]_0 [D]_0}{k_{-1} + k_2 [D]_0}.$$

If $k_{-1} >> k_2[D]_0$, the second step of the reaction is rate determining, and we must place the second transition state higher than the first. If $k_2[D]_0 >> k_{-1}$, the first step is rate determining and the first transition state must be placed higher than the second.

The relative magnitude of k_{-1} and $k_2[D]_0$ is obviously the determinant of the identity of the rate determining step. Neglecting activity coefficients for simplicity, we have:

$$RT \ln k_{-1} = RT \ln \left(\frac{\kappa T}{h}\right) + RT \ln K_{-1}^{\ddagger}$$
$$= \mu_C^\circ - \mu_{TS1}^\circ + RT \ln \left(\frac{\kappa T}{h}\right),$$

and

$$RT \ln k_2 + RT \ln[D]_0 = RT \ln \left(\frac{\kappa T}{h}\right) + RT \ln K_2^{\ddagger} + RT \ln[D]_0$$
$$= \mu_C^\circ + \mu_D^\circ - \mu_{TS2}^\circ + RT \ln[D]_0 + RT \ln \left(\frac{\kappa T}{h}\right).$$

Therefore

$$RT \ln \left(\frac{k_2[D]_0}{k_{-1}}\right) = \mu_{TS1}^\circ - \mu_{TS2}^\circ + \mu_D^\circ + RT \ln[D]_0.$$

Thus, the vertical placement of the second transition state relative to the first depends on the chemical potential of D, which depends on the concentration of D.

If we include activity coefficients in the above development, we will find that the relative heights on our 'reaction coordinate diagram' also depend on these. It is possible, therefore, to get somewhat confused even when all steps of a mechanism are unimolecular if we change solvent or ionic strength.

2.6 Salt and Solvent Effects on Reaction Rates

Since transition state theory postulates thermodynamic equilibrium between transition states and reactants, any change in rate constant for a reaction can occur only if activity coefficients of reactants and/or transition states change, as shown in Eq. (2.7). Remember that k_f° is an honest-to-goodness thermodynamic **constant** once we have chosen our standard state.

Eq. (2.7) is used in two different ways. If we know something about the activity coefficients, we can use Eq. (2.7), along with a measured k_f°, to predict values for k_f in states other than the standard state. Or, if we have measured values for k_f°, k_f, and the activity coefficients of the reactants, we can obtain the activity coefficient of the transition state.

The first type of usage is frequently applied to *correcting* rate constants measured at different ionic strengths to a single standard state, usually infinite dilution. For example, if we wish to study the mass law effect of added chloride ion on an SN1 reaction of an alkyl chloride, we must somehow 'correct' for the effect of the added chloride salt on activity coefficients of reactants and transition state. Most frequently, for aqueous solution, this involves the use of Debye-Hückel theory [7]. Two limiting forms of the equations are most common:

$$\text{(a)} \quad \log \gamma_i = -\mathcal{A} z_i^2 \mathcal{I}^{1/2}, \quad \text{or (b)} \quad \log \gamma_i = \frac{-\mathcal{A} z_i^2 \mathcal{I}^{1/2}}{1 + \mathcal{B}\mathcal{I}^{1/2}}, \qquad (2.13)$$

where z_i is the ionic charge of species i, and $\mathcal{I}$ is the ionic strength of the solution, defined as

$$\mathcal{I} = \frac{1}{2} \sum_j m_i z_j^2,$$

where the indicated sum is over all ionic species in solution. The value of $\mathcal{A} = 1.875 \times 10^6 (\epsilon T)^{-3/2}$ is 0.51 for aqueous solution at 25°C. The constant $\mathcal{B}$ in the latter equation depends on the ionic radius of the ionic species i. Numerical values of the constants for various ionic radii and temperatures are given in many handbooks such as Lange's [8]. These equations are reasonably accurate for aqueous solutions and ionic strengths up to approximately 0.1M.

Within the applicability of Eq. (2.13a), for a reaction:

$$A^{+n} + B^{-m} \quad \overset{K^{\ddagger}}{\rightleftarrows} \quad [TS^{+(n-m)}]^{\ddagger} \quad \longrightarrow \quad P^{+(n-m)},$$

we have

$$\log \frac{k_f}{k_f^\circ} = \log \frac{\gamma_A \gamma_B}{\gamma_{\ddagger}} = -\mathcal{A}\mathcal{I}^{1/2}(n^2 + m^2 - (n-m)^2 = -\mathcal{A}\mathcal{I}^{1/2}(2mn),$$

or a similar equation in which $\mathcal{I}^{1/2}$ is replaced by $\mathcal{I}^{1/2}/(1 + \mathcal{B}\mathcal{I}^{1/2})$ if Eq. (2.13b) is applicable. If the reaction is indeed an elementary step, then the transition state must have the charge which is the algebraic sum of the reactant charges, and, within the range of the Debye-Hückel equations, the activity coefficient of the transition state is set. There have been innumerable studies which accord with the predicted behavior in reasonably dilute aqueous solutions.

Once one leaves dilute aqueous solutions it is no longer certain, or even likely, that the activity coefficient of the transition state can be predicted even if the activity coefficients of the reactant species are known. The usual approach in these situations is to study the salt effects on rate constants and then attempt an interpretation of the data in terms of the ratio of activity coefficients of reactants and transition state. Only in unusually thorough studies are the activity coefficients of the reactants likely to be determined separately. A fairly typical set of results of measurement of salt effects on rate constants of a reaction in non-dilute aqueous solution is shown in Table 2.1. The partial reaction mechanism which had been assigned from other studies was:

$$
\begin{aligned}
&p\text{-}O_2NC_6H_4SO_2CH_2OClO_3 + H_2O \\
&\xrightleftharpoons{slow} p\text{-}O_2NC_6H_4SO_2CH^{\ominus}OClO_3 + H_3O^{\oplus} \qquad (2.14) \\
&\xrightarrow{fast} p\text{-}O_2NC_6H_4SO_2H + HCOOH + ClO_3^{\ominus}
\end{aligned}
$$

TABLE 2.1

Salt Effects on Reaction (2.14) at 25°C

In water containing $10^{-3}M$ HCl. $k_{\psi}^{\circ} = 32.5 \times 10^{-4} s^{-1}$.

Data from Reference [9]

Salt	M	10^4k_{ψ}	Salt	M	10^4k_{ψ}
KBr	0.50	29.5	Na_2SO_4	0.50	93.1
	1.00	26.3	$(n\text{-}Bu)_4NBr$	0.05	36.1
$NaClO_4$	0.25	28.9		0.10	36.7
	0.50	25.4		0.20	42.9
	1.00	20.5		0.30	53.4
$(CH_3)_4NBr$	0.20	33.1		0.50	87.0
	0.80	36.3		0.80	164.
$CaCl_2$	0.50	25.1			

There are several things worth noting. Most obviously, the rate constant depends on the specific identity of the salt as well as the salt concentration and charge type. Some of the salts cause increases, while others cause decreases, in the rate constant. The effect of moderate salt concentrations on rate constant can be quite large. At least two factors complicate salt effects at moderate to high salt concentrations. First, the formation of ionic aggregates, such as ion-pairs, triplets, etc., becomes significant at much above 0.1M in aqueous solutions, and at much lower concentrations in most non-aqueous solvents. Second, salt effects on the activity coefficients of non-electrolytes [10] become appreciable. In the case of reaction (2.14), the effect of the salts on the activity of water must also be considered since the solvent is a formal reactant.

The most extensive studies of salt effects on reaction rates in non-aqueous solvents have dealt with solvolysis reactions. In many of these reactions, it is found that there is a linear relationship between the observed pseudo-first-order rate constants and concentration of salt as expressed by the equation [11]

$$k_\psi = k_\psi^\circ(1 + b[salt]). \tag{2.15}$$

Some of the values reported for the constant b are shown in Table 2.2 for the solvolysis of p-methoxyneophyl tosylate,

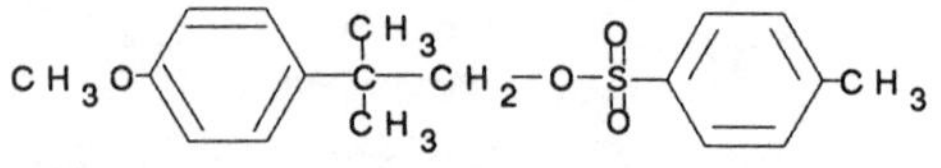

p-Methoxyneophyl Tosylate

TABLE 2.2
Salt Effects on Solvolyses of p-Methoxyneophyl Tosylate
at 75°C, b values for Eq.(2.15) from [12]

Solvent[a]	Salt	b	Solvent	Salt	b
DMSO	$LiClO_4$	0.	THF	$LiClO_4$	482.
DMF	$LiClO_4$	1.	Acetone	$LiClO_4$	47.
Acetic acid	$LiClO_4$	12.	Acetone	$NaClO_4$	17.
Ac_2O	$LiClO_4$	47.	Acetone	$(n\text{-}Bu)_4NClO_4$	3.

[a] DMSO, dimethyl sulfoxide; DMF, dimethylformamide; THF, tetrahydrofuran; Ac_2O, acetic anhydride.

Although all of the b values shown in Table 2.2 are positive, some negative values have been reported. For example, in the solvolysis of neophyl tosylate in aqueous dioxane, values of $b = -1$ for sodium hydroxide, and $b = -0.3$ for sodium chloride have been reported [12]. The fact that the b values depend on the identity of the salt makes it difficult, or impossible in some cases, to separate such things as mass law effects or kinetic order of the reactions with respect to concentrations of added ionic nucleophiles from the salt effects on activity coefficients. Considerable controversy has arisen from interpretations of such data [13].

An example of the use of Eq. (2.7) to infer information concerning the structure of transition states from knowledge of their activity coefficients is given in a study [14] of solvolyses of t-butyl chloride in a number of solvents. The activity coefficients for the transition states were determined, relative to a standard state in dilute methanol solution, by measuring rate constants for the solvolysis reaction and activity coefficients of the reactant in various solvents. Standard free energies of transfer of the transition state, $\Delta G_t^{\circ\ddagger} = -RT \ln \gamma_\ddagger^s$, were then compared with standard free energies of transfer of tetramethylammonium chloride free ions and ion-pairs. Some of the data are shown in Table 2.3.

TABLE 2.3
Comparison of ΔG_t° for t-BuCl Solvolysis TS‡ and Tetramethylammonium Chloride
Methanol solution is the standard state. Values in Kcal/mole. data from reference [14]

Solvent	ΔG_t°(ion-pair)	ΔG_t°(free ions)	$\Delta G_t^{\circ\ddagger}$(t-BuCl)
Water	-3.0	-3.7	-1.65
Methanol	0	0	0
Ethanol	1.5	3.0	1.06
DMF	3.5	4.3	2.71
Acetonitrile	4.5	5.6	3.20
Acetone	6.6	8.9	4.55
Diethyl ether	15.1	28.	8.11

A plot of the transfer free energies for the transition state vs those for the ion-pair is linear with a slope of approximately 0.7, leading to the conclusion that the transition state structure is similar to that of an ion-pair, but with a charge of only 0.7 units on the two counterions.

Some of the most extensive data concerning activity coefficients of transition states is available for SN2 type reactions involving nucleophiles other than the solvents. If any of the reactants or transition state are ionic species, the problem of obtaining single ion activity coefficients arises. There are several approaches to these which are reasonably consistent [15], and some of the reported data are shown, expressed as free energies of transfer, in Tables 2.4, and, as $\log \gamma$'s, in Table 2.5.

TABLE 2.4
Free Energies of Transfer From Water to Other Solvents
Kcal/mole at 25°C, data from [15]

	Solvent[a]					
Ion	MeOH	DMF	DMSO	MeCN	PC	TMS
Li^+	0.9	-2.3	-3.5	-	5.7	-
Na^+	2.0	-2.5	-3.3	3.3	3.6	-0.7
K^+	2.4	-2.3	-2.9	1.9	1.4	-1.0
Et_4N^+	0.2	-2.0	-3.0	-2.1	-	-
Cl^-	3.0	11.0	9.2	-	9.0	12.6
N_3^-	2.6	8.2	6.1	7.3	6.9	9.5
Br^-	2.7	7.2	6.1	7.6	7.1	9.5
I^-	1.6	4.5	2.2	4.5	4.2	4.9
ClO_4^-	1.4	0.1	-0.3	-	-	-
Ph_4B^-	-5.6	-9.1	-8.8	-7.8	-8.5	-8.5

[a] MeOH= methanol; DMF= dimethylformamide; DMSO= dimethyl sulfoxide; MeCN= acetonitrile; PC= propylene carbonate; TMS= tetramethylenesulfone.

The values shown for transition states in Table 2.5 are, of course, obtained from the rate constants of the SN2 reactions in methanol and dimethylformamide solutions and the activity coefficients for the two reactant species for each reaction. One may compare the values for transition states with the values for the various solutes shown to get some idea of the charge distributions of the transition states. For the attacks of negatively charged nucleophiles on neutral methyl halides or tosylates, the activity coefficients indicate that the transition states have

TABLE 2.5
Logarithms of Activity Coefficients in DMF Solution Relative to Methanol Standard State [16]

Solute	$\log \gamma^{DMF}$	Solute	$\log \gamma^{DMF}$
$CH_3CO_2^-$	9.2	CH_3Cl	-0.4
Cl^-	6.5	CH_3Br	-0.3
Br^-	4.9	CH_3I	-0.5
N_3^-	4.9	t-Butylchloride	-0.2
I^-	2.6	Benzylchloride	0.1
SCN^-	2.7	p-$NO_2C_6H_4I$	-1.2
Picrate$^-$	-0.4	$(C_6H_5)_4C$	-1.6
p-$CH_3C_6H_4SO_3^-$	3.5	$[N_3^- \cdot CH_3 \cdot {}^+SMe_2]^\ddagger$	-1.3
$(C_6H_5)_4B^-$	-2.7	$[N_3^{\partial-} \cdot CH_3 \cdot I^{\partial-}]^\ddagger$	-0.2
Na^+	-3.9	$[N_3^{\partial-} \cdot CH_3 \cdot OTs^{\partial-}]^\ddagger$	2.3
K^+	-3.7	$[N_3^{\partial-} \cdot CH_3 \cdot Cl^{\partial-}]^\ddagger$	1.2
Cs^+	-3.3	$[Cl^{\partial-} \cdot CH_3 \cdot Br^{\partial-}]^\ddagger$	1.4
$(C_6H_5)_4As^+$	-2.7	$[AcO^{\partial-} \cdot CH_3 \cdot I^{\partial-}]^\ddagger$	1.8

the negative charge spread over a large region since values are much smaller than those for the attacking anion.

It is fairly certain that one of the major factors determining activity coefficients, or free energies of transfer, of anionic species in the solvents such as dimethylformamide, or the other non-hydroxylic solvents listed in these tables, relative to water or methanol as standard state is hydrogen bonding stabilization of the anions in water or methanol. Except for the very largest anions with shielded charge, such as the tetraphenylboride ion, the free energies of transfer of anions from the hydroxylic to the non-hydroxylic solvents is positive. Those anions having the larger ionic radii, or the more delocalization of negative charge such as in picrate or perchlorate ions, have least hydrogen bond stabilization in methanol or water, and therefore less positive free energies of transfer to the non-hydroxylic solvents.

There are also significant contributions to transfer free energies from solvent-solvent interactions. Such interactions must be overcome in order to make the 'holes' in which the solute species are situated, and the larger the solute, the larger the hole and the greater the energy required. Such solvent-solvent interactions include hydrogen-bonding, and the transfer free energies for large solutes, such as tetraphenylmethane and p-nitrophenyl iodide shown in Table 2.5, from methanol to the polar non-hydroxylic solvents are negative as a result.

Dipole moments of solutes and solvents are also important in affecting free energies of transfer. To the extent that dipole moments are proportional to dielectric constants, they are very similar for methanol, dimethylformamide, dimethyl sulfoxide, acetonitrile, propylene carbonate, and tetramethylenesulfone, which all have dielectric constants between 35 and 50. Water, however, with a dielectric constant of over 80 is quite different from the other solvents. Free energies of transfer of all of the reasonably small ions, both cations and anions, from water to methanol are positive. Only very large ions, such as tetraphenylboride, show negative free energies for transfer from water to methanol; presumably because of the greater energy required for 'hole' formation in water than in methanol.

The negative free energies for transfer of cations from water or methanol to dimethylformamide or dimethyl sulfoxide may be attributed to the Lewis acid-base interactions of the cations with the lone-pairs of electrons on the partially negative oxygens of these latter solvents. The smaller the cation, the greater the interaction with these solvents.

2.7 Reactions With Solvent as a Formal Reactant

The types of studies involving determinations and interpretations of activity coefficients for transition states cannot be applied to reactions in which the solvent is a formal reactant in the rate-determining step, and, in fact, could be invalid even in the cases which we have discussed above. For an ordinary solute species, even one strongly solvated and intimately associated with solvent molecules, everything works out fine with the formalism of chemical potential neglecting the solvation, so long as the interconversion of various solvates is rapid relative to any reaction being studied. This situation is just another example, in fact, of the principle discussed earlier in Section 2.5. Hammett has a particularly informative discussion of *primitive* and *sophisticated* chemical potentials pertaining to ordinary solute species [17]. As mentioned earlier, transition states are nearly, but not exactly, like ordinary species. If solvent molecules are intimately involved in the bonding changes that occur in a reaction , or, more precisely [18], if the solvent is involved in the motion along the reaction coordinate, we have to treat the transition states in different solvents as entirely different species.

The SN2 solvolysis reactions of alkyl halides and tosylates in various solvents have received particular attention and can serve as examples of approaches to this problem of reactions in which the solvent is a reactant. In reactions of this sort, there is believed to be both charge separation due to partial breaking of the carbon-halogen or carbon-tosylate bond, and partial bond formation between carbon and solvent at the transition states. At least two different properties of the solvent, then, are expected to be operative.

The ability of the solvent to support the separation of charge in SN2 reactions might be expected to be similar to that in SN1 reactions. A totally empirical approach to evaluate this *solvent ionizing power* was taken in a study of solvent effects on the rates of solvolysis of t-butyl chloride [19]. At the time of the first study, it was thought that t-butyl chloride underwent a 'pure' SN1 reaction with no specific involvement of solvent. The rate constants for solvolysis in various solvents, k^s_{t-BuCl}, were used to define a parameter, Y_s:

$$Y_s = \log k^s_{t-BuCl} - \log k^o_{t-BuCl} \quad \text{where} \quad k^o_{t-BuCl} = k^{80\%ethanol}_{t-BuCl}.$$

The solvent 80% ethanol, 20% water was taken as the standard solvent for which Y_s is defined to be zero. Other alkyl halides and tosylates which undergo SN1 solvolyses were expected, and sometimes found, to have rate constants which would be correlated by the equation:

$$\log k^s_{RX} - \log k^o_{RX} = mY_s,$$

in which m might be interpreted as a measure of the extent of charge separation at the transition state relative to that for the transition state for t-butyl chloride. For those compounds which are believed to react by the SN1 mechanism, values of m are in fact between 0.8 and 1.2, indicating that the extent of ionization at the transition states for SN1 reactions does not vary greatly.

Over the years since the first definition of Y_s values, there have been several modifications of the concept. These modifications have all recognized that solvents are more-or-less intimately involved at the transition states of the reactions, at least to the extent of having started to hydrogen-bond to the anionic leaving group. Since different leaving groups require different extents of hydrogen-bonding, relative to the more general electrostatic or 'polar', stabilization of solvents, different leaving groups should require different Y_s values. Particular attention

TABLE 2.6
Nucleophilicities and Ionizing Powers of Solvents
data from references [20]-SBS ,[21]-PEP, and [22]-KAF.

	N_{OTs}		N_{Cl}		N_{neut}	Y_{OTs}		Y_{Cl}	
Solvent	SBS	PEP	SBS	PEP	KAF	SBS	PEP	SBS	PEP
H_2O	—	—	-0.26	0.00	-1.15	—	—	3.49	3.82
MeOH	-0.04	-0.23	0.01	-0.27	0.62	-1.09	-0.67	-1.09	-0.85
EtOH	0.00	-0.06	0.09	-0.08	0.56	-1.75	-1.60	-2.03	-1.79
i-PrOH	—	—	0.09	0.14	—	—	—	-2.73	-2.36
CH_3COOH	-2.35	-1.49	-2.05	-1.73	-1.46	-0.61	-0.90	-1.64	-1.61
HCOOH	-2.35	-1.49	-2.05	-1.73	-1.73	3.04	2.75	2.05	2.38
CF_3COOH	-5.56	-4.09	-4.74	-4.75	—	4.57	4.72	1.84	3.48
CF_3CH_2OH	-3.00	—	-2.78	—	-2.42	1.80	—	1.05	—
DMF	—	1.94	—	2.25	—	—	-2.29	—	-1.74
DMSO	—	2.11	—	2.46	—	—	-1.31	—	-0.60
CH_3CN	—	1.60	—	1.85	—	—	-2.05	—	-1.62

has been given the tosylate and chloride leaving groups [20]; some data are available for neutral leaving groups which should be nearly free of the hydrogen-bonding contributions. Some of the various values of Y_s for a few representative solvents are given in Table 2.6.

There have been a number of attempts to relate the Y_s values to other measures of solvent polarity. The earliest of these other measures was the determination of solvent effects on the wavelength of the charge transfer absorption of N-methyl, 4-carbomethoxypyridinium iodide [23]. The absorption of a photon of light by this ion-pair is believed to lead to charge transfer forming the radical pair, so that more energy should be required in those solvents furnishing greater stabilization of the ion-pair ground state. Since the energy for the charge transfer is directly related to the wavelength of light absorbed, $E_{ct} = h\nu = hc/\lambda$, the measure of solvent effects is simply a matter of running a spectrum in various solvents. There have been several other spectral measurements suggested for this same purpose and these have come to be known as 'solvatochromic' scales. Considerable effort has been directed toward expressing both the solvatochromic and solvolysis-based measurements of solvent ionizing power in terms of more fundamental solvent properties such as hydrogen-bond donating ability, hydrogen-bond accepting ability, and polarity [24].

The treatments of SN2 reactions have all taken the form of the four parameter equation first used for reactions of methyl iodide [25]:

$$\log\left(\frac{k^s_{RX}}{k^o_{RX}}\right) = SN_s + mY_s,$$

where S and m depend only on the particular alkyl halide or tosylate concerned, N_s is a measure of solvent nucleophilicity, and Y_s is usually the same measure of solvent ionizing power discussed above. The separation of the two solvent properties is non-trivial and is the subject of continuing studies. Some of the N_s values are shown with corresponding Y_s values resulting from different approaches to this separation in Table 2.6.

The simplest approach to the separation of N_s and Y_s effects is that used by Bentley and co-workers [20]. They define a set of Y_s values based on the solvolyses of 2-adamantyltosylate,

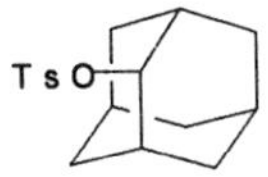

2-adamantyltosylate

which are certainly free of any backside nucleophilic solvent involvement, and then define a set of N_s based on these Y_s values and rate constants for solvolyses of methyltosylate by the equation;

$$N_s \equiv \log\left(\frac{k^s_{MeOTs}}{k^o_{MeOTs}}\right) - 0.3Y_s.$$

Peterson [21] uses a more statistical approach which does not involve one particular choice of reference compounds. The different approaches sometimes... note particularly the N_{OTs} and Y_{Cl} for CF_3COOH in Table 2.6... lead to quite significant differences in values for the solvent parameters.

2.8 Determinations of Stereochemistry of Transition States

The studies of salt and solvent effects on reaction rates are primarily useful in providing information relative to the ionic charges or charge separations of transition states. We would like to know a great deal more about the structures of transition states. For reactions involving chiral species, one of the least ambiguous features of transition state structures that can sometimes be obtained is stereochemistry. In any general case of reaction at a chiral center, it is possible to start with a pure enantiomer and then to determine the configuration at the chiral center of the product. If one observes, for example, the same absolute configuration for product as for reactant of an elementary step of reaction, then the transition state must have had a geometry corresponding to this same configuration; if the elementary step was a substitution at the chiral center, then the incoming group must have attached to the same side of the chiral center as the leaving group at the transition state.

These studies of stereochemistry are generally very difficult to carry out since one must start with enantiomerically pure starting material, and must know the stereochemistry of product relative to reactant. Of course, if chiral reactant produces only racemic product, the task is somewhat simpler in that enantiomerically pure reactant would not be required. The studies can also be made much easier in the cases of substitution reactions in which the entering and/or leaving group can be radioactively *tagged.* As an example, let's consider nucleophilic substitutions of alkyl iodides.

The general technique involves a study of an *identity reaction*, a reaction in which the reactant and product are the same, such as the attack of iodide ion on an alkyl iodide. The reaction is made observable by using a radioactively labelled reactant iodide ion:

$$\text{R–I} + {}^*\text{I}^- \underset{k}{\overset{k^*}{\rightleftharpoons}} \text{R–}{}^*\text{I} + \text{I}^-$$

The reaction is obviously reversible, and, neglecting the extremely small isotope effect, the rate constant for the forward reaction, k^*, is equal to that for the reverse reaction, k, regardless of mechanism. Tracer levels of the radioactive isotope are used; that is, the amount of labelled iodide is extremely small in comparison with the total amount of iodide.

It is convenient to define two quantities, X and Y:

$$X = \frac{[{}^*I^-]}{[I^-] + [{}^*I^-]} \simeq \frac{[{}^*I^-]}{[I^-]} \quad \text{and} \quad Y = \frac{[R^*I]}{[R^*I] + [RI]} \simeq \frac{[R^*I]}{[RI]}$$

where the last $\simeq$ in each definition arises from the fact that we are using tracer levels. The reaction is studied by starting with unlabelled RI and a large excess of the trace labelled iodide ion such that the quantity X remains essentially constant throughout the reaction, and the quantity Y is equal to zero at the start of reaction and becomes equal to X at equilibrium. The reaction is followed by isolating the alkyl iodide at various times and determining Y by a radiation counting technique. We can set up the rate equation as

$$\begin{aligned}
\frac{dY}{dt} = \frac{1}{[RI]}\frac{d[R^*I]}{dt} &= \frac{1}{[RI]}\left\{k[RI][^*I^-] - k[R^*I][I^-]\right\} \\
&= \frac{1}{[RI]}k[I^-]\left\{[RI]\frac{[^*I^-]}{[I^-]} - [R^*I]\right\} \\
&= k[I^-]\left\{\frac{[^*I^-]}{[I^-]} - \frac{[R^*I]}{[RI]}\right\} \\
&= k[I^-]\{X - Y\} \\
\frac{dY}{\{X-Y\}} &= k[I^-]dt \\
\ln\frac{\{X-Y\}}{X} &= k[I^-]t \equiv k_{ex}t,
\end{aligned}$$

where the equivalence sign in the last line defines k_{ex} as the pseudo-first-order rate constant for exchange.

Now consider the reaction of a chiral RI:

$$(R)\text{-RI} + \text{I}^- \underset{k'}{\overset{k'}{\rightleftharpoons}} (S)\text{-RI} + \text{I}^-,$$

where the process may, or may not, be the same as the exchange above. Note again that this reaction is reversible with the same rate constant for forward and reverse directions; the reaction will proceed until there is complete racemization. The actual attack of iodide on RI might produce either retention or inversion, but the process we are now discussing is only for inversion. Let α be the angle of rotation of plane polarized light by the reaction solution in a cell in a polarimeter, and let α_M be the molar rotation by the pure (R)-RI. Then

$$\alpha = \alpha_M\left([(R)\text{-RI}] - [(S)\text{-RI}]\right)$$

We'll see that we don't really need to know α_M. If we look at α as a function of time, we have

$$\begin{aligned}
\frac{d\alpha}{dt} &= \alpha_M\left\{\frac{d([(R)\text{-RI}] - [(S)\text{-RI}])}{dt}\right\} = 2\alpha_M\left\{\frac{d[(R)\text{-RI}]}{dt}\right\} \\
&= -2k'[I^-]\alpha_M\left\{[(R)\text{-RI}] - [(S)\text{-RI}]\right\} \\
&= -2k'[I^-]\alpha \\
\ln\frac{\alpha}{\alpha_0} &= -2k'[I^-]t \equiv -k_{ract},
\end{aligned}$$

where α_o is the angle of rotation of light by the solution at $t = 0$, and the last $\equiv$ sign defines the pseudo-first-order rate constant for racemization.

If the measurements of isotopic exchange and racemization are carried out on the same solution to obtain k_{ex} and k_{rac}, we can easily deduce the stereochemical outcome of each substitution. If every iodide exchange gives inversion of configuration, then the k for the exchange process is the same as the k' for the inversion reaction, and we must have $k_{rac} = 2k_{ex}$. Reversing the argument, if $k_{rac} = 2k_{ex}$, each exchange must have produced inversion. A little thought produces the following possibilities:

- If $k_{rac} = 2k_{ex}$, exchange gives 100% inversion.
- If $k_{rac} = k_{ex}$, exchange gives 100% racemization.
- If $k_{rac} < k_{ex}$, exchange gives dominant retention.
- If $k_{ex} < k_{rac} < 2k_{ex}$, exchange gives dominant inversion.

Nearly every nucleophilic substitution reaction of alkyl derivatives studied which shows second order kinetics also shows complete inversion of configuration. This, in fact, has come to be regarded as one of the distinguishing features of second-order nucleophilic substitution reactions. Even in cases loaded to favor front-side attack of nucleophiles, such as the attack of anionic nucleophiles on alkyl trimethylammonium ions or alkyl dimethylsulfonium ions [26], 100% inversion of configuration is observed. Thus, the transition states for these reactions are believed to have a geometry represented by a trigonal bipyramid in which the entering nucleophile and the leaving group occupy the two opposite apical positions.

The use of isotopic exchange and racemization is only one of many clever ways of getting at the stereochemistry of reactions. Another of the fairly general techniques that have been used involves the reactions of diastereomeric reactants, or the production of diastereomeric products. The study of reactions of diastereomers has become particularly prevalent since the advent of high resolution n.m.r. spectrometry which sometimes allows very easy stereochemical analysis.

An example of the study of diastereomeric reactants is provided in the acetolysis reactions of the diastereomeric 3-phenyl, 2-butyltosylates [27] in which the stereochemical results provide evidence for the intermediacy of the phenonium ion shown, and for the back-side attack both of the phenyl group on the neighboring carbon as the tosylate departs and of the solvent on the back-side of the carbon as the phenyl group departs to give a double inversion:

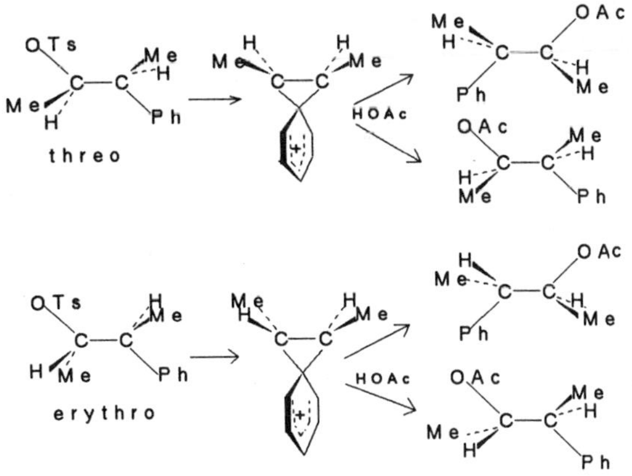

Acetolyses of 3-Phenyl, 2-Butyltosylates

The threo-tosylate produces only the threo-acetate, while the erythro-tosylate produces only the erythro-acetate. The optically active *L*-threo-tosylate produces racemic threo-acetate, while the optically active *L*-erythro-tosylate produces optically active *L*-erythro-acetate (Note that the last two structures at the bottom right of the reaction scheme are identical... **not** enantiomers.).

An even simpler example of stereochemical determination is provided by the bromination of *cis*- and *trans*-2-butenes. The fact that reaction of the *trans*-isomer produces *meso*-2,3- dibromobutane, while reaction of the *cis*-isomer produces *D,L*-2,3-dibromobutane [28], is indicative of the formation of a bromonium ion intermediate which undergoes back-side attack by bromide ion to give the final product.

Normally achiral reaction centers can sometimes be made chiral by the use of isotopes so that the three dimensional character of transition states can be revealed. A primary carbon can, for example, be made a chiral center by substituting one hydrogen by deuterium.

Some particularly impressive studies, which illustrate several worthwhile general points, have been carried out with phosphate esters [29]. The studies have utilized optically pure phosphate esters containing all three stable isotopes of oxygen, ^{16}O, ^{17}O, and ^{18}O. For example, the rearrangement of phosphate monoester derived from propanediol to its isomer without intermediacy of the cyclic diester was shown to proceed with retention of configuration at phosphorous [30]. The mechanism inferred by the result is shown:

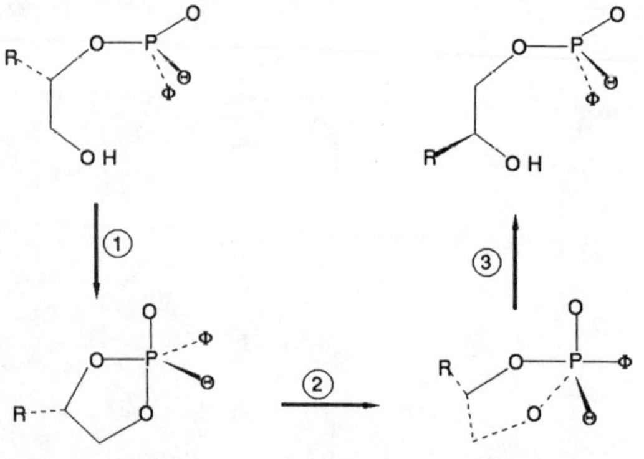

Rearrangement of Chirally Labelled Phosphate Ester

In step 1, the OH group attacks at an apical position to give the trigonal bipyramidal intermediate, which pseudo-rotates in step 2 to place the other 'ether' oxygen in the apical position, from which it leaves in step 3. Note that if the attack of OH is at the apical position, then microscopic reversability requires that that OH leave from the apical position. Since the two groups in the present example are so similar, it is virtually required that the leaving of the other OH is from the apical position.

In a later chapter, we'll look more closely at some stereochemical studies which have been combined with other studies to provide even more detailed information about mechanisms and structures of transition states for some solvolysis reactions.

2.9 Partial Molar Volumes of Transition States

The partial molar volumes of solute species, $\overline{V}_i$

$$\left(\frac{\partial \mathcal{V}}{\partial m_i}\right)_{T,P,m_j} \equiv \overline{V}_i,$$

where $\mathcal{V}$ is the volume of the system, determine the pressure effects on equilibrium constants through the relationship [1]:

$$\left(\frac{\partial \ln K}{\partial P}\right)_T = -\frac{\Delta \overline{V}^{\circ}}{RT} + \Delta m \left(\frac{\partial \ln \rho}{\partial P}\right)_T,$$

where $\Delta \overline{V}^{\circ}$ is the difference in molar volumes between products and reactants, ρ is the density of the solution, and Δm is the difference in the stoichiometric coefficients of products and reactants. This last term, in Δm, arises because of the use of the molarity concentration scale since molarity changes with density of the solution; the term is usually small and frequently neglected or avoided by using the molal concentration scale.

There is a completely analogous expression:

$$\left(\frac{\partial \ln k}{\partial P}\right) = -\frac{\Delta V^{\ddagger}}{RT} + \Delta m^{\ddagger}\left(\frac{\partial \ln \rho}{\partial P}\right)_T,$$

for the effect of pressure on rate constants [31]. If $\Delta V^{\ddagger}$ is determined by measuring the effect of pressure on rate constant, and the partial molar volumes of the reactants are known or determined, the partial molar volume of the transition state for the reaction can be evaluated. For $|\Delta V^{\ddagger}| \approx 25 cm^3/mole$ each kilobar of pressure will change the rate by roughly a factor of two. Studies frequently employ pressures up to 10 kilobar, and some employ even higher pressures.

Reactions not involving ions are simplest to interpret. Diels-Alder reactions have received particular attention [32]. A typical result is that the reaction of maleic anhydride with isoprene in acetone solution at $35°C$ has $\Delta V^{\ddagger} = -39 cm^3/mole$ which is slightly more negative than the $\Delta \overline{V}° = -36 cm^3/mole$ for the reaction. This means that the transition state actually has a smaller partial molar volume than does the product of the reaction. The authors interpret this to indicate 'secondary bonding' in the transition state; it could also result from a polar character for the transition state.

The partial molar volumes of ions have large contributions from electrostriction of solvent around the ions. For example, the reaction

$$CO_2 + H_2O \rightleftharpoons H^+ + HCO_3^-$$

has $\Delta \overline{V}° = -28 cm^3/mole$ [33]. For reactions having transition states with large amounts of charge separation, we can expect large negative contributions to $\Delta V^{\ddagger}$. The solvolysis of *t*-butyl chloride in aqueous ethanol , for example [33], has $\Delta V^{\ddagger} = -20 cm^3/mole$, and the reaction of methyl iodide with pyridine in acetone [33] has $\Delta V^{\ddagger} = -30 cm^3/mole$.

Since compactness of a species, and therefore small partial molar volume, usually implies considerable order or few degrees of freedom, the ΔV's of reactions, both equilibrium and activation, are usually related to the ΔS's, entropies, of the reactions.

2.10 REFERENCES

[1]. L. P. Hammett, **Physical Organic Chemistry**, 2nd Ed., McGraw–Hill Publishers, New York, **1970**, Chapter 2.

[2]. J. Hine, **J. Org. Chem., 31**, 1236 (**1966**).

[3]. C. D. Ritchie and D. J. Wright, **J. Amer. Chem. Soc., 93**, 6574 (**1971**).

[4]. L. P. Hammett, *loc. cit.*, pp. 216–218.

[5]. L. P. Hammett, *loc. cit.*, pp. 142–144.

[6]. L. P. Hammett, *loc. cit.*, pg. 119.

[7]. L. P. Hammett, *loc. cit.*, pg. 188–192.

[8]. J. A. Dean, **Lange's Handbook of Chemistry**, 12'th Ed., McGraw–Hill Publishers, New York, **1979**, pp. 5.3–5.6.

[9]. L. Menninga and J. B. F. N. Engberts, **J. Amer. Chem. Soc., 98**, 7652 (**1976**).

[10]. F. A. Long and W. F. McDevit, **Chem. Rev., 51**, 119 (**1952**).

[11]. L. P. Hammett, *loc. cit.*, pg. 215.

[12]. D. J. Raber, J. M. Harris, and P. v. R. Schleyer in **Ions and Ion-Pairs in Organic Reactions, Vol. 2**, Ed. by M. Szwarc, Wiley-Interscience Publishers, New York, **1974**, pp. 247–374.

[13]. D. J. Raber, J. M. Harris, R. E. Hall, and P. v. R. Schleyer, **J. Amer. Chem. Soc., 93**, 4821 (**1971**); R. A. Sneen and H. M. Robbins, *Ibid.*, **94**, 7868 (**1972**); See reference 12 also.

[14]. M. H. Abraham, **Prog. Phys. Org. Chem., 11**, 1 (**1974**).

[15]. B. G. Cox, G. R. Hedwig, A. J. Parker, and D. W. Watts, **Austr. J. Chem., 27**, 477–501 (**1974**).

[16]. A. J. Parker, **Chem. Rev., 69**, 1 (**1969**).

[17]. L. P. Hammett, *loc. cit.*, pp. 16–19.

[18]. C. D. Ritchie, **Pure and Applied Chem., 51**, 153 (**1979**).

[19]. E. Grunwald and S. Winstein, **J. Amer. Chem. Soc., 70**, 846 (**1948**).

[20]. F. L. Schadt, T. W. Bentley, and P. v. R. Schleyer, **J. Amer. Chem. Soc., 98**, 7667 (**1976**); T. W. Bentley in **Nucleophilicity**, Ed. by J. M. Harris and S. P. Mcmanus, #215 in **ACS Advances in Chemistry Series**, Washington, D. C., **1987**.

[21]. P. E. Peterson in **Nucleophilicity**, *loc. cit.*, pp. 299–314.

[22]. D. N. Kevill, S. W. Anderson, and E. K. Fujimoto in **Nucleophilicity**, *loc. cit.*, pp. 269–283.

[23]. E. M. Kosower, **J. Amer. Chem. Soc., 80**, 3253 (**1958**).

[24]. J. L. M. Abboud, M. J. Kamlet, and R. J. Taft, **Prog. Phys. Org. Chem., 13**, 485–630 (**1981**). M. H. Abraham, P. L. Grellier, A. Nasehzadeh, and R. A. C. Walker, **J. Chem. Soc., Perkin II, 1988**, 1717.

[25]. C. G. Swain, R. B. Moseley, and D. E. Bown, **J. Amer. Chem. Soc., 77**, 3731 (**1955**).

[26]. H. R. Snyder and J. R. Brewster, **J. Amer. Chem. Soc., 71**, 291 (**1949**); S. Siegel and A. F. Graefe, *ibid.*, **75**, 4521 (**1953**).

[27]. D. J. Cram, **J. Amer. Chem. Soc., 71**, 3863 (**1949**); S. Winstein and K. C. Schreiber, *ibid.*, **74**, 2165 (**1952**). Additional references and examples are given by T. H. Lowry and K. S. Richardson, **Mechanism and Theory in Organic Chemistry**, 3rd Ed., Harper and Row Publishers, New York, **1987**, pp. 434–444.

[28]. See T. H. Lowry and K. S. Richardson, **Mechanism and...**", *loc. cit.*, pp. 579–584, for a discussion and primary references.

[29]. G. Lowe, **Accts. Chem. Res., 16**, 244–251 (**1983**).

[30]. S. L. Buchwald, D. H. Pliura, and J. R. Knowles, **J. Amer. Chem. Soc., 104**, 845 (**1982**).

[31]. N. S. Isaacs, **Liquid Phase High Pressure Chemistry**, Wiley, New York, **1981**.

[32]. J. R. McCabe and C. A. Eckert, **Accts. Chem. Res., 7**, 251 (**1974**).

[33]. W. J. LeNoble, **Prog. Phys. Org. Chem., 5**, 207 (**1967**).

2.11 PROBLEMS

Problem 2.1

The pseudo-first-order rate constant for solvolysis of a compound RX in aqueous solution was found to vary with temperature as follows:

k_ψ(sec$^-$1)	T°C	k_ψ(sec$^-$1)	T°C
1.20×10^{-4}	5.0	4.70×10^{-4}	30.0
1.64×10^{-4}	10.0	7.90×10^{-4}	40.0
2.72×10^{-4}	20.0	1.23×10^{-3}	50.0

Calculate E_a, $\ln Z$, $\Delta G^{\circ\ddagger}$, $\Delta H^{\circ\ddagger}$, and $\Delta S^{\circ\ddagger}$ for the reaction at $25°C$.

Problem 2.2

An isomerization reaction is reported to have $\Delta H^{\circ\ddagger} = 21.3$ kcal/M, and $k_\psi = 2.3 \times 10^{-4} sec^{-1}$, both at $20°C$. Calculate the rate constant expected for $50°C$.

Problem 2.3

The reaction of p-tolyldiazonium ion with 2-napthylamine-6-sulfonate in aqueous solution at $10°C$, with potassium chloride added to maintain constant ionic strength during each individual measurement of rate, gave the following second-order rate constants at the various ionic strengths studied:

$\mathcal{I}, molar$	k_2, $M^{-1}sec^{-1}$	$\mathcal{I}, molar$	k_2, $M^{-1}sec^{-1}$
0.05	0.189	0.20	0.144
0.10	0.165	0.30	0.135
0.15	0.153	0.50	0.130

See if the data are in accord with either Eq. (2.13)(a) or (2.13)(b). Use reference [7] to obtain values for the constants in Eq. (2.13)(b); try using ionic radii of 2 Å, 4 Å, and 8 Å for the reactant ions. Pay particular attention to whether or not the dependence of rate on ionic strength is in accord with the expected charge of zero for the transition state. Assuming that the lowest ionic strength used, 0.05 M, is low enough to give behavior in accord with the equations, calculate values for k_2 at zero ionic strength by each of the equations and for each of the ionic radii tried. What conclusion do you reach from this exercise?

Problem 2.4

The reaction of a carbocation, R^+, with sulfite ion was studied in aqueous solution at 25°C. The solutions were buffered to maintain constant pH throughout each rate measurement, and the total concentration of sulfite plus hydrosulfite was in large excess over the concentration of the carbocation so that pseudo-first-order kinetics were observed. The buffer solutions were prepared so that the ionic strength was known for the final solutions and the pH of each solution was measured with a pH meter. A weighed portion of sodium bisulfite was then added to a known volume of buffer solution so that the concentration of sulfite ion plus bisulfite ion is known. The carbocation was added, and the rates were determined. The following data resulted:

$\mathcal{I}, M$	pH	$[NaHSO_3]$ added, M	k_ψ, sec^{-1}
0.01	7.50	1.00×10^{-3}	0.962
0.05	7.50	2.00×10^{-3}	1.41
0.10	7.50	3.00×10^{-3}	1.75
0.15	8.00	1.00×10^{-3}	0.588
0.20	8.00	2.00×10^{-3}	1.06
0.25	8.00	3.00×10^{-3}	1.46

Show that the data are consistent with a reaction of $SO_3^=$ with the cation and the expected charge, -1, for the transition state. Use a pK_a value of 7.21 for bisulfite ion at infinite dilution in aqueous solution. Use Eq. [2.13](b) to calculate activity coefficients. Values of $\mathcal{A} = 0.511$ and of $\mathcal{B} = 1.32$ for HSO_3^- and $\mathcal{B} = 1.48$ for $SO_3^=$ are taken from reference [7]. Note that the measurement of pH gives the **activity** of the hydrogen ion.

Hint: Write the equation for the acid ionization of bisulfite ion, complete with activity coefficients, and solve for the **concentration** *ratio of sulfite to bisulfite ions at each pH and ionic strength. Then make a table of data with concentrations and activities of sulfite ion, and the calculated values for the second-order rate constants. There are some short-cuts that could be used, but the above procedure is the simplest way to proceed. You should find* $k_2^\circ = 2.0 \times 10^3 M^{-1} sec^{-1}$.

Problem 2.5

At very low salt concentrations in acetic acid solvent, threo-3-p-anisyl-2-butyl-p-bromobenzene- sulfonate solvolyzes with a rate constant of $7.8 \times 10^{-5} sec^{-1}$ at 25°C. Use the data in Table 2.2 to calculate the rate constant in the presence of 0.3 M $LiClO_4$.

Problem 2.6

The rate constant for the solvolysis of 2-adamantyl tosylate in formic acid at 25°C is $2.6 \times 10^{-5} sec^{-1}$. Use the data in Table 2.6 to calculate the rate constant for solvolysis in acetic acid.

Problem 2.7

The second-order rate constant for the reaction of azide ion with methyl chloride in methanol is $7.9 \times 10^{-5} M^{-1} sec^{-1}$ at 25°C. Use the data in Table 2.5 to calculate the rate constant in dimethylformamide.

Problem 2.8

The reaction of p-nitrobenzyl bromide with azide ion at 25°C has a rate constant of $4.0 \times 10^{-3} M^{-1} sec^{-1}$ in methanol and of $3.3 \times 10^{1} M^{-1} sec^{-1}$ in dimethylformamide. The activity coefficient of p-nitrobenzyl bromide in dimethylformamide is 0.070, relative to a standard state in methanol solution. Calculate the activity coefficient for the transition state in dimethylformamide relative to methanol standard state.

Problem 2.9

The rate constant for solvolysis of 2-adamantyl tosylate at 25°C is $4.7 \times 10^{-7} sec^{-1}$ in 50% ethanol-water and is $2.4 \times 10^{-8} sec^{-1}$ in 80% ethanol-water. At the same temperature, the rate constant for solvolysis of methyl tosylate is $4.41 \times 10^{-5} sec^{-1}$ in 50% ethanol-water and is $2.22 \times 10^{-5} sec^{-1}$ in 80% ethanol-water. Use Bentley's approach, outlined on pg 69, to calculate Y_{OTs} and N_{OTs} values for 50% ethanol-water.

Problem 2.10

The solvolysis of benzhydryl chloride in 80% acetone-water at 0°C was studied in the presence of varying concentrations of added potassium chloride. Ionic strength was kept at 0.10 M in all solutions by the addition of the required amounts of potassium bromide. The following pseudo-first-order rate constants were determined:

[KCl], M	k_ψ, sec^{-1}	[KCl], M	k_ψ, sec^{-1}
0.01	2.56×10^{-6}	0.07	1.66×10^{-6}
0.02	2.35×10^{-6}	0.085	1.53×10^{-6}
0.04	2.01×10^{-6}		

Calculate the ratio k_{-1}/k_2 for the reaction scheme:

$$\text{R-Cl} \underset{k_{-1}}{\overset{k_1}{\rightleftharpoons}} \text{R}^+ + \text{Cl}^-, \qquad \text{R}^+ + \text{solvent} \xrightarrow{k_2} \text{product}$$

where R^+ is a very reactive intermediate.
Hint: Use the Bodenstein approximation to obtain an expression for the pseudo-first-order rate constant, k_ψ. Look closely at the expression for $1/k_\psi$.

Problem 2.11

The reaction of ethylene oxide in water is believed to involve a rapid equilibrium protonation of the oxygen of the ethylene oxide, followed by a slow, rate-determining, SN2 attack of hydroxide ion on the protonated ethylene oxide. The protonated ethylene oxide, at reasonable pH, is a reactive intermediate to which the Bodenstein approximation may be applied. How do you expect the pseudo-first-order rate constant in buffered solutions to depend on pH? What salt effects would you expect to observe?
Hint: Read through Section 2.5 again.

Problem 2.12

Aniline and 2,4-dinitrochlorobenzene in ethanol solution react to form 2,4-dinitrodiphenylamine and hydrochloric acid. The reaction was followed under pseudo-first-order conditions with aniline present in large excess. The following data were obtained at 25°C:

$[PhNH_2]$, M	k_ψ, sec^{-1}	$[PhNH_2]$, M	k_ψ, sec^{-1}
0.200	1.47×10^{-5}	0.800	4.65×10^{-5}
0.300	2.12×10^{-5}	1.00	5.44×10^{-5}
0.400	2.72×10^{-5}	1.50	6.98×10^{-5}
0.600	3.76×10^{-5}	2.00	8.15×10^{-5}

Suggest a mechanism for the reaction and evaluate as many of the rate and equilibrium constants in your scheme as possible.
Hint: Consider the possibility of a rapid equilibrium reaction of the aniline with the 2,4-dinitrochlorobenzene to form an adduct which is **not** *a reactive intermediate.*

Problem 2.13

Transition states are always less stable than reactants or products of any step of a reaction, so both $\Delta G^{\circ\ddagger}$ and $\Delta H^{\circ\ddagger}$ must be positive for any elementary step of a reaction. There have been cases, however, in which the measured second-order rate constants for some reactions $A + B \rightleftharpoons C$ actually decreased with temperature.

How can you rationalize this result?

Hint: What would happen if A and B were in some rapid equilibrium to form something other than C?

Problem 2.14

The reaction of an optically pure α-bromocarboxylic acid in methanol solution produces the corresponding α-methoxycarboxylic acid. The reaction was followed by observed by following the production of bromide ion. Reactions were carried out in buffered solutions under pseudo-first-order conditions, so that the concentration of H^+ was constant throughout any one of the rate measurements. The α-methoxycarboxylic acid was isolated at the end of each kinetic run, and its optical purity was determined. The following data were obtained:

$[H^+]$, M	k_ψ, sec^{-1}	Product stereochemistry
1.00×10^{-3}	1.01×10^{-8}	100% inversion
1.00×10^{-4}	1.11×10^{-8}	82% net inversion
1.00×10^{-5}	2.10×10^{-8}	4% net inversion
1.00×10^{-6}	1.20×10^{-7}	66% net retention
1.00×10^{-7}	1.10×10^{-6}	80% net retention
1.00×10^{-8}	1.00×10^{-5}	82.6% net retention
1.00×10^{-9}	5.50×10^{-5}	82.6% net retention
1.00×10^{-10}	1.00×10^{-4}	80% net retention
1.00×10^{-11}	1.20×10^{-4}	66% net retention
1.00×10^{-12}	2.10×10^{-4}	4% net inversion
1.00×10^{-13}	1.00×10^{-3}	82% net inversion

The α-bromocarboxylic acid has an ionization constant of $1.00 \times 10^{-9} M$ in methanol, and the ion-product constant of methanol, $[H^+][CH_3O^-]$, is $1.00 \times 10^{-17} M^2$ at the temperature of the experiments. By X% net inversion, we mean that (100-X)% of the product is racemic; therefore, that there is X + $\frac{1}{2}$(100-X) percent inverted product, and $\frac{1}{2}$(100-X) percent retained product.
Propose a mechanism for the reaction which is consistent with the data, and calculate as many rate constants for your scheme as possible.
Hint: This is a long problem which must be approached systematically. First, note that that both the α-bromocarboxylic acid and the α-bromocarboxylate ion may be reactants. These two species are in very rapid equilibrium. Make another column for the above table which shows the fraction of the acid and the fraction of the carboxylate present at each pH.
Next, note that k_ψ could be written as a sum of a rate constant giving retained product and one giving inverted product; $k_\psi = k_I + k_R$. Then, the ratio %inversion/%retention is equal to k_I/k_R. You can add another two columns to the table, then, containing the values for $k_I = k_\psi \times$(fraction inverted), and $k_R = k_\psi \times$(fraction retained). Now, begin to consider the reaction scheme:

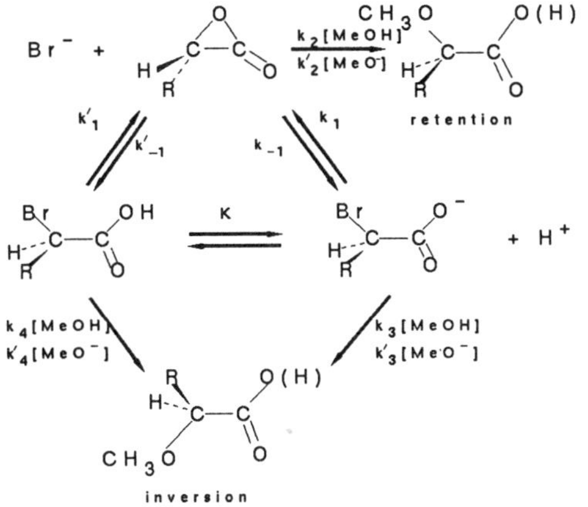

Reaction Scheme

in which the α-lactone is a reactive intermediate. You should now be able to evaluate some of the rate constants in the scheme by just scanning your table of data. As you evaluate them, subtract their contributions from the total k_ψ *to get the remaining ones. Not all of the rate constants in the scheme contribute to the observed rates, and there are some cases where contributions could have arisen from either of two paths. In the latter case, choose the chemically more reasonable possibility.*

Problem 2.15

Show that the equation between Eqs. (2.4) and (2.5) follows from the assumption that the number of transition states formed from products is proportional to the activity of the product, and the number formed from reactants is proportional to the product of activities of reactants.

Hint: First, define 'net fraction' to be the number of transition states formed from reactants minus the number formed from products, divided by the number from reactants. Find the ratio of the two proportionality constants from the fact that equal numbers of transition states are formed from reactants and products at equilibrium.

CHAPTER 3
Structural Effects on Reactivity

3.1 Introduction

The ultimate goal of the physical organic chemist is to be able to predict the course, rate, and equilibrium of any organic reaction under any set of conditions from simple first principles. There is ample opportunity left for future generations to contribute to attaining the goal, but some progress has been made.

At the simplest, most approximate, level, we can recognize that organic chemistry is a tractable subject largely because the concept of the chemistry of functional groups is useful. Every student of sophomore organic chemistry is familiar with the fact that, for example, all aliphatic alcohols react with sodium metal to form alkoxides, or that nearly all alkenes undergo hydroboration. In fact, things are usually even more well-behaved than indicated by these qualitative statements. For example, virtually all unsubstituted alkane carboxylic acids have acid ionization constants which are nearly the same. The studies of bond lengths, bond dissociation energies, heats of hydrogenation, and heats of formation in elementary organic chemistry were meant to establish the fact that particular structural features in a molecule are associated at least semi-quantitatively with physical and chemical properties of the molecule.

You may also have learned in elementary organic chemistry that certain substituents on a molecule are *electron-withdrawing* and that others are *electron-donating*. Electron-withdrawing substituents on an alcohol, for example, such as in 2,2,2-trifluoroethanol, cause the OH group to be more acidic than in the unsubstituted alcohol. In the present chapter, you will learn that these ideas can be made more quantitative.

3.2 Additivity Schemes for Atomic, Bond, and Group Properties

The concepts of covalent and ionic radii of atoms should be familiar from elementary courses. These concepts are useful because we are able to assign values for these radii that accurately give the bond lengths of bonds in molecules or contact distances in ionic crystals.

You should recognize that in actual practice it is the bond lengths or contact distances that are known from experiment. If we want to assign a covalent radius to carbon, for example, we might look at the experimental values of the C-C bond lengths in a number of hydrocarbons, then take half of the average length as the covalent radius of carbon. We would find that the C-C bond length in virtually all saturated hydrocarbons is 1.53 Å within a few hundredths of a unit. We would then assign a covalent radius of 0.77 Å to saturated carbon. We could assign a covalent radius of 1.33 Å to iodine from the bond length of 2.66 Å in I_2. Now, we can estimate the bond length of the C-I bond in methyl iodide to be $0.77 + 1.33 = 2.10$Å, which turns out to be in perfect agreement with the measured value. Such perfect agreement is not always found, and, in general, covalent radii are assigned by considering all known bond lengths involving the particular atom, then setting the covalent radius such that the best agreement with experiment is obtained. Some values for covalent radii are shown in Table 3.1.

TABLE 3.1
Covalent Radii of Some Common Atoms
Data from [1].

Atom	Single Bond	Double Bond	Triple Bond
C	0.77	0.67	0.60
N	0.74	0.62	0.55
O	0.74	0.62	
H	0.37		
F	0.72		
Cl	0.99		
Br	1.14		
I	1.33		

The inclusion of different values for multiple bonds to the atoms allows for the fact that multiple bonds are generally shorter than single bonds, and is actually a correction to simple atomic additivities. If we are willing to accept a little more 'correction', by allowing different covalent radii for different ligancies or hybridizations of the atoms, even cases such as the C-C single bond in butadiene, with a bond length of 1.47 Å, or the C-C single bond in butadiyne, with a bond length of 1.37 Å can be handled (See Problem 3.8 at the end of this chapter.).

It is quite remarkable that atomic contributions to thermodynamic quantities can be assigned in the same way as the contributions to bond lengths. These assignments allow reasonable estimates of entropies and heat capacities of molecules. Some values of atomic contributions to the thermodynamic quantities, again with different values for different ligancies, are shown in Table 3.2

TABLE 3.2
Atomic Contributions to Entropies and Heat Capacities
Cal/deg-mole at 25°C
Data from [2].

Atom	C_p^o	Standard Entropy			
		Ligancy			
		1	2	3	4
H	0.85	21.0			
C	3.75		5.3	-13.5	-32.6
N	3.40	22.9	5.2	-12.1	
O	3.40	25.5	8.8		
F	2.40	25.5			
Cl	3.70	28.4			
Br	4.20	31.3			

As an example of the use of data from Table 3.2, consider 2-butene, consisting of eight hydrogens, two carbons with ligancy of three, and two carbons with ligancy of four. Thus, the standard C_p^o is estimated as $8 \times 0.85 + 4 \times 3.75 = 21.8$ e.u., and the standard entropy as $8 \times 21.0 - 2 \times 13.5 - 2 \times 32.6 = 75.8$ e.u. The experimental values are $C_p^o = 19.0$ e.u., $S^o = 72.1$ e.u. for cis- 2-butene, and $C_p^o = 21$ e.u., $S^o = 70.9$ e.u. for trans-2-butene. Cyclic compounds and ionic species cannot be accomodated by this simple atomic additivity scheme, and attempts to use a similar scheme for enthalpies or free energies fail badly for any but the simplest molecules.

It should be noted that if any property of molecules is given by an atomic additivity scheme allowing only for differences in ligancies, then that property will show no change for any chemical reaction which does not involve changes in ligancies. For example, if we have any reaction that can be represented as:

$$\text{A-X} \; + \; \text{B-Y} \quad \xrightleftharpoons{\Delta R_{ex}} \quad \text{A-Y} \; + \; \text{B-X}, \tag{3.1}$$

where R is any quantity given by atomic additivity, then ΔR_{ex} must equal zero for the reaction. We'll see shortly that this is the type of reaction which we need to consider first when we wish to go beyond additivity schemes.

In order to find an additivity scheme that is at all useful for enthalpies or free energies, we must go considerably beyond atomic additivity. We could either abandon the simple principle of additivity altogether, or attempt to assign additive contributions for larger moieties of molecules than the atoms. The first step in this latter direction is to attempt bond additivity schemes. Bond additivity schemes have been heavily used by organic chemists, and for many years served as the basis for evaluating other phenomena such as resonance stabilizations. Some bond additivity parameters are given in Table 3.3 for a few very common bonds.

TABLE 3.3
Bond Additivity Parameters for Some Common Bonds
Data from [3].

Bond	Bond Length Å	$\Delta H^{\circ}_{dissociation}$ kcal/mole	Dipole Moment Debyes
C-C	1.53	83.1	0
C=C	1.34	145.0	0
C≡C	1.21	198.0	0
C-H	1.12	98.8	0.30
C-O	1.43	84.0	0.86
C=O	1.20	179.0	2.40
N-H	1.03	93.4	1.31
O-H	0.97	117.5	1.53
C-N	1.51	62.0	0.40
C=N	1.29	121.0	0.90

As an example of how bond additivity has been used to evaluate other phenomena, we can consider the heats of hydrogenation of ethylene and of propene. The hydrogenation of either molecule involves converting one H-H bond and one C=C bond into two C-H bonds and one C-C bond. According to bond additivity, the heats of hydrogenation should be the same. In fact, the heat of hydrogenation of ethylene is 32.8 kcal/mole, and that of propene is 30.1 kcal/mole. The fact that the value for propene is less than that for ethylene by 2.7 kcal/mole was originally attributed to a stabilization of propene by a resonance interaction of the methyl group with the double bond, called *hyperconjugation* or *double-bond-no-bond resonance.*

More modern considerations [4] place less emphasis on the contribution of resonance stabilization to the energy, and recognize the possibility of steric interactions of the methyl group with the methylene group in the hydrogenated product.

One may go even further in this direction of considering larger moieties in additivity schemes by defining group contributions to molecular properties. For example, group contributions of -CH_3, -CH_2-, -OH, etc. to thermodynamic properties of organic molecules have been assigned [1], [2] that allow heats of formation of mono-functional organic compounds to be calculated with an accuracy of a few kcals/mole. The most accurate of these additivity schemes become quite complex in allowing 'corrections' for gauche conformations, cis-interactions in alkenes, and other such structural features [5].

The basic conclusion to be drawn from the successes of these additivity schemes is that the properties of many molecules are largely determined by nearest-neighbor interactions of the atoms. When one begins to consider organic compounds containing two or more polar groups, however, it becomes obvious that non-nearest-neighbor interactions make significant contributions to the thermodynamic properties of the molecules. None of the additivity schemes are accurate enough for practical application to enthalpies or free energies of reactions of such multiply substituted organic compounds. There is simply no way that additivity schemes can account for the repulsions or attractions of two dipolar groups within a molecule, or for the strong steric interactions of two very large groups that are forced into close proximity by the geometry of a molecule, such as in the case of a 1,3-diaxially substituted cyclohexane. There are, nevertheless, several other simple empirical approaches to handling these non-nearest-neighbor interaction effects that are highly successful within reasonable limits.

One of the very early and well known concepts which moves away from additivity schemes is that of *electronegativity*, which attributed the discrepancies in bond energies from atomic additivity to a sort of ionic-covalent resonance stabilization resulting from differences in electron-attracting characters of the two atoms involved in the bonding. The fact that the bond dissociation of an *A-B* bond is, in general, greater than the mean of the bond dissociation energies of *A-A* and *B-B* is attributed to a difference in electronegativities of *A* and *B*. Quantitatively, the difference in electronegativities of *A* and *B* is defined by:

$$[\chi_A - \chi_B]^2 \equiv \Delta H^\circ_d(AB) - \frac{\Delta H^\circ_d(AA) + \Delta H^\circ_d(BB)}{2} \tag{3.2}$$

where the dissociation energies are expressed in units of e.v./mole (1 e.v.= 23.06 kcal). The individual atomic electronegativities, χ_A and χ_B, are given absolute values by assigning a value of 4.0 to the most electronegative atom, fluorine. A little algebra will show you that this definition of electronegativity leads to a 'predicted' value of

$$\begin{aligned} \Delta H^\circ_{ex} &= 2[\chi_A - \chi_B][\chi_X - \chi_Y]\text{e.v./mole} \\ &= 46.1[\chi_A - \chi_B][\chi_X - \chi_Y]\text{kcal/mole} \end{aligned} \tag{3.3}$$

for the exchange reaction of Eq. (3.1), above. A little thought should convince you that the simplest mathematical function involving any atomic property that leads to non-zero ΔH°_{ex} for reactions representable by Eq. (3.1) is a square function such as that in Eq. (3.2). This fact will perhaps make it less surprising that several of the best known empirical relations dealing with substituent effects on reactivities of organic molecules have the same form as Eq. (3.3).

The original formulation of electronegativities as atomic properties does not accomodate the general problem of substituent effects on organic reactions, such as the one mentioned in the introduction of 2,2,2-trifluoroethanol being a considerably stronger acid than ethanol. The effect of the fluorines on the acidity can be expressed most conveniently by considering the two reactions:

$$CF_3CH_2OH \overset{K_{TFE}}{\rightleftharpoons} CF_3CH_2O^- + H^+$$

$$CH_3CH_2OH \overset{K_{Et}}{\rightleftharpoons} CH_3CH_2O^- + H^+$$

which may be combined to give the exchange reaction:

$$CF_3CH_2OH \quad + \quad CH_3CH_2O^- \quad \xrightleftharpoons{K_{ex}} \quad CF_3CH_2O^- \quad + \quad CH_3CH_2OH$$

with the effect of the change from methyl to trifluoromethyl group given by

$$\log K_{ex} = \log K_{TFE} - \log K_{Et} \quad \text{and} \quad \Delta G^\circ_{ex} = \Delta G^\circ_{TFE} - \Delta G^\circ_{Et}.$$

Since all of the atom-atom bonds on the left of this exchange are also present on the right, the electronegativity concept 'predicts' $\Delta G^\circ_{ex} = 0$. The actual value of ΔG°_{ex} in aqueous solution at $25°C$ is approximately 8 kcals/mole.

One way of accomodating such experimental facts is to extend the concept of electronegativity to groups. For example, we could try to assign electronegativities to groups such as CF_3-, CH_3-, $HOCH_2$-, and $-OCH_2-$. This approach has been considered [6] and shown to lead to results which are indistinguishable from the Hammett and Taft approaches to substituent effects which we shall examine now.

3.3 The Hammett Equation

Regularities in the changes in reactivities with changes in structures of organic molecules were first noted for *meta*- and *para*-substituted benzene derivatives. At a qualitative level, it was observed that in any reaction where a methyl group, for example, substituted for a hydrogen at the *para*-position caused an increase in reactivity, then a nitro group substituted at the same position caused a decrease in reactivity, and *vice versa*. At the quantitative level, it was frequently found that plots of $\log K_X$'s for one reaction vs. those for another approximated straight lines, where the K_X's are the equilibrium constants for the various X-substituted benzene derivatives.

As examples of this behavior, data for the ionization constants of several substituted benzoic acids:

$$\text{X-C}_6\text{H}_4\text{COOH} \xrightleftharpoons{K_a} \text{X-C}_6\text{H}_4\text{COO}^- + \text{H}^+$$

and for rate and equilibrium constants of reactions of similarly substituted benzenediazonium ions with sulfinate ions:

$$\text{X-C}_6\text{H}_4\text{N}_2^+ + \text{C}_6\text{H}_5\text{SO}_2^- \xrightleftharpoons{k_f, K_{eq}} \text{X-C}_6\text{H}_4\text{N=NSO}_2\text{C}_6\text{H}_5$$

TABLE 3.4
Logarithms of Rate and Equilibrium Constants, Ionizations of Benzoic Acids in Water (Log K_a), and Reactions of ArN_2^+ with $C_6H_5SO_2^-$ in MeOH, (Log K_{eq} and Log k_f) at 25° [7], [8].

X	Log K_a	Log K_{eq}	Log k_f
H	-4.20	5.16	2.51
p-CH_3	-4.37	4.64	2.20
p-Cl	-3.99	6.04	2.99
m-CF_3	-3.77	7.14	3.67
m-Cl	-3.82	6.89	3.56
p-CN	-3.55	7.68	4.17
p-NO_2	-3.44	8.01	4.28

are given in Table 3.4, and the log-log plots are shown in Fig. 3.1.

Since $\Delta G^\circ = -RT \ln K$, the Log-Log plots are equivalent to plots of ΔG°'s for one reaction vs those for another, and the relationships are frequently called *linear free energy relationships.*

The linear plots can be expressed in linear equation form; for example:

$$\log K_{eq}^X - \log K_{eq}^H = 3.8(\log K_a^X - \log K_a^H)$$

is the equation for the line drawn in Fig. 3.1 for the equilibria of the diazonium ion reactions. It is convenient to choose some particular substituent and some particular reaction of substituted benzene derivatives to be the standards with which all others are compared. The standard substituent is chosen to be hydrogen atom, making the unsubstituted benzene derivative the one with which others are compared. The standard reaction should involve easily obtainable reactants and have easily measurable equilibrium or rate constants. A particularly convenient reaction is, in fact, just the ionization of substituted benzoic acids which we have been discussing. The pK_a's of a large number of these acids have been measured in water at 25°C. Some of the data are shown in Table 3.5.

Once the standard reaction is chosen, we may abbreviate the appearance of the linear equations by defining:

$$\sigma_X \equiv \log K_a^X - \log K_a^H \qquad (3.4)$$

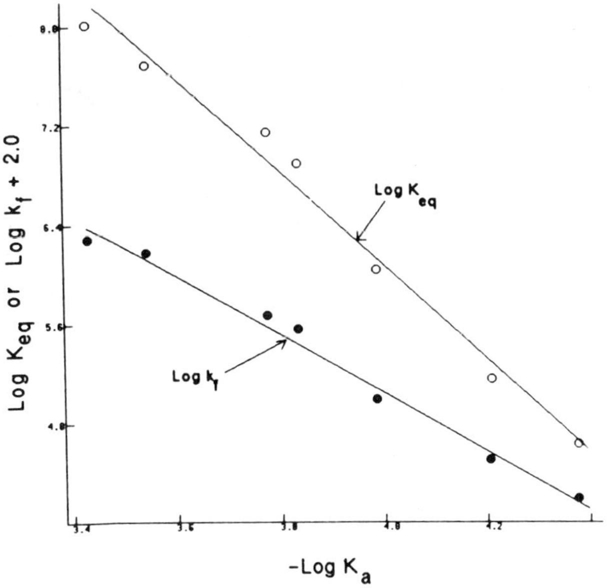

Figure 3.1: *Plot of Log K_{eq} and Log k_f vs Log K_a from Table 3.4*

and denoting the slope of the plots as ρ. The equation for the line describing the K_{eq} data in Fig. 3.1 then becomes:

$$\log K_{eq}^{X} - \log K_{eq}^{H} = \log\left(\frac{K_{eq}^{X}}{K_{eq}^{H}}\right) = \sigma_X \rho, \tag{3.5}$$

with ρ=3.8. An equation such as Eq. (3.5) is called a *Hammett equation* in honor of L. P. Hammett, who first wrote the correlations in this form [11]. The σ_X parameters are called Hammett substituent constants of the substituents X. Each set of rate constants or equilibrium constants for a reaction series, such as the equilibrium constants of the variously substituted benzenediazonium ions reacting with sulfinate ion in methanol, which is correlated by a Hammett equation will have some particular value of ρ. The series of rate constants for the benzenediazonium ion reactions, for example, has $\rho = 2.3$.

There are several commonly used terms that can be defined at this point. First, those substituents which cause the substituted benzoic acid to be stronger than the hydrogen substituted acid have positive σ

TABLE 3.5
pK_a's of Some Substituted Benzene Derivatives in Water* at 25°C
Ar=X-C_6H_4-

X	ArCOOH ref. [8]	ArCH_2COOH ref. [9]*	ArOH ref. [10]	ArNH_3^+ ref. [10]
H	4.20	4.50	9.90	4.59
m-CH_3	4.25	4.56	10.09	4.68
p-CH_3	4.37	4.57	10.27	5.09
m-CH_3O	4.09	4.49	9.64	4.23
p-CH_3O	4.49	4.55	10.19	5.33
m-CF_3	3.77	-	8.90	-
m-F	3.86	4.34	9.28	3.58
p-F	4.14	4.43	9.88	4.64
m-Cl	3.82	4.33	9.11	3.52
p-Cl	3.99	4.37	9.42	3.98
m-Br	3.81	4.33	9.02	3.52
p-Br	4.00	4.36	9.33	3.87
m-HO	4.08	4.49	-	-
p-HO	4.58	4.59	-	-
m-NC	3.60	4.18	8.56	-
p-NC	3.55	4.11	7.96	-
m-O_2N	3.46	4.14	8.35	2.45
p-O_2N	3.44	4.05	7.15	1.00
m-$(CH_3)_3N^+$	3.16	3.80	8.05	-
p-$(CH_3)_3N^+$	3.24	3.89	8.34	-
p-OHC	3.98	-	7.64	-
p-EtO_2C	3.75	-	8.54	-

*Data for Arylacetic acids are for 10% ethanol-90%water solution.

values, and those which cause the substituted benzoic acid to be weaker than the hydrogen substituted acid have negative σ values. Notice that $\sigma_H = 0$ by Eq. (3.4). We speak of those substituents having positive σ values as being *electron withdrawing* and those with negative σ values as being *electron donating*. If a reaction series has a positive ρ value, the reaction is *aided by electron withdrawing groups*, and if it has a negative

ρ value, it is ***aided by electron donating groups.*** Both the equilibria and the rates of the benzenediazonium ion reactions discussed above are ***aided by electron withdrawing groups*** since both series have positive ρ values.

The definition of σ values in Eq. (3.4) has also **defined** the ρ for the ionization of benzoic acids in water at 25° to be +1.00. Since this reaction involves a neutral benzoic acid being converted to a negatively charged benzoate ion, we may expect any reaction series which involves conversion of a neutral reactant to a negatively charged product, **or a positively charged reactant to a neutral product**, to have a positive ρ. Conversely, any reaction series which involves conversion of a neutral reactant to a positively charged product, or a negatively charged reactant to a neutral product, will have a negative ρ. We can also expect that the magnitude of the ρ for a reaction will depend on the amount of change in charge between reactant and product, and **on the distance of the charge from the substituted benzene ring.** For example, we should expect that the ρ for the acid-dissociation of anilinium ions:

$$ArNH_3^+ \xrightleftharpoons{K_{dis}} ArNH_2 + H^+$$

is positive and greater than 1.00 since the charge is on the nitrogen directly attached to the benzene ring, whereas the charge of benzoate ions is on the oxygens which are one atom removed from the ring. The experimental ρ for ionization of anilinium ions in water is 2.8. Similarly, the ρ for the rates of an electrophilic aromatic substitution reaction:

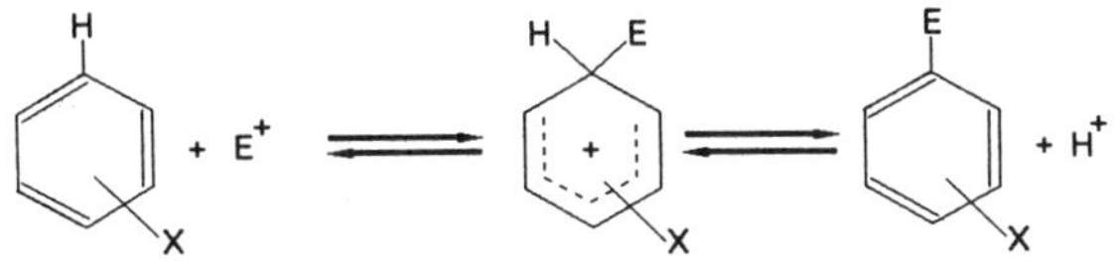

having a transition state for the rate-determining step which resembles the intermediate, should be negative in sign and of magnitude considerably greater than 1.0 because the charge is actually on the benzene ring. Typical experimental values [12] are -6.2 for nitration in nitromethane and -12. for bromination in acetic acid solution. Values of ρ are solvent dependent; for the ionizations of benzoic acids in dimethyl sulfoxide, for example, ρ=2.6. Any solvent-solute interaction that stabilizes ionic species will generally decrease the magnitude of the ρ for reactions involving such species.

In practice, we turn all of the above arguments around and use the signs and magnitudes of ρ values for rate constant series to infer the charge types and developments of the transition states of reactions. For example, the rate constants for solvolyses of a series of methylarenesulfonates in ethanol at 70°C are correlated by the Hammett equation with ρ=1.3 [13]. In order to interpret this value, we must account for the fact that ρ values are temperature and solvent dependent. Usually, ρ is proportional to 1/T, where T is the absolute temperature [14]. Therefore, the ρ of 1.3 at 70° would be approximately 1.5 at 25°. The most direct way to handle the solvent effect on the ρ is to note that the ionization constants for benzoic acids in ethanol at 25° is 1.96 [15]. Since the location of charge on the benzoate ions and sulfonate ions are comparable, we would expect that the full development of the sulfonate charge would produce ρ=1.96. The temperature-corrected observed value for the solvolyses then indicates that the charge development is approximately 76% complete (100% × 1.5/1.96) at the transition state.

The most direct interpretations of ρ values for rate series utilize a comparison with the ρ for the equilibrium constants of the same reactions under the same set of conditions. The ratio of the ρ's, ρ_{rate}/ρ_{eq}, gives the extent of progress of charge development toward product at the transition state. One must be careful that the ρ's used are those for an elementary step of the overall reaction if the reaction is complex.

Substituents on a benzene ring are believed to exert their effects on reactivities primarily by way of two distinguishable modes of action. The inductive, or polar, mode of action arises from interactions of dipoles or charges in the substituent and in the reaction site. The resonance mode of interaction arises from resonance interactions of the substituents with the benzene ring placing partial charges on the *ortho-* and *para-* carbons which interact with the reaction site. Those substituents having a lone pair of electrons on the atom attached to the benzene ring can have resonance forms placing negative charge on the *ortho* and *para-* carbons, while those with an unsaturated atom attached to the ring place positive charge at the same positions. For example:

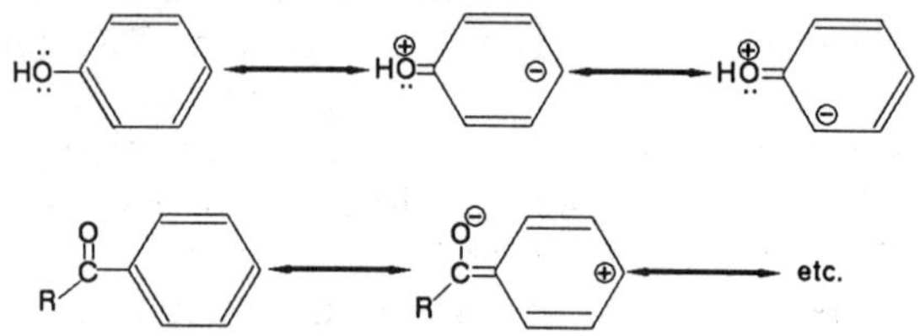

The success of the Hammett equation is due to the remoteness of the substituents to the reaction sites, which prevents any direct steric interactions between the two sites. *Ortho*-substituents generally fail to

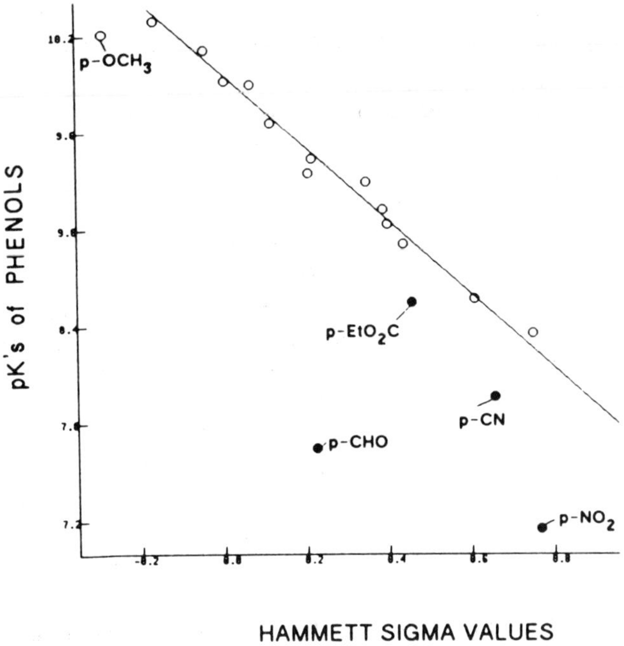

Figure 3.2: *Hammett Plot of pK's of Phenols from Table 3.5.*

obey the Hammett equation, although limited reaction series can be found in which *ortho*-substituents can be included [16]. Even in *para*-substituted compounds, deviations from the simple Hammett equation are observed for cases in which resonance forms involving direct interactions between reaction site and substituent are possible. For example, the data in Table 3.5 for the ionizations of phenols are shown plotted vs the Hammett σ's in Fig. 3.2.

The p-substituted phenols for which the resonance forms of the phenoxides can be written such as:

are significantly more acidic than expected from the corresponding benzoic acids. Special substituent constants, denoted as σ^-, have been defined for those substituents capable of such resonance to be used in reaction series involving a reaction site of either reactant or product which bears a lone pair of electrons on the atom attached to the benzene ring. The σ^- values are obtained, essentially, by moving the points in Fig. 3.2 horizontally onto the line drawn through the points for *meta*-substituents [16a].

Notice in Fig. 3.3 that the p-nitro substituent influences the acidity of anilinium ions in the same way as it does the phenols, indicating that the σ^- values are not limited to use with negatively charged reaction sites. One might have expected that the charge separated resonance forms which can be written for the anilines would be less important than the charge delocalized forms written for the phenoxides.

The feature of the reactions requiring these enhanced substituent constants is a difference in the resonance stabilization of the reactant and product of the reaction series. The phenoxide ions are stabilized more than are the phenols, and the anilines are stabilized more than are the anilinium ions, by substituents capable of the direct resonance electron withdrawal. This *differential* nature of the effect is probably responsible for the similar substituent effects on the ionizations of the phenols and anilines. Although the phenoxides probably have more such resonance stabilization than do the anilines, the phenols also have more than do the anilinium ions.

There is also a possibility of direct resonance interactions between substituent and reaction site for those p-substituents bearing a lone pair of electrons on the atom attached to the benzene ring with reaction sites having an unsaturated atom attached to the ring. A set of substituent constants, σ^+, have been defined [17] based on the rates of solvolysis of t-cumyl chlorides believed to have transition states closely resembling the carbocation intermediates, where resonance forms such as illustrated for the p-methoxy group:

are important.

Some of the various σ constants are shown in Table 3.6.

Both σ^- and σ^+ are the same as the ordinary σ values for all *meta*-substituents and for those *para*-substituents not capable of the direct resonance interactions of the types shown. These direct resonance interactions of *para*-substituents having lone-pairs of electrons on the atom attached to the benzene ring are believed to 'contaminate' even the ordinary σ values, since resonance forms, such as illustrated for the p-hydroxy group:

are possible which are more important than the corresponding forms for the benzoate ions. In order to obtain a set of σ's free of these 'contaminations' from direct interactions, reaction series having the reaction site 'insulated' from the benzene ring, such as the ionizations of substituted phenylacetic acids reported in Table 3.5, have been used to define sets of σ values, denoted as σ°, or σ^{n} [9], [18]. Again, these differ from the ordinary σ values only for those p-substituents having lone pairs on the atom attached to the benzene ring, notably the p-OR, p-NH_2, and p-NR_2 substituents. Unfortunately, reaction series such as the ionizations of the phenylacetic acids have relatively small substituent effects because of the greater separation of substituent and reaction site than in the benzoic acids. This means that small experimental errors in the determinations of the equilibrium constants, or small contributions from

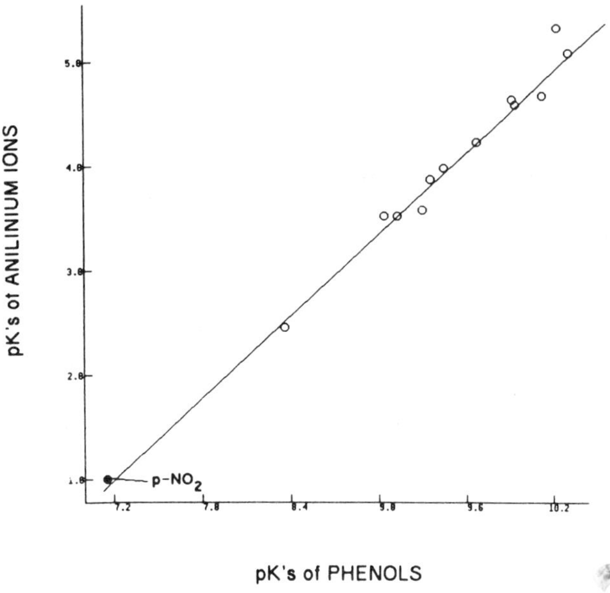

Figure 3.3: *pK's of Phenols and Anilinium Ions. Data from Table 3.5.*

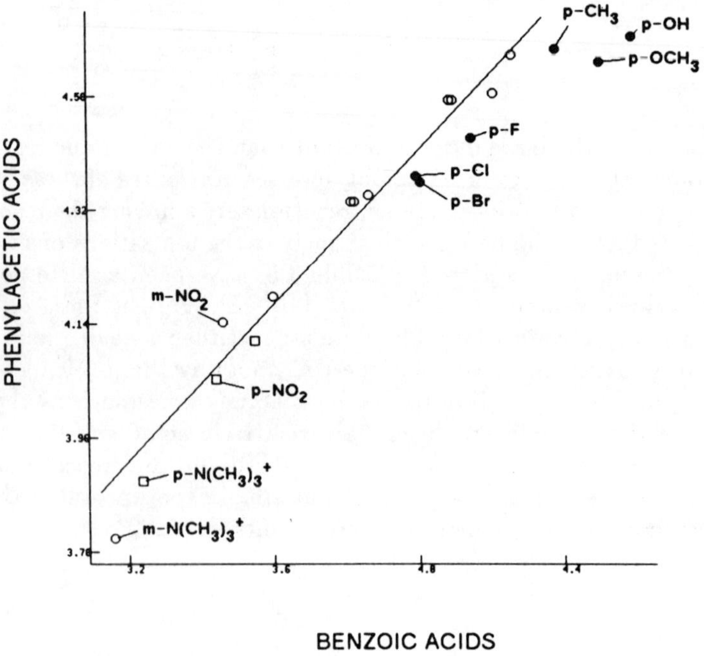

Figure 3.4: *Plot of pK's of Phenylacetic Acids vs Those For Benzoic Acids. Data from Table 3.5.*

'non-Hammett' effects, lead to appreciable uncertainties in the derived σ° values. Thus, as seen from the plot of pK's of phenylacetic acids vs those for benzoic acids in Fig. 3.4, there is considerable scatter in the plot even for the *meta*-substituents and *para*-substituents where the direct resonance interactions do not occur.

Most of the more recent attempts to account for enhanced substituent effects have used an approach which attempts to separate these direct resonance interactions in a more fundamental way. We will return to this topic after discussing the more general separation of polar and resonance modes of substituent effects.

3.4 The Separation of Inductive and Resonance Effects

As already stated above, the substituent effects in benzene reactivities can be attributed to two different modes of operation; the polar or inductive mode and the resonance mode. The attempts to evaluate the separate contributions of these two modes have focussed on substituent effects in saturated systems. Two different approaches have led to essentially the same results.

The most straightforward approach was to work with a system which has a similar spatial isolation of substituent and reaction site, a similar relative orientation of substituents and reaction sites, and a similar geometric rigidity, to those in the benzene derivatives, but having a saturated group to which the substituents and reaction sites are attached. The ideal system on all these points, designed by Roberts and Moreland [19], was that of the 4-substituted, bicyclo[2.2.2]octane-1-carboxylic acids:

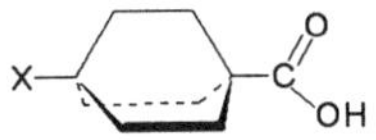

and their derivatives. The compounds with X=H, COOEt, OH, CN, COOH, and Br were synthesized and studied.

A set of substituent constants, denoted as σ^I, were defined from the pK_a's of the acids measured in 50% aqueous ethanol:

$$\sigma_X^I = \frac{1}{1.45}(\log K_{a(X)} - \log K_{a(H)}), \tag{3.6}$$

where the factor 1.45 is intended to place the σ^I values on the same scale as the Hammett σ's, and is equal to the Hammett ρ for the ionization of benzoic acids in 50% ethanol solvent. Several reactions of the 4-X-bicyclo[2.2.2]octane-1-carboxylic acid derivatives, for example the base catalyzed hydrolyses of the ethyl esters, were shown to obey an equation of the same form as the Hammett equation, but using the σ^I substituent constants.

Most interesting, it was observed that

$$\sigma_{p-X} - \sigma_X^I \approx 3.0(\sigma_{m-X} - \sigma_X^I)$$

for the substituents studied. This is an indication that the polar effect of substituents, measured by the σ^I values, are the same in the benzene series and the aliphatic series, and that the resonance effects of the substituents in the benzene series are approximately three times greater at the *para-* than at the *meta-* positions. Assuming that the Hammett substituent constants for a group, X, can be written as a sum of a polar substituent constant, σ_X^I, which is the same for the *para-* and *meta-*X, and a resonance substituent constant which differs for the two positions:

$$\sigma_{m-X} = \sigma_X^I + \sigma_{m-X}^R \quad \text{and} \quad \sigma_{p-X} = \sigma_X^I + \sigma_{p-X}^R,$$

the data for the limited number of substituents indicated that $\sigma^R_{p-X} \approx 3\sigma^R_{m-X}$.

Considerably later, ionizations of quinuclidinium ions:

having all of the desirable features of the bicyclo[2.2.2]octane series, and with the reaction site closer to the substituents, were used to obtain a fair variety of σ^I values [20].

A completely different approach to the evaluation of polar and resonance effects [21] was taken at about the same time as that of the Roberts approach. It was assumed that polar, resonance, and steric effects are always additive, and that, as a result, any substituent effect on any reactivity could be expressed as:

$$\log K_X - \log K_H = P_X + R_X + S_X,$$

where P_X, R_X, and S_X are the polar, resonance, and steric effects, respectively, of substituent X relative to H. The general idea, now, to evaluate any particular effect is to somehow eliminate the other two.

The rates of hydrolyses of a large number of substituted acetate esters, under acidic and basic conditions, had been reported in the literature, and the mechanisms of the acid-catalyzed and base-catalyzed reactions had been pretty well established. For the acid-catalyzed reactions, the transition state for the rate-determining step is believed to closely resemble the intermediate:

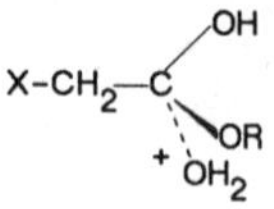

Structure I

and for the base-catalyzed reactions, the transition state for the rate-determining step is believed to closely resemble the intermediate:

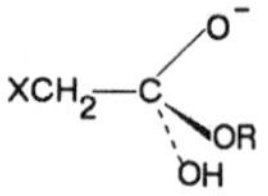

Structure II

Structures I and II differ only in charge and the hydrogens attached to oxygens. In particular, on going from the ester to either transition state, the carbonyl carbon becomes 'four-coordinate'; any change in steric effect or resonance effect should be nearly the same on going from the ester to either of the two structures. The polar effects, however, should be quite different since structure I bears a positive charge while structure II has a negative charge. Using superscripts A and B for the acid and base catalyzed reactions, respectively, we may write:

$$(\log k_X^A - \log k_H^A) = P_X^A + R_X^A + S_X^A$$

and

$$(\log k_X^B - \log k_H^B) = P_X^B + R_X^B + S_X^B,$$

Since $R_X^A = R_X^B$ and $S_X^A = S_X^B$, we obtain:

$$(\log k_X^A - \log k_H^A) - (\log k_X^B - \log k_H^B) = P_X^A - P_X^B.$$

The left-hand-side of this equation was then used to evaluate polar substituent constants, originally denoted $\sigma^*_{CH_2X}$, but later shown to be proportional to the σ_X^I as defined above. We have, therefore:

$$\sigma_X^I = \Upsilon(\log k_X^A - \log k_H^A) - (\log k_X^B - \log k_H^B), \tag{3.7}$$

where Υ is simply a constant chosen to make the two definitions, Eqs. (3.6) and (3.7), the same.

In still later work, it was found that the ionizations of α–substituted acetic acids in water accurately obey the equation[1]:

$$\log K_a^X - \log K_a^H = \sigma_X^I \rho_I. \tag{3.8}$$

Most of the currently known σ^I values are derived from either the rates of hydrolysis of acetate esters or from the ionizations of acetic acids [8]. Some values are given in Table 3.6.

Having the polar substituent constants, one may now attempt to evaluate the resonance effects of substituents in various ways. Taft and his co-workers [22] have advocated a fairly general approach through the equation:

$$\log \frac{K^X}{K^H} = \sigma_X^I \rho_I + \sigma_X^R \rho_R, \tag{3.9}$$

and have tabulated four distinct sets of resonance substituent constants, obtained from 'ordinary' (benzoic acid ionization) σ, 'pure' (ie; phenylacetic acid ionizations) σ°, σ^+, and σ^- values.

Another form of separation of resonance effects in benzene reactivities has given more practically interpretable results. Shortly after the

---

1 This form of equation for aliphatic reactivities is called the Taft equation

definition of the σ^+ constants, the equation, known as the Yukawa-Tsuno Equation:

$$\log \frac{k_X}{k_H} = \rho\left[\sigma_X^\circ + r(\sigma_X^+ - \sigma_X^\circ)\right],$$

where r is an adjustable parameter, was shown to give better correlations of some reaction series than the Hammett equation using either ordinary σ° or σ^+ values [23]. The parameter r was interpreted as a measure of the extent of direct resonance between electron donating substituents and electron withdrawing reaction sites, ranging from zero for 'insulated' reactions to unity for the cumyl chloride solvolyses defining σ^+. A slightly different form of the Yukawa-Tsuno equation [24]:

$$\log \frac{k_X}{k_H} = \sigma_X^\circ \rho + (\sigma_X^+ - \sigma_X^\circ)\rho_R, \tag{3.10}$$

shows more clearly the relative magnitudes of the direct resonance effect and the polar plus indirect resonance effect, measured by ρ_R and ρ, respectively.

The use of Eq. (3.10) for gaining information concerning the progress of transition states toward product structures has been particularly interesting. This information is obtained by comparing ρ's for series of rates with those for the corresponding equilibria, as already mentioned above for the simple ρ values. For example [24], the rates and equilibria of reactions of a series of substituted acetophenones with bisulfite ion were studied and the rate and equilibrium constants for the elementary step:

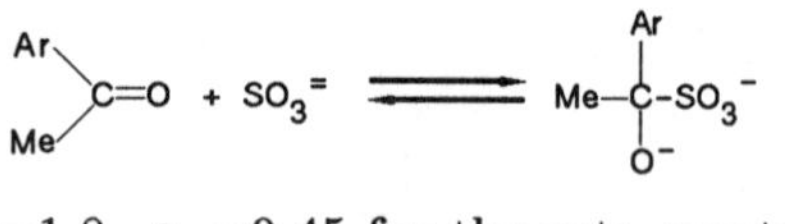

were found to give ρ=1.8, ρ_R=0.45 for the rate constants, and ρ=2.3, ρ_R=0.95 for the equilibrium constants. The ratio of the ρ's for rates and equilibria,1.8/2.3 $\approx$ 0.8, indicates that charge development at the transition state for the addition step is 80% complete, while the ratio of the ρ_R's, 0.45/0.95 $\approx$ 0.5, indicates that the loss of resonance interaction of the substituents with the carbonyl group is only 50% complete at the transition state.

It was also shown that the use of Eq. (3.10) for the rates of reactions of substituted benzyl halides with nucleophiles gave more reasonably interpreted correlations than other equations. These reactions show highly curved, or even U-shaped, plots of log k's against ordinary Hammett σ's. By the use of Eq. (3.10), one obtains positive ρ values, but **negative** ρ_R values, for reactions of anionic nucleophiles. These are interpreted

TABLE 3.6
List of Substituent Constants
σ^I_X, σ_{m-X}, and σ_{p-X} from ref. [8]. σ^+_X and σ^-_X from Ref. [25].

X	σ^I_X	σ_{m-X}	σ_{p-X}	σ^-_{p-X}	σ^+_{p-X}
$-CH_3$	0(-0.01)*	-0.06	-0.17	-	-0.31
$-CH_2CH_3$	0(-0.01)*	-0.05	-0.15	-	-0.30
$i-C_3H_7$	0(-0.01)*	-0.04	-0.15	-	-
$-C{\equiv}CH$	0.29	0.29	0.25	-	-
$-CH{=}CH_2$	0.11	0.06	-0.04	-	-
$-C_6H_5$	0.12	0.09	0.01	-	-0.17
$-CCl_3$	0.36	0.40	0.44	-	-
$-CF_3$	0.40	0.46	0.51	-	-
$-CH_2Cl$	0.17	0.15	0.09	-	-0.01
$-CH_2OH$	0.11	0.10	0.04	-	-
$-CH_2OCH_3$	0.11	0.08	0.01	-	-
$-CH_2OAc$	0.15	0.13	0.06	-	-
$-CH_2OC_6H_5$	0.12	0.09	0.01	-	-
$-CH_2CN$	0.20	0.20	0.16	-	-
$-COOH$	0.30	0.36	0.41	-	-
$-CONH_2$	0.28	-	-	0.62	-
$-COOCH_3$	0.32	0.38	0.43	0.68	-
$-COCH_3$	0.30	0.38	0.50	0.87	-
$-NH_2$	0.17	-0.21	-0.63	-	-1.3
$-NHAc$	0.28	0.16	-0.07	-	0.00
$-N(CH_3)_2$	0.17	-0.16	-0.71	-	-1.7
$-NHC_6H_5$	0.30	-0.02	-0.56	-	-
$-OH$	0.24	0.13	-0.38	-	-0.92
$-OCH_3$	0.30	0.11	-0.28	-	-0.78
$-OSO_2CH_3$	0.55	0.47	0.31	-	-
$-OOCCH_3$	0.38	0.30	0.15	-	-
$-OC_6H_5$	0.40	0.23	-0.08	-	-0.50
$-SH$	0.27	0.25	0.15	-	-
$-SCF_3$	0.45	0.46	0.44	-	-
$-SCH_3$	0.30	0.17	-0.08	-	-0.60
$-F$	0.54	0.34	0.06	-	-0.07
$-Cl$	0.47	0.37	0.22	-	0.11
$-Br$	0.47	0.34	0.22	-	0.15
$-I$	0.40	0.35	0.24	-	0.13
$-N_3$	0.43	0.32	0.12	-	-
$-CN$	0.57	0.61	0.65	0.90	-
$-NO_2$	0.67	0.74	0.77	1.24	-
$-CO_2^-$	-0.19	-0.09	0.04	-	-0.03
$-SO_3^-$	0.15	-	0.48	-	-
$-N(CH_3)_3^+$	1.07	1.04	0.97	-	-
$-S(CH_3)_2^+$	0.90	1.11	1.14	-	-

*The σ^I values for all saturated alkyl groups must be zero for logical consistency [26].

as indicating that the charge of the transition state is negative, but that electron-donating substituents can stabilize the transition state by direct-resonance interactions of the same type as that for the cationic intermediates of the cumyl chloride solvolyses.

These examples of the use of Eq. (3.10) are among the many cases of reactions in which there has been found to be an incomplete synchronization of various interactions as one proceeds from reactant to product of a reaction [27]. The comparisons of ρ's for rates and equilibria are the most common ways of measuring these 'extents of reactions' at transition states.

A very small selection of various types of ρ from the hundreds available in the literature are reported in Table 3.7.

TABLE 3.7
Examples of Rho Values

Reaction	Solvent	T°C	Log	σ-Type	Value
XCH_2COOH ionization [21]	water	25	K	σ^I	3.8
XCH_2CH_2OH ionization [21]	i-PrOH	27	K	σ^I	3.0
4-X-quinuclidinium ionization [20]	water	25	K	σ^I	5.1
$ArCH_2COOH$ ionization [28]	water	25	K	σ°	0.56
$ArCH_2CH_2COOH$ ionization [28]	water	25	K	σ°	0.24
ArCOOH ionization [28]	MeOH	25	K	σ	1.54
ArCOOH ionization [28]	EtOH	25	K	σ	1.65
ArCOOH ionization [15]	EtOH	25	K	σ	1.96
ArCOOH ionization [29]	DMSO	25	K	σ	2.6
ArCOOH ionization [29]	CH_3CN	25	K	σ	2.8
X-pyridinium ionization [20]	water	25	K	σ	5.2
ArOH ionization [16a]	water	25	K	σ^-	2.1
$ArNH_3^+$ ionization [16a]	water	25	K	σ^-	2.8
$Ar_3COH + H^+$ ionization [15]	water	25	K	σ^+	-3.6
$XCH_2CH(OEt)_2 + H^+$ hydrolysis [21]	mixed¹	25	k	σ^I	-8.1
1-X, 2-PrOTs solvolysis [21]	HOAc	30	k	σ^I	-7.7
1-X, 2-Me, 2-PrCl solvolysis [21]	mixed²	25	k	σ^I	-7.3
XCH_2CH_2OTs solvolysis [21]	EtOH	100	k	σ^I	-1.6
$XCH_2CH_2Br + C_6H_5S^-$ [21]	MeOH	20	k	σ^I	-1.4
ArCOCl hydrolysis [16a]	mixed³	25	k	σ	1.8
t-cumylchloride solvolysis [17]	mixed⁴	25	k	σ^+	-4.5
t-cumylchloride solvolysis [17]	EtOH	25	k	σ^+	-4.7
$ArSO_3Me$ solvolysis [13]	EtOH	70	k	σ	1.3
$ArSO_3Et$ solvolysis [13]	EtOH	70	k	σ	1.3
$ArSO_3CH(CH_3)_2$ solvolysis [13]	EtOH	70	k	σ	1.6

[1] Water:dioxane(50:50). [2] Water:ethanol(20:80). [3] Water:acetone(5:95). [4] Water:acetone(10:90).

3.5 Rho Values, Fall-off Factors, and Symmetry

As already pointed out in Section 3.2, above, the effect of substituents on a reaction can be formulated in terms of an exchange reaction. The types of reactions that we have discussed in connection with the Hammett and Taft equations can be written:

$$X - G - R \quad \overset{K_X^{RP}}{\rightleftharpoons} \quad X - G - P$$

and

$$Y - G - R \quad \overset{K_Y^{RP}}{\rightleftharpoons} \quad Y - G - P,$$

with G being a benzene ring for Hammett series or a saturated fragment for Taft series. The two reactions can be combined into the exchange reaction:

$$X - G - R \; + \; Y - G - P \quad \overset{K_{XY}^{RP}}{\rightleftharpoons} \quad X - G - P \; + \; Y - G - R$$

with $K_{XY}^{RP} = K_X^{RP}/K_Y^{RP}$. Using σ's and ρ's to represent either Taft or Hammett parameters, the equilibrium constants should be given by:

$$\log K_{XY}^{RP} = \log \frac{K_X^{RP}}{K_Y^{RP}} = (\sigma_X - \sigma_Y)\rho_{RP}. \tag{3.11}$$

We could have written the same exchange reaction, however, if we had considered the two reactions:

$$R - G - X \quad \overset{K_R^{XY}}{\rightleftharpoons} \quad R - G - Y$$

and

$$P - G - X \quad \overset{K_P^{XY}}{\rightleftharpoons} \quad P - G - Y,$$

the equilibrium constants should be given by:

$$\log K_{XY}^{RP} = \log \frac{K_R^{XY}}{K_P^{XY}} = (\sigma_R - \sigma_P)\rho_{XY}. \tag{3.12}$$

If the Taft or Hammett equations apply to both reaction series, that in which R is converted to P, with X and Y as substituents, and that in which X is converted to Y with R and P as substituents, the two equations must be equal [30]:

$$(\sigma_X - \sigma_Y)\rho_{RP} = (\sigma_R - \sigma_P)\rho_{XY},$$

$$\frac{(\sigma_X - \sigma_Y)}{\rho_{XY}} = \frac{(\sigma_R - \sigma_P)}{\rho_{RP}},$$

and, since R and P may be varied independently of X and Y, we must have:

$$\frac{(\sigma_X - \sigma_Y)}{\rho_{XY}} = \Omega \text{ (a constant).} \tag{3.13}$$

This equation shows that rho values and sigma values are directly related to one another, and furnishes a foundation for the interpretations of the rho values presented above. For the exchange reaction, then, we can write:

$$\log K_{XY}^{RP} = \frac{-\Delta G^{\circ}_{ex}}{2.303RT} = \Omega(\sigma_X - \sigma_Y)(\sigma_R - \sigma_P),$$

which has exactly the same form as Eq. (3.3) obtained from the concept of electronegativity applied to the exchange reaction of Eq. (3.1). The σ's given in Table 3.6 for charged substituents could be used to evaluate the constant, Ω ($\approx$0.4 in water), but this turns out not to be very useful, partly because of the extreme solvent dependences of the values for such groups [8].

It is also possible to show that the σ_X^I and $\sigma_{CH_2X}^I$ values for all X are proportional to one another by considering reaction series in which the substituents are the X groups and G is -CH_2CH_2- or in which the substituents are XCH_2 groups and G is -CH_2-. This is left as an exercise at the end of the chapter.

3.6 Treatment of Steric and Conformational Effects

Although attempts have been made to correlate steric effects on reactivities by approaches similar to those used for polar and resonance effects discussed above, they have not been very successful. It has become increasingly apparent that the things typically associated with the term 'steric effects' actually have a number of different sources and modes of operation. Conformational energies, bond angle deformation energies, bond length deformation energies, and Van der Waals attractions or repulsions are only some of the contributions.

TABLE 3.8
Some Data for Cyclic Compounds [31]

Ring size	$\Delta H^{o}_{combustion}$, kcal/$CH_2$ group	relative acetolysis rate constant of tosylate at 60°C
3	166.6	2. x 10^{-5}
4	164.0	14
5	158.7	16
6	157.4	(1.0)
7	158.3	31
8	158.6	285
9	158.8	266
10	158.6	539
11	158.4	67
12	157.7	-
14	157.4	-

The data in Table 3.8, dealing with cyclic compounds, can serve to illustrate some of the different types of steric effects and how these interact with one another.

The variations in the heats of combustion per CH_2 group for the cyclic C_3 to C_6 systems can be rationalized on the basis of ring strain, or bond angle distortion, effects. The value drops quite regularly from cyclopropane with a bond angle of 60° to cyclohexane which has the normal tetrahedral bond angle for all C-C-C bonds. The increase in value as one proceeds from cyclohexane to the larger rings results from compromises between conformational and bond angle factors. Cyclohexane can adopt the well-known chair conformation, which gives tetrahedral angles for all bonds and perfectly staggered conformations of the groups attached to each C-C bond.

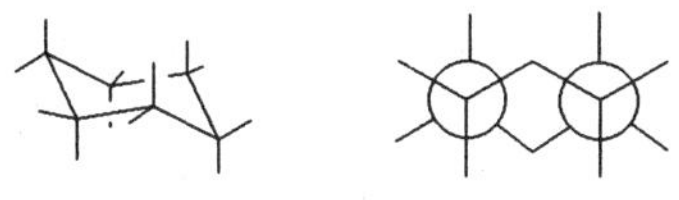

None of the other medium sized rings can satisfy both of these conditions simultaneously. The relative acetolysis rate constants of the cycloalkyl tosylates are difficult to rationalize consistently, but several effects can probably be identified. The acetolyses are believed to occur by the SN1 mechanism in which the transition state is similar in structure to the carbocation intermediate. Thus, the carbon initially attached to the tosylate becomes similar to a trigonal carbon at the transition state. For the small rings, this should produce more angle strain in the transition state, with a preferred angle of 120°, than in the reactant, leading to smaller rate constants. This is probably one of the factors contributing to the slow rate for the cyclopropyl tosylate, but obviously something else is going on with the cyclobutyl tosylate leading to a greater rate than for cyclohexyl tosylate. One factor which will increase rate is relief of eclipsing interactions on going from the four-coordinate reactant carbon to the three-coordinate carbon of the transition state. This is probably the major factor causing both cyclopentyl and cycloheptyl tosylates to have larger rate constants than cyclohexyl tosylate. It should be evident that one needs to consider a large number of factors in rationalizing the data for the medium sized rings... again, eclipsing, bond-angle distortions, bond-length distortions, etc. will interact to give compromise structures for both reactants and transition states that will be very difficult to rationalize even qualitatively.

The entire problem of predicting the properties of complex molecules is tied into the problem of predicting geometries because of compromises of the sort discussed above. A large amount of progress in this area has been made by way of what are called *molecular mechanics* calculations [32]. These calculations use parameterized force fields for bending, twisting, and stretching of individual bonds in the molecules. For example, one might approximate bond-bending energies as quadratic functions of bond-angle with force constants chosen to give best fits to some set of experimental data. The geometries and resulting energies of molecules would be calculated by minimization of energy with respect to all bond-angles, eclipsing angles, etc. Such calculations require a tremendous amount of initial work to establish the force field parameters, such as the bending force constants, which will reproduce experimental data with the greatest accuracy. The multi-parameter, multi-coordinate minimizations to obtain molecular geometries and energies are difficult and tricky because of the existence of local minima and the coupling of variables. Nevertheless, computer programs are now available [33] which can be run on moderate sized computers and which can handle quite complex systems. These, together with computer graphics software which gives photo-quality displays of molecular models which may be oriented and highlighted to show any desired region of the molecule, are becoming routine tools in attempting to relate molecular structures and properties.

3.7 REFERENCES

[1]. S. W. Benson and J. H. Buss, **J. Chem. Phys., 29**, 546 (**1958**).

[2]. S. W. Benson, F. R. Cruickshank, D. M. Golden, G. R. Haugen, H. E. O'Neal, A. S. Rodgers, R. Shaw, and R. Walsh, **Chem. Revs., 69**, 279 (**1969**).

[3]. L. N. Ferguson, **The Modern Structural Theory of Organic Chemistry**, Prentice-Hall Publ., Englewood Cliffs, N. J., **1963**.

[4]. M. J. S. Dewar and H. Schmeising, **Tetrahedron, 11**, 96 (**1960**).

[5]. (a). J. Hine, **Structural Effects on Equilibria in Organic Chemistry**, Wiley-Interscience Publ., N.Y., N. Y., **1975**, Chap. 1; (b). T. H. Lowry and K. S. Richardson, **Mechanism and Theory in Organic Chemistry**, 3rd Ed., Harper and Row Publ., N. Y., N. Y., **1987**, pp159–177.

[6]. J. Hinze, M. A. Whitehead, and H. H. Jaffe, **J. Amer. Chem. Soc., 85**, 148 (**1963**); P. R. Wells, **Prog. Phys. Org. Chem., 6**, 111 (**1968**).

[7]. C. D. Ritchie, J. Saltiel, and E. S. Lewis, **J. Amer. Chem. Soc., 83**, 4601 (**1961**).

[8]. M. Charton, **Prog. Phys. Org. Chem., 13**, 119 (**1981**).

[9]. A. J. Hoefnagel and B. M. Wepster, **J. Amer. Chem. Soc., 95**, 5357 (**1973**).

[10]. J. W. Larson and L. G. Hepler in **Solvent-Solute Interactions, Vol 1**, Ed. by J. F. Coetzee and C. D. Ritchie, Marcel Dekker Publ., N. Y., N. Y., **1969**, Chap. 1.

[11]. L. P. Hammett, **Chem. Revs., 17**, 125 (**1935**).

[12]. J. Hine, **Physical Organic Chemistry**, McGraw-Hill Publ., N. Y., N. Y., **1962**, pg. 372.

[13]. D. N. Kevill, K. C. Kolwyck, D. M. Shold, and C. B. Kim, **J. Amer. Chem. Soc., 95**, 6022 (**1973**).

[14]. This would be trivial if $\Delta S°$ were constant for the reaction series; usually, it is not, but ρ is still proportional to 1/T. See C. D. Ritchie and W. F. Sager, **Prog. Phys. Org. Chem., 2**, 323 (**1964**), particularly pgs. 353–378, and O. Exner, **Prog. Phys. Org. Chem., 10**, 411 (**1973**).

[15]. Ref. [5a], pg. 162.

[16]. (a). L. P. Hammett, **Physical Organic Chemistry**, 1st Ed., McGraw-Hill Publ., N. Y., N. Y., **1940**, Chap. 7; (b). M. Charton, **Prog. Phys. Org. Chem., 8**, 235 **(1971)**.

[17]. H. C. Brown and Y. Okamoto, **J. Amer. Chem. Soc., 80**, 4979 **(1958)**.

[18]. R. W. Taft, S. Ehrenson, I. C. Lewis, and R. E. Glick, **J. Amer. Chem. Soc., 81**, 5352 **(1959)**.

[19]. J. D. Roberts and W. T. Moreland, **J. Amer. Chem. Soc., 75**, 2167 **(1953)**.

[20]. R. W. Taft and C. A. Grob, **J. Amer. Chem. Soc., 96**, 1236 **(1974)**; C. A. Grob and M. G. Schlageter, **Helv. Chim. Acta, 59**, 264 **(1976)**.

[21]. R. W. Taft in **Steric Effects in Organic Chemistry**, Edited by M. S. Newman, John Wiley and Sons Publ., N. Y., N. Y., **1956**, Chap. 13.

[22]. S. Ehrenson, R. T. C. Brownlee, and R. W. Taft, **Prog. Phys. Org. Chem., 10**, 1, **(1973)**.

[23]. Y. Yukawa and Y. Tsuno, **Bull. Chem. Soc., Japan, 32**, 971 **(1959)**.

[24]. P. R. Young and W. P. Jencks, **J. Amer. Chem. Soc., 101**, 3288 **(1979)**.

[25]. C. D. Ritchie and W. F. Sager, **Prog. Phys. Org. Chem., 2**, 323 **(1964)**.

[26]. C. D. Ritchie, **J. Phys. Chem., 65**, 2091 **(1961)**.

[27]. C. F. Bernasconi, **Accts. Chem. Res., 20**, 301 **(1987)**.

[28]. R. W. Taft and I. C. Lewis, **J. Amer. Chem. Soc., 81**, 5343 **(1959)**.

[29]. C. D. Ritchie and R. E. Uschold, **J. Amer. Chem. Soc., 90**, 2821 **(1968)**.

[30]. J. Hine, **J. Amer. Chem. Soc., 81**, 1126 **(1959)**; C. D. Ritchie and W. F. Sager, **J. Amer. Chem. Soc., 83**, 3498 **(1961)**.

[31]. E. L. Eliel, N. L. Allinger, S. J. Angyal, and G. A. Morrison, **Conformational Analysis**, John Wiley and Sons Publ., N. Y., N. Y., **1965**.

[32]. U. Burkert and N. L. Allinger, **Molecular Mechanics**, American Chemical Society Publ., Wash., D. C., **1982**; J. C. Tai and N. L. Allinger, **J. Amer. Chem. Soc., 110**, 2050 (**1988**).

[33]. The program titled **MM2(85)** is available from Quantum Chemistry Program Exchange, Department of Chemistry, University of Indiana, Bloomington, Ind., 47405.

3.8 PROBLEMS

Problem 3.1

Without looking at Table 3.7, estimate the signs and magnitudes of the ρ's for the following reaction series and state which of the various σ's should be used for the correlation:

$$\mathrm{ArOH} \underset{H_2O\ 25^\circ}{\overset{K_a}{\rightleftharpoons}} \mathrm{ArO^-} + \mathrm{H^+}$$

$$\mathrm{Ar_3COH} + \mathrm{H^+} \underset{H_2O\ 25^\circ}{\overset{K}{\rightleftharpoons}} \mathrm{Ar_3C^+} + \mathrm{H_2O}$$

$$\mathrm{ArB(OH)_2} + \mathrm{H_2O} \underset{H_2O\ 25^\circ}{\overset{K}{\rightleftharpoons}} \mathrm{ArB(OH)_3^-} + \mathrm{H^+}$$

$$\mathrm{ArCH_2CH_2NH_2} + \mathrm{H^+} \underset{H_2O\ 25^\circ}{\overset{K_b}{\rightleftharpoons}} \mathrm{ArCH_2CH_2NH_3^+}$$

$$\mathrm{ArNH_3^+} \underset{DMSO\ 25^\circ}{\overset{K_a}{\rightleftharpoons}} \mathrm{ArNH_2} + \mathrm{H^+}$$

Problem 3.2

The rates of hydrolyses of the acetals X-$CH_2CH(OEt)_2$ in water at 25°C are first order with respect to concentrations of acetal and first order with respect to concentration of H^+. ρ_I=-8.1. Write a mechanism for the reaction which is consistent with this information.

Problem 3.3

A study of the elimination reactions of β-arylethyl derivatives was reported by C. H. DePuy and C. A. Bishop, **J. Amer. Chem. Soc., 82**, 2532 **(1960)**. The rates of the reactions were first order with respect to the concentrations of the β-arylethyl compounds, and first order with respect to concentrations of alkoxide ion in alcohol solutions. Rate constants for the reactions:

$$ArCH_2CH_2I \; + \; EtO^- \xrightarrow{k} ArCH = CH_2 \; + \; EtOH$$

in ethanol at 30°C were as follows: Ar=C_6H_5, $k = 1.3 \times 10^{-2}$; Ar=p-ClC_6H_4, $k = 4.66 \times 10^{-2}$; Ar=p-$CH_3OC_6H_4$, $k = 4.3 \times 10^{-3}$; Ar=m-BrC_6H_4, $k = 9.4 \times 10^{-2}$, all in units of $M^{-1}sec^{-1}$. Make a Hammett plot of the data to determine the ρ for the reaction. What does the sign and magnitude of ρ tell you about the transition state for these reactions?

Problem 3.4

Assume that $\Delta S°$ has the same value for each substituted derivative in a Hammett reaction series and show that ρ is proportional to 1/T.

Hint: Write ΔG°_{ex} for an exchange reaction in terms of $\sigma\rho$, then note that $d\Delta G^{\circ}_{ex}/dT = -\Delta S^{\circ}_{ex} = 0$.

Problem 3.5

The pK_a of phenylacetic acid is 4.31 and that of acetic acid is 4.76, both in water at 25°C. The ρ_I value for the ionizations of substituted acetic acids under these conditions is 3.8. Without looking at Table 3.6, find the σ^I value for the phenyl group.

Problem 3.6

From the ρ values in Table 3.7 for the ionizations of substituted acetic acids and of substituted phenylacetic acids, together with the σ^I value for the phenyl group which you obtained in the last problem, derive an expression relating σ^n_X and $\sigma^I_{X-C_6H_4}$ values for all X groups.

Hint: Write both the Taft equation and the Hammett equation for the ionizations of the substituted phenylacetic acids.

Problem 3.7

Show that if the Taft equation applies both to reactions of the type

$$X - CH_2A \rightleftharpoons X - CH_2B$$

and

$$X - CH_2CH_2A \rightleftharpoons X - CH_2CH_2B,$$

then $\sigma^I_X = \tau\sigma^I_{XCH_2}$ for all X. From the data given in Table 3.6, evaluate τ.

Hint: follow the lead given in the last paragraph of section 3.7 on pg. 70.

Problem 3.8

The length of the C-C bond in ethane is 1.53Å, that of the methyl-vinyl C-C= bond in propene is 1.51Å, and that of the methyl-acetylenic C-C≡ bond in propyne is 1.46Å. The central (ie. the C_2-C_3) C-C bond in 1,3-butadiene is 1.47Å, in 1-butene,3-yne is 1.43Å, and in 1,3-butadiyne is 1.37Å. The bond lengths shorter than that in ethane have been sometimes attributed to resonance effects in the unsaturated compounds. Propose an alternate explanation.

Hint: Suppose sp^3, sp^2, and sp hybridized carbons had different covalent radii.

Problem 3.9

The ΔG° for the conversion of axial-cyclohexyltosylate to the equatorial conformer is -0.7 kcal/mole. That for the conversion of axial-t-butyl cyclohexane to the equatorial conformer is approximately -7 kcal/mole. Assuming that the solvolyses of 4-t-butylcyclohexyl tosylates proceed by an SN1 mechanism having a transition state with a structure essentially that of the carbocation intermediate, estimate the relative rate constants for solvolyses of cis- and trans-4-t-butylcyclohexyl tosylate, k_{cis}/k_{trans}.

Hint: Draw the preferred conformations of the two different compounds and the transition states. How much energy would be required to convert the cis-isomer into the trans-isomer? How much energy would be required to convert one transition state to the other? Write k_{cis}/k_{trans} in terms of these energies.

Problem 3.10

Which set of σ's would be needed to handle substituent effects on the ionizations of 2,6-dimethyl, N,N-dimethylanilinium ions?

Hint: Consider how the 2,6-dimethyl groups might influence the conformation of the dimethylamino group.

CHAPTER 4
More About Transition States and Intermediates

4.1 Introduction

The concepts of transition state theory were introduced in Chapter 2, and we have now discussed the structures of transition states for several reactions. In most cases, we have compared the transition states to either reactants, products, or, more frequently, intermediates of the reactions considered. Transition states, however, differ fundamentally from ordinary species in one respect, and, in order to understand structural effects on reactivity beyond the simplest cases, we need to understand this difference. This involves some understanding of potential energy surfaces. Some of the qualitative characteristics of such surfaces and their pertinence to structural effects on reactivities will be considered in the present chapter.

Consideration of the precise nature of transition states and of ordinary chemical species will lead to ideas of limitations on the magnitudes of rate constants for reactions, of how changes in structure can lead to changes in mechanisms of reactions, of complications that can occur in mechanisms as we begin to reach limits in reactivities of intermediates, and of some ideas and facts regarding selectivities of some reactions. We will continue to use very common reactions, such as the nucleophilic substitution and elimination reactions of alkyl halides, as examples of the applications of these ideas.

4.2 The Distinction Between Transition States and Intermediates

Ordinary chemical species, including reactive intermediates of reactions, are at local minima of the many-dimensional potential energy surfaces representing energy as a function of relative atomic positions. That rather long-winded, but exact, statement means that any distortion of bond lengths or bond angles of normal species requires an input of energy... the species sits in a depression or hole in the potential energy surface. Any elementary step of a reaction involves moving from one hole, representing the reactant, on a potential energy surface to another hole, representing the product of the elementary step. If you can visualize such a surface, imagine now that you start pouring water into one of the holes. . . at some point, you will have added enough water that it overflows from the hole into which you are pouring it to the other hole. There will be some particular point on the surface that the water level must reach before it can flow into the other hole. This point represents the transition state for the elementary step. Such a point, in general, will be a *saddle point* in the surface. Energy will be at a maximum with respect to motion along the direction of flow at the point, but will be at a minimum with respect to motion in any perpendicular direction. For a three-dimensional surface, the saddle point will look like the highest point in a pass through a mountain range.

On our imaginary watered surface, the direction of flow of the water at the saddle point is called the *reaction coordinate.* Distortion of the bond angles or bond lengths of a transition state in a manner corresponding to motion along the reaction coordinate leads to a **decrease** in energy of the system. Distortion along any perpendicular coordinate leads to an increase in energy, just as for an ordinary species. This is the distinguishing feature between transition states and intermediates or other ordinary species involved in reactions. No matter how 'reactive' an intermediate may be, if it is really an intermediate, it is in a potential energy hole, and some amount of energy must be supplied to transform it to another species. A transition state, however, is at a saddle point and will be transformed into either reactant or product of the elementary step without any input of energy.

In more precise mathematical language, we can specify the complete geometry and position of any system of N nuclei and associated electrons by specifying three positional coordinates for each nucleus. We can, therefore, write the potential energy of such a system as a function of $n = 3N$ coordinates, x_1, x_2, $\ldots x_n$:

$$E = f(x_1, \ldots, x_n).$$

In general, there will be a number of *stationary points* for the system which are defined as points in the n dimensional space where $\partial E/\partial x_i = 0$ for all i. At any such point, a Taylor expansion of energy as a function of displacements from the point will begin with quadratic terms. That is, if the point $x_1^\circ, \ldots, x_n^\circ$ is a stationary point where $E = E_\circ$, then, defining $q_i = (x_i - x_i^\circ)$, we have

$$\frac{\partial E}{\partial q_i} = 0 \quad \text{for all } i,$$

and

$$\Delta E = E - E_o = \sum_{i=1}^{n} \sum_{j-i}^{n} c_{ij} q_i q_j + \textit{higher order terms.}$$

If we consider a small enough region about the stationary point, the higher order terms in ΔE can be neglected, and ΔE is a quadratic function of the coordinates, q. This is just the situation which was discussed on pg. 24 in Section 1.5, where it was shown that ΔE can be written as a matrix equation:

$$\Delta E = \begin{pmatrix} q_1 & q_2 & \cdots & q_n \end{pmatrix} \begin{pmatrix} \frac{\partial^2 E}{2\partial q_1^2} & \frac{\partial^2 E}{2\partial q_1 \partial q_2} & \cdots & \frac{\partial^2 E}{2\partial q_1 \partial q_n} \\ \frac{\partial^2 E}{2\partial q_2 \partial q_1} & \frac{\partial^2 E}{2\partial q_2^2} & \cdots & \frac{\partial^2 E}{2\partial q_2 \partial q_n} \\ \vdots & \vdots & \ddots & \vdots \\ \frac{\partial^2 E}{2\partial q_n \partial q_1} & \frac{\partial^2 E}{2\partial q_n \partial q_2} & \cdots & \frac{\partial^2 E}{2\partial q_n^2} \end{pmatrix} \begin{pmatrix} q_1 \\ q_2 \\ \vdots \\ q_n \end{pmatrix},$$

or, in matrix notation:

$$E = \vec{q}\mathcal{D}\vec{q}^{\dagger}.$$

The matrix of second derivatives can be diagonalized, as pointed out in Section 1.5, and the equation rewritten in terms of rotated coordinates and a diagonal matrix:

$$E = \begin{pmatrix} q_1' & \cdots & q_n' \end{pmatrix} \begin{pmatrix} \lambda_1 & 0 & \cdots & 0 \\ 0 & \lambda_2 & \cdots & 0 \\ \vdots & \vdots & \ddots & \vdots \\ 0 & 0 & \cdots & \lambda_n \end{pmatrix} \begin{pmatrix} q_1' \\ q_2' \\ \vdots \\ q_n' \end{pmatrix},$$

or, in matrix notation:

$$E = \vec{q}'\Lambda\vec{q}'^{\dagger}, \text{ where } \mathcal{U}^{\dagger}\mathcal{D}\mathcal{U} = \Lambda, \text{ and } \vec{q}' = \vec{q}\mathcal{U}.$$

A negative eigenvalue, λ_i, of the $\mathcal{D}$ matrix means that the energy is at a maximum with respect to motion along the coordinate q_i', and a positive eigenvalue means that the energy is at a maximum with respect to motion along the corresponding coordinate. For the sake of completeness, we should note that six of the eigenvalues of the $\mathcal{D}$ matrix will be zero for a non-linear system. The q' coordinates corresponding to these zero eigenvalues are the three cartesian coordinates for translations of the entire system, and the three coordinates corresponding to rotations of the entire system, without changing relative locations of the nuclei.

We can now state precisely the definitions of ordinary chemical species, including 'reactive' intermediates, and of transition states. Any stationary points for which there are no negative eigenvalues of $\mathcal{D}$ correspond to configurations of ordinary chemical species. **Any stationary points for which there is one, and only one, negative eigenvalue correspond to configurations of transition states.** Other stationary points may exist which are local maxima, or various other types of 'bumps' on the surface.

4.3 Simplified Potential Energy Diagrams for Reactions

Reacting systems generally involve a large enough number of atoms that it is humanly impossible to visualize the many-dimensional potential energy surfaces of the systems; for example, the reaction of methyl chloride with fluoride ion would require the ability to conceptualize in 13 dimensions...the 12 relative coordinates and potential energy. Since we are used to seeing things in three dimensions, it is natural to attempt to visualize the surfaces generated by potential energy as a function of some two chosen coordinates. The use of such diagrams involving two coordinates was suggested for considering the several possible mechanisms of elimination reactions [1], and this type of use was further refined in connection with general acid-base catalysis [2], and considerations of reactive intermediates [3] in a number of types of reactions.

In elimination reactions of alkyl halides or tosylates in the presence of a base, B, we can choose the C-X and the B-H bond lengths as the two coordinates in which we might be primarily interested since these bonds undergo the greatest change in the reactions.

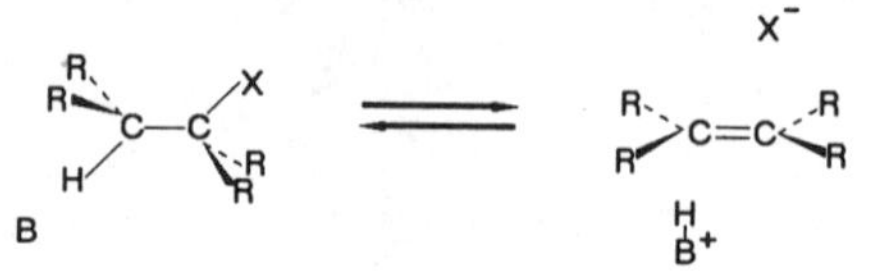

A great deal of care is necessary in defining exactly what we mean, however, in writing potential energy as a function of these two coordinates alone. Obviously, for any given pair of values for these coordinates, other coordinates could take on any values, and, since potential energy depends on those values, it is indeterminate. We get around this difficulty by choosing the pair of values for our two coordinates, and then minimizing the potential energy of the system with respect to all other coordinates. This almost always leads to a well-defined surface, but we must keep in mind what is really involved in motion on the surface. In the elimination reaction, for example, moving from one point to another of the two chosen coordinates almost certainly involves changes in the

C-C distance, the R-C-R bond angles, the C-B distance, or even rotational coordinates within the R groups. These changes are not generally proportional to the changes in the coordinates shown, and it is even possible that discontinuous changes occur in the coordinates not shown.

A hypothetical potential energy surface for the elimination reaction under discussion is shown in Fig. 4.1 by means of a perspective drawing and a contour diagram.

The lines in the contour diagram follow points of constant energy on the surface. The numbers shown are the energies, in arbitrary units, of the points on the lines. The product of the elimination reaction is in the depression at the lower-left of the figures, and, as shown in the contour diagram, has a potential energy of somewhat less than 1 unit. The reactant alkyl halide and base is in the depression at the upper-right of the figures, and has a potential energy of somewhat less than 3 units. The depression at the upper-left of the figures corresponds to a carbanion and the protonated base, with a potential energy <8 units. The depression at the lower-right corresponds to a carbocation and the halide ion, with a potential energy <9 units. There are five points, marked with ‡'s on the contour diagram, corresponding to transition states. The lowest energy transition state is at the center of the diagram, the 'pass between the two mountain peaks', and has a potential energy between 7 and 8 units. The other four transition states are along lines connecting reactant or product with the depressions corresponding to carbocation and carbanion species.[1]

Notice that even though carbanion and carbocation 'intermediates' exist for this system, the lowest energy path from reactant to product does not pass through these species, and a concerted one step mechanism would be followed. The *reaction coordinate* for the concerted mechanism

[1] There is an interesting, but presently unanswerable, question associated with the hypothetical surface in Fig. 4.1. Suppose we had some way of generating the [carbanion, BH^+] species as a reactant and that we wish to know the mechanism of conversion of this species to the alkene. The reasonably direct path from the carbanion to alkene has a transition state with an energy of a little more than 10 units. **There are, however, lower energy paths available for the reaction which pass through the transition state just to the right of the carbanion in the Figure.** The interesting, presently unanswerable, question, is whether or not the alkyl halide is an intermediate in the reaction. Once the system crosses the transition state just to the right of the carbanion, there are an infinite number of paths to the alkene which never exceed the energy of that transition state and which do not pass through the alkyl halide 'depression'. There are similar conundrums associated with reactions of the carbocation on this surface.

A.

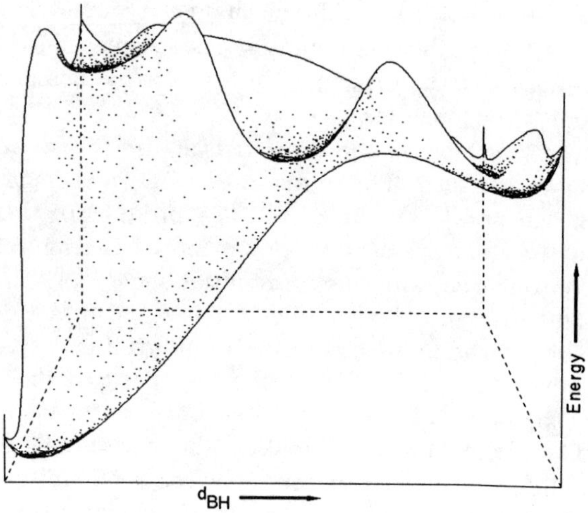

B.

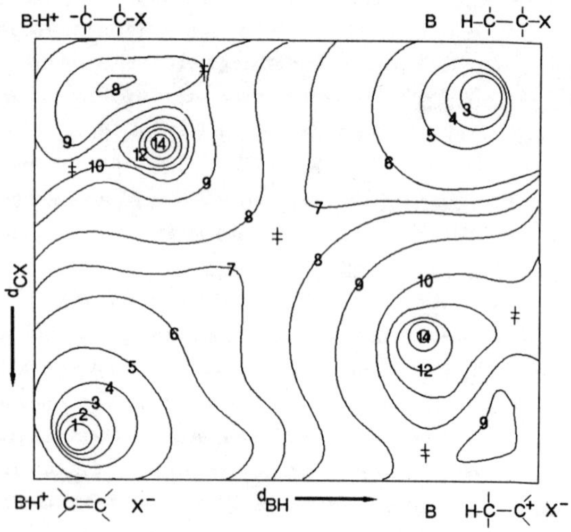

Figure 4.1: *Hypothetical Perspective and Contour Potential Energy Diagrams for Elimination Reactions.*

is the direction of negative curvature at the central transition state, and is a coordinate directed from the upper-right to the lower-left corner of the diagram. More precisely, in this diagram, and for the concerted pathway, the reaction coordinate is defined by: $rc = (1/\sqrt{2})(d_{CX} - d_{BH})$.

It is now possible to look a little more closely at some of the interpretations of structural effects on reactivities that we discussed in the last chapter. There, we interpreted the effect of structural changes on rates and equilibria of the reaction to say something about the structure of the transition state. If the effect of structural change on $\log k$ was, for example, one-half the effect on $\log K$, we interpreted this to mean that the transition state was half-way in structure between reactants and products. It should be obvious from a consideration of Fig. 4.1 that this type of interpretation depends on an assumption that changing the height of one point relative to a second point on the potential energy surface produces a change in height of points between the two that is proportional to the distance of the point along the line connecting the two. Obviously, there are relative energies other than those of reactants and products which might be changed by a change in structure. We could, for example, easily think of changes in structure of an alkyl halide which might primarily affect the relative energies of the carbanion and carbocation species. If the change in structure changed the height of the carbanion species upward and the height of the carbocation species downward by the same amounts relative to reactants and products, the change in height of a perfectly central transition state might be expected to be unchanged relative to reactant and product. Using the same assumption that change is proportional to the distances, we could interpret the effect of such a change in relative energies of carbanion and carbocation on $\log k$ in terms of the location of the transition state along a line between the upper-left and lower-right corners of the diagram.

Note, however, that the assumption that change in relative energy of any two points produces proportional changes in energies of all intervening points requires that these changes also affect the **location** of the transition state. If we were to change the structure of the reactant in such a manner that the carbocation became less stable relative to all the other species, raising the lower-right corner of the surface, we would shift the location of the transition state toward the upper-left of the diagram... away from the high energy corner. If we raised the energy of the product with respect to all other species, we would shift the location of the transition state **toward** the product... **toward** the high energy corner. The magnitudes of the resulting shifts in locations of saddle points depend on the curvatures of the surface at the saddle points and the location of the saddle points. The shift in location of the central transition state in Fig. 4.1 which would be expected as a result of raising the energy of the carbanion or carbocation corners is a shift perpendicular to the reaction coordinate (which runs between the lower-left and upper-right of the diagram) and is a shift **away from** the raised corner. The shift in location of central transition state expected from raising the reactant or product corner is a shift parallel to the reaction coordinate and is a shift **toward** the raised corner. These are frequently referred to as *perpendicular effects* and *parallel effects*, respectively. If the shifts in location of a transition state were appreciable, we would expect curvature in Hammett or Taft plots for the reactions. This is very rarely observed, and is taken to mean that the curvatures of the potential energy surfaces at the transition states are so great that ordinary Hammett or Taft structural changes are not large enough to have appreciable effects on location of transition states for most reactions.

4.4 Upper Limits on Rate Constants for Reactions

In Section 2.4, the idea that all transition states are transformed to other species with a universal rate constant of $\kappa T/h$ was presented as an essential part of transition state theory. We will see later that this idea can actually be derived from quantum mechanical considerations. The rate constant means that the half-life of transition states is given by $t_{1/2} = h \ln 2/\kappa T$, which is approximately 10^{-13} secs. near room temperature. In slightly looser language, we speak of the *lifetime* of a species as being the reciprocal of the first-order rate constant for its decomposition. The *lifetime* of a transition state, then, is *ca.* $1.6 \times 10^{-13} sec.$ at $25°C$. Any ordinary chemical species must have a lifetime greater than that of a transition state. Normal vibrational frequencies of bonds are on the order of $10^{14} sec^{-1}$, so ordinary chemical species have a lifetime of at least several molecular vibrations.

These considerations place an upper limit of $\kappa T/h \approx 6 \times 10^{12} sec^{-1}$ on the rate constant for any first order reaction at ordinary temperatures. What about other types of reactions?

It seems obvious that two species reacting in a second order reaction must come into contact with each other before reaction can occur. We can write such reactions, then, as occurring in two elementary steps:

$$\mathrm{A + B} \underset{k_{-d}}{\overset{k_d}{\rightleftharpoons}} [\mathrm{A \cdots B}] \underset{k_{-1}}{\overset{k_1}{\rightleftharpoons}} \text{Product},$$

where, for the moment, we can think of the complex, [A···B], in rather imprecise terms, as an *encounter complex*. Assuming that the concentration of the complex is never significant in comparison with concentrations of reactants and products, we may use the Bodenstein approximation to obtain the expression:

$$k_f = \frac{k_d k_1}{k_{-d} + k_1}$$

for the observed forward, second order, rate constant k_f. Either the formation of the encounter complex, or its decomposition to product, may be rate determining depending on the relative magnitudes of k_{-d} and k_1. For our immediate purpose, it is only necessary to note that:

$$k_f \leq k_d \quad \text{since} \quad \frac{k_1}{k_1 + k_{-d}} \leq 1.$$

The rate constant k_f cannot exceed the rate constant for diffusion of the two species together in solution. In solution, solute species diffuse essentially by moving from one 'hole' in the solvent to another. Stated differently, the solute moves by separating solvent molecules surrounding it. The transition state for the diffusion process will look something like an irregularly solvated solute. Rates of diffusion of solutes in solution can be measured or can be calculated fairly well from hydrodynamic theory [4]. The rates depend somewhat on the sizes of the solute particles and strongly on the viscosity of the solvent. The rate constant for diffusion of two solute particles together is dependent upon the rates of diffusion, sizes, charges and concentrations of the two solutes. In aqueous solution at room temperature for ordinary sized species, $k_d \approx 10^9 - 10^{10} M^{-1} sec^{-1}$. The larger values are found for oppositely charged solute ions, and the smaller values are found for similarly charged solutes. The largest value found for k_d in aqueous solution is $\approx 10^{11} M^{-1} sec^{-1}$ for $H^+ + OH^-$ where both ions have abnormally high diffusion rates due to so-called *Grotthus chain* mechanisms in which the proton or hydroxide are transferred along a chain of hydrogen-bonded solvent molecules by bond changes [5].

One point needs to be kept in mind to avoid confusion as we proceed. Our discussion of these reactions has assumed that the encounter complex is a reactive intermediate *whose concentration must be small in comparison with the total concentration of reactants and products throughout the reaction.* If the complex is formed in significant concentrations, then we cannot use the Bodenstein approximation and the reaction becomes two consecutive reactions that must be treated as such.

With the Bodenstein approximation, the question of rate-determining step, as already pointed out, is a question of the relative magnitudes of k_{-d} and k_1. The magnitude of k_{-d} is related to that of the diffusion rate constant, k_d, through the equilibrium constant, K_{assoc}, for the formation of the encounter complex from the two reactants:

$$k_{-d} = \frac{k_d}{K_{assoc}}$$

Let's try to get some ideas of the magnitudes of the equilibrium constants that might be appropriate. Suppose the species, A, is a moderate-sized organic molecule, roughly spherical in shape and having a radius of 3Å. Taking the diameter of a solvent molecule as 2Å, we can calculate the volume around an A molecule within a solvent diameter as:

$$V = \frac{4}{3}\pi[(3+2)^3 - 3^3]\mathring{A}^3 = 4.1 \times 10^{-22} cm^3,$$

and, multiplying by 6.02×10^{23}, we obtain a volume of $247 cm^3/mole$ of A. The probability of finding a B molecule within this volume in a total volume of 1 *liter*, then is $\approx 0.25M^{-1}$, which we can take as the equilibrium constant for formation of an encounter complex of A and B in the absence of any attractive or repulsive interactions of A and B. Values between $0.017M^{-1}$ [6] and $0.3M^{-1}$ [7] have been considered in the literature as reasonable estimates of this quantity. Using $K_{assoc} = 0.25M^{-1}$ and $k_d = 5 \times 10^9 M^{-1} sec^{-1}$, we can calculate $k_{-d} = 2 \times 10^{10} sec^{-1}$. Any factors, such as electrostatic attractions of ions or hydrogen-bonding, which would stabilize the encounter complex will increase K_{assoc} and decrease k_{-d}.

Even with $k_{-d} = 2 \times 10^{10} sec^{-1}$, k_1 values well below the limit of $6.3 \times 10^{12} sec^{-1}$ are possible which allow the diffusion step to be rate-determining. Such reactions are called *diffusion-controlled* reactions, and are such that any change in structures of A or B to increase their reactivities do not lead to increases in the observed rate constants since such structural changes usually have insignificant effects on diffusion rates.

Some of the best examples of what can happen when $k_1 \geq k_{-d}$ are provided by looking at the reverse type of reaction to that which we wrote above. If we start with an A-B molecule that dissociates into an [A···B] complex which can recombine to form a new molecule, or dissociate to produce other products, some interesting chemistry can result.

4.5 Ion-Pair Intermediates in SN1 Reactions

The solvolysis of 3-methyl-3-chloro-1-butene in acetic acid solution buffered with acetate ion produces the allylic isomer, 3-methyl-1-chloro-2-butene, simultaneously with a mixture of the two isomeric acetates,

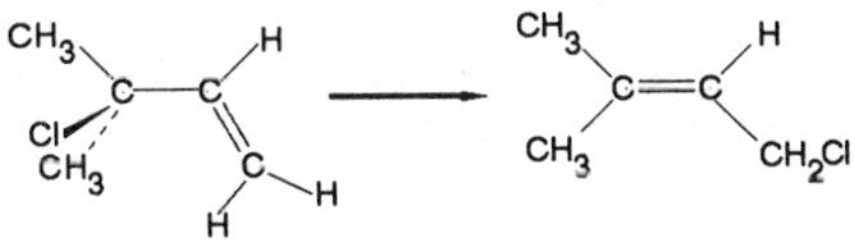

and the amount of 3-methyl-1-chloro-2-butene formed is not changed by adding chloride ion to the solution [8]. This last fact means that the isomerization does not occur by formation of an allylic cation which is captured by chloride ion competitively with capture by solvent and acetate ion. It was postulated that the reactions were proceeding through an intermediate in which the carbon-chlorine bond is broken, but in which the chloride ion is still held in close proximity to the carbocation... an ion-pair intermediate rather than free ions. The collapse of the ion-pair, with the chloride ion bonding to the 1-carbon to produce the allylic isomer of the reactant, was termed *internal return*. The observations must mean that chloride ion reacts with the carbocation in the ion-pair at least as fast as it can diffuse away to produce free ions which would react with solvent or acetate ion. The rate constant, k_{-d}, for diffusion of the two ions apart in acetic acid is probably considerably smaller than the value we estimated in Section 4.3 for an encounter complex because the low dielectric constant of acetic acid would be expected to allow a large electrostatic attraction of the ions, leading to a large K_{assoc}.

The first study to begin to probe the details of ion-pair behaviors involved the acetolysis of optically active threo-3-p-anisyl-2-butyl p-bromobenzenesulfonate in the presence of added salts [9]. It was mentioned in Section 2.8, pg. 73, that 3-phenyl-2-butyl p-toluenesulfonate undergoes acetolysis through a phenonium ion intermediate, and that the threo-reactant produces exclusively the d,l-threo-acetate. There is even stronger evidence [10] that the 3-p-anisyl compound undergoes acetolysis exclusively through a symmetrical phenonium ion. The important point here is that once the phenonium ion is formed, whether free or ion-paired, optical activity is lost; therefore, the rate of the bond heterolysis step can be followed by following the rate of loss of optical activity. A first-order rate constant for racemization, k_α, can be obtained by polarimetry, and a first-order rate constant for formation of acetate product, k_t, can be obtained by titrating the p-bromobenzenesulfonic acid in aliquots of the solution taken at various times. As expected for a reaction involving an ion-pair intermediate which undergoes internal return, k_α was greater than k_t under all reaction conditions. Addition of lithium brosylate to the solution caused only a normal salt effect increase in both k_α and k_t, showing that there is no appreciable trapping of a free carbocation by brosylate ion. The surprise came when the reaction was studied in the presence of various concentrations of added lithium perchlorate. The data are shown in Table 4.1.

TABLE 4.1
Solvolysis of Threo-3-p-anisyl-2-butyl brosylate in Acetic Acid at 25°*C*
Data from [9]

$[LiClO_4]$, M	$10^5 k_\alpha$, sec^{-1}	$10^5 k_t$, sec^{-1}	k_α/k_t
0	8.06	1.96	4.1
1.0×10^{-3}	-	2.68	-
5.0×10^{-3}	-	4.69	-
1.0×10^{-2}	8.73	5.70	1.54
3.0×10^{-2}	11.1	8.40	1.32
6.0×10^{-2}	15.6	11.5	1.36
1.0×10^{-1}	19.8	16.1	1.23

The effect of the lithium perchlorate on k_α is the normal salt effect in such solvents (Section 2.6, pg. 62) and is correlated by the equation

$$k_\alpha = k^\circ_\alpha(1 + 14[LiClO_4]).$$

The effect on k_t is quite different, however, giving a sharp increase in k_t for quite low salt concentration, and then a more normal increase, essentially the same as that on k_α, at higher salt concentrations.

The sharp increase in k_t at low salt concentrations may be understood as resulting from a trapping of the initially formed carbocation-brosylate ion-pair by perchlorate ion to form a carbocation-perchlorate ion-pair which produces only acetate product because any collapsed perchlorate product would solvolyze very rapidly. The perplexing part of the data is that k_α remains greater than k_t even at high concentrations of lithium perchlorate where any extrapolation of the trapping at low salt concentration requires that the trapping is essentially complete. The constancy of the ratio k_α/k_t at the high salt concentrations was interpreted as requiring that the transition states for the racemization of starting material and formation of acetate product be very similar, and, since product is clearly formed from ion-pairs, that the racemization of starting material also occurs through an ion-pair[2]. This must, however, be a different ion-pair from the one that is trapped by lithium perchlorate. The existence of two types of ion-pairs, an *intimate* ion-pair and a *solvent-separated* ion-pair, was postulated. Only the solvent-separated ion-pair can be trapped by perchlorate ion. The full reaction scheme proposed can be written:

$$\text{ROBs} \underset{k_{-i}}{\overset{k_i}{\rightleftharpoons}} [\text{R}^{+-}\text{OBs}] \text{ intimate ion-pair}$$

$$[\text{R}^{+-}\text{OBs}] \underset{k_{-ip}}{\overset{k_{ip}}{\rightleftharpoons}} [\text{R}^{+}//^{-}\text{OBs}] \text{ solvent-separated ion-pair}$$

$$[\text{R}^{+}//^{-}\text{OBs}] + \text{ClO}_4^- \underset{k_{-ex}}{\overset{k_{ex}}{\rightleftharpoons}} [\text{R}^{+}//^{-}\text{O}_4\text{Cl}] + {}^{-}\text{OBs} \text{ anion exchange}$$

$$[\text{R}^{+}//^{-}\text{OBs}] + \text{HOAc} \xrightarrow{k_s} \text{ROAc} + \text{HOBs}$$

$$[\text{R}^{+}//^{-}\text{O}_4\text{Cl}] + \text{HOAc} \xrightarrow{k_s'} \text{ROAc} + \text{HClO}_4$$

Two additional studies of this system strongly confirm the intermediacy of a solvent-separated ion-pair and its anion exchange with other ions. First, was the observation of a salt-induced mass law effect. It was already noted that the acetolysis reaction, with no lithium perchlorate present, fails to give a mass law effect. When lithium perchlorate is present, however, a mass law effect is observed [11], as expected from the postulated k_{-ex} step of the above scheme. Second, was the observation of product formation from added anions expected to give derivatives which solvolyze more slowly than brosylates. The addition of lithium bromide to the solution gave rate effects on k_α and k_t similar to those produced by lithium perchlorate, and the bromide product was observed to be formed [12].

It should be noted that the polarimetric rate measurements actually detect the return of the brosylate ion to the original 3-carbon while the anisyl group moves to the original 2-carbon to produce the enantiomer of the starting reactant. If the brosylate returned to the original 2-carbon more frequently than it did to the 3-carbon, a process which has been called *hidden return*, k_α would underestimate the rate of ion-pair formation. It was originally hoped that this hidden return would prove to be unimportant, and that the patterns of ion-pair reactivities could

---

2 There is, however, room to question this interpretation since only 20–30% of the observed k_α at high salt concentrations gives return to starting material. Extreme precision in rate measurements would be necessary to detect appreciable differences in response to salt effects. It seems odd that the intimate ion-pair cannot be trapped even in other solvolysis reaction systems with very good trapping agents, such as azide ion. Nevertheless, the interpretation given is generally accepted. There are some completely different types of systems, to be discussed later, which provide stronger evidence for the two types of ion-pairs proposed here.

be established using the relatively simple polarimetric and titrimetric measurements described above.

In order to check this point, the technique of isotopic labelling was used as another way of detecting the ion-pair return process. Optically active p-chlorobenzhydryl p-nitrobenzoate, specifically labelled with ^{18}O at the carbonyl oxygen:

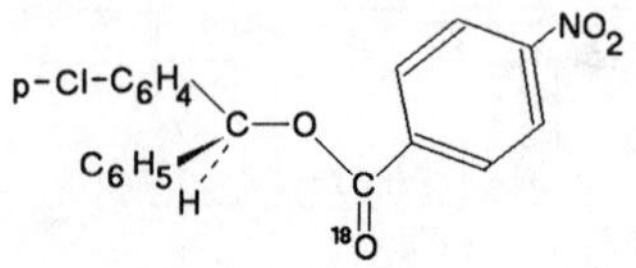

was prepared and its solvolysis in aqueous acetone was studied [13]. Three processes could be followed to produce three rate constants: k_t for the rate of production of p-nitrobenzoic acid detected titrimetrically; k_α for the rate of loss of optical activity detected polarimetrically; and k_{eq} for the rate of equilibration of the isotopic label between the ether and carbonyl positions of the ester detected by isolation and specific degradation of ester followed by mass analysis of the degradation products. No incorporation of ^{14}C into the ester was detected when the solvolysis was carried out in the presence of ^{14}C-labelled p-nitrobenzoic acid...ruling out the existence of any return from free carbocation.

The process measured by k_α involves both racemization of starting ester and formation of racemic solvolysis product. The rate constant for racemization of the starting ester, k_{rac}, can be obtained as $k_{rac} = k_\alpha - k_t$. Some of the values obtained are shown in Table 4.2.

TABLE 4.2
Solvolysis of p-Chlorobenzhydryl-p-nitrobenzoate
in aqueous acetone at 99.6°C
Data from [13]

Rate Constant	90% Acetone	80% Acetone
k_α, sec^{-1}	2.7×10^{-6}	2.0×10^{-5}
k_t, sec^{-1}	1.4×10^{-6}	1.2×10^{-5}
k_{eq}, sec^{-1}	3.5×10^{-6}	1.8×10^{-5}
k_{rac}, sec^{-1}	1.3×10^{-6}	0.8×10^{-5}

If one is willing to assume that the ether and carbonyl oxygens are equivalent in the ion-pair, the ratio $k_{eq}/(k_{eq} + k_t)$ gives the fraction of the ion pairs that return to ester, and the ratio $k_{rac}/2k_{eq}$ gives the fraction inversion/retention for the return. In 90% acetone, 71% of the ion-pairs return to ester, and, of the returns, 19% give inversion while 81% give retention according to this assumption. Most interestingly, when the solvolysis in 80% acetone was carried out in the presence of 0.141M sodium azide, it was found that $k_\alpha = k_t$ but that scrambling of the isotope still occurred in the ester. This result indicated that racemization does not occur in the intimate ion-pair although ^{18}O scrambling does.

Unfortunately, this latter conclusion cannot be generalized. There are several cases known for which $k_{rac} > k_{eq}$, and there are some cases where added azide ion does not stop racemization of the starting ester. Some illustrative data are as follows:

- For 1-anisylethyl p-nitrobenzoate in aqueous acetone with no azide ion added, $k_{eq} = 2k_{rac} = k_t$. With 0.14 M sodium azide present, $k_{eq} = 2k_{rac} = 0.5k_t$. [14]
- For p-methylbenzhydryl p-nitrobenzoate in aqueous acetone, $k_t = 1.0 \times 10^{-7} sec^{-1}$, $k_{eq} = 2.9 \times 10^{-7} sec^{-1}$, and $k_{rac} = 5.0 \times 10^{-8} sec^{-1}$. [15]
- For 3-phenyl-2-butyl tosylate labelled at the sulfonyl oxygen, solvolysis in acetic acid gives $k_{rac} = 2k_{eq}$. [16]

The last-mentioned result indicates that the oxygens of the p-toluenesulfonate anion are not equivalent in the ion-pair. In this study [16], it was also possible to resolve the enantiomers of the starting ester and to determine the ^{18}O distribution in each enantiomer. The enantiomer of the starting ester was found to have 17% of the ^{18}O in the C-O-S bond. The data show that as the sulfonate anion moves from the 2- to the 3-carbon, there is a preference for the original oxygen to rebond. The indication is that the rates of migration from one carbon to the other, rotation of the sulfonate group relative to the carbocation, collapse of the ion-pair, and reaction of the ion-pair with solvent are closely competitive. Since the rates of migration and rotation must be extremely fast, so must be the rate of collapse of the ion-pair.

We might begin to be concerned about just how fast the collapses of some ion-pairs really are. We might also begin to wonder whether or not there is a real distinction between SN1 and SN2 mechanisms. In particular, is it possible that an SN2 mechanism involves two steps:

$$RX \underset{k_{-i}}{\overset{k_i}{\rightleftharpoons}} [R^{+-}X]$$

$$[R^{+-}X] + Y^- \xrightarrow{k_Y} RY + X^-$$

with $k_Y[Y^-] \leq k_{-i}$. This would mean that the two mechanisms differ only in the matter of relative rates of reactions of the ion-pair intermediates.

4.6 Potential Energy Diagrams for SN1 and SN2 Reactions

There are a few more points regarding the connection between potential energy surfaces and rates of reactions which need to be cleared up before we proceed much further. It may already have occurred to you that our discussions of Fig. 4.1 were in terms of pathways governed by the relative potential energies of the various transition states, while the expression for a rate constant is in terms of the **free energy** of the transition state. In fact, there is a slight problem here. Transition state theory defines transition states as saddle points on the **potential** energy surface, but concludes that it is the **free** energy of the transition state that determines rate constant. Thus, when we discuss competition between two or more paths on a potential energy surface, as we did in Section 4.3 for elimination reactions, we should consider the relative **free** energies of the transition states. This becomes a particularly important point when the stoichiometries of the different transition states are not the same. Entropy differences are then present which are associated with the differences in the degrees of freedom of a system, and free energy differences between species having different compositions will depend on activities (and, therefore, concentrations) of some of the reactants or products.

When we discussed the formation of 'encounter complexes' above, the equilibrium constant was estimated using an assumption that no interaction energies of the two particles existed, and that only probability (and, therefore, only entropy) need be considered. Experimentally, the standard entropy observed in solution for association of two species is frequently somewhere around -18 e.u., giving a contribution of ca. +5.4 kcal/mole to ΔG°, and a factor of 10^{-4} to the equilibrium constant. The observed cases all involve strong interaction energies between the two species, which certainly lead to specific orientations within the complexes, which were not considered in our discussion of 'encounter complexes'. The question of exactly how much orientation we should consider in encounter complexes is not clearly answerable, and this uncertainty is the primary factor leading to the widely different estimates of association constants ($\sim 0.4 > K_{assoc} > \sim 0.017$) mentioned in the earlier discussion.

Differences in stoichiometry of different transition states are obviously involved in the reactions of ion-pairs or free ions in the SN1 reactions that we have been discussing, and in the distinction between SN1 and SN2 reactions. In these reactions, there are clearly two bonds that undergo primary change; the one from carbon to the leaving group, and the one from carbon to the entering group. A reasonable diagram for the reactions can be constructed using coordinates corresponding to these two bonds, and having regions with different stoichiometries separated from one another [3] as shown in Fig. 4.2. In order to keep things as simple as possible, only one type of ion-pair is shown.

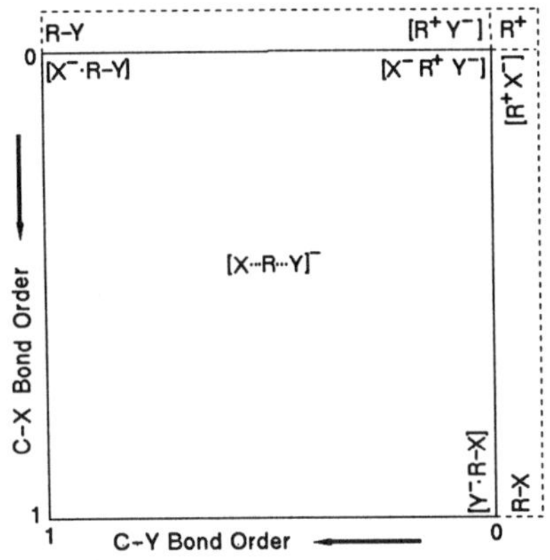

Figure 4.2: *Diagram Showing Locations of SN Intermediates and Transition States in Terms of Two Bond Coordinates as Discussed in Ref. [3].*

The region enclosed by dashed lines at the top of the diagram contains only the R and Y moieties; the region enclosed by dashed lines at the right contains only the R and X moieties; the small region enclosed in the dashed lines at the upper-right contains only the R moiety; and, the region enclosed in the solid lines contains all three, R, X, and Y, moieties. Crossing any line involves a diffusion of at least two species together or apart. Any reaction having a transition state for the rate-determining step within the solid boxed region would be first order with respect to concentration of attacking anion, and, therefore, classified as SN2. Reactions following a path around the region enclosed by the dashed lines, and through the free carbocation, would be classified as 'normal' SN1. Which path is followed depends, of course, on the relative free energies of the transition states for the paths.

Assuming that all of the species shown in the diagram actually exist as ordinary species, we can write a general reaction scheme:

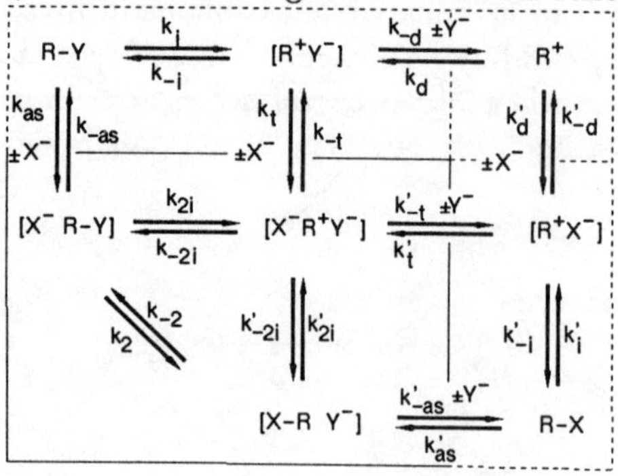

Generalized SN Reaction Scheme

where the lines are those in Fig. 4.2. The barriers to reactions across the lines are expected to be just the barriers to diffusion of the species. In good ionizing solvents these diffusion barriers are probably not much more than 2–4 kcal/mole.

We'll continue to keep things as simple as possible through the following discussion by assuming that the concentrations of X^- and Y^- are equal and not so high that associations are forced. Reaction coordinate diagrams will help to visualize some of the differences in mechanisms, but it must be kept in mind that relative heights of points on these diagrams may be concentration dependent as was pointed out in Section 2.5, pg. 58.

Let's begin the consideration of the full scheme with a reaction of a species such as benzhydryl chloride, which has a good leaving anion and forms a reasonably stable carbocation, with azide ion, which is a very good nucleophile. For this system, the conversion of alkyl halide to the ion pair by the k_i path is expected to be fairly easy. In good ionizing solvents, there should be no appreciable stabilization of the $[R^+Y^-]$ ion-pair by association with X^- in the $[X^-R^+Y^-]$ triple-ion, and only a moderate diffusion barrier for the k_{-d} path to free carbocation. Therefore, there is no reason for the reaction to take the k_t path with its entropy loss requirement. Once R^+ has formed, azide ion diffuses in to form the $[R^+X^-]$ ion-pair, which then collapses to product. Since azide ion is a better nucleophile than chloride ion, $k'_{-i} > k_{-i}$, and, assuming that the diffusion steps are faster than the collapse of the ion-pairs, the reaction follows a classical SN1 mechanism. A reaction coordinate

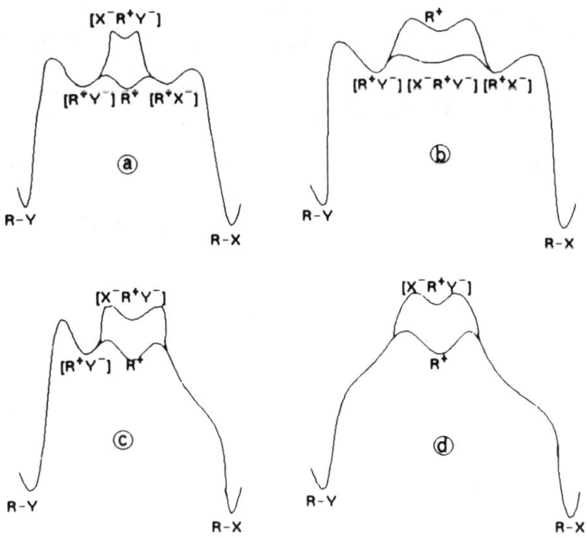

Figure 4.3: *Possible Reaction Coordinate Diagrams for SN1 Reactions. (a) Ion-pairs and free carbocation intermediates exist. (b) Dissociation of ion-pairs is energetically unfavorable. (c) Barrier for collapse of ion-pair to product disappears. (d) Barriers to collapse of both ion-pairs disappear.*

diagram would look something like Fig. 4.3a.

If the same reaction is run in a poor ionizing solvent, the dissociations of ion-pairs to free carbocation become energetically unfavorable. The free energy of the carbocation is raised to a higher level than that of the ion-triplet $[X^-R^+Y^-]$, and the reaction path is forced through the boxed area of the diagram regardless of whether the ion-triplet is an ordinary species or a transition state. A reaction coordinate diagram would look something like Fig. 4.3b. The reaction would still show essentially all of the characteristics of a classical SN1 reaction, including a possible mass-law effect if the barriers for reversion of $[R^+X^-]$ to R-Y are not so high relative to the barrier for collapse to R-X that they cannot be reversed by increasing concentrations of Y^-. The reactions discussed in the last section were mostly carried out in relatively poor ionizing solvents, and are probably of this basic type.

If we change to an alkyl halide which forms a carbocation so reactive that the barrier to collapse of the ion-pair $[R^+X^-]$ to product disappears, that ion-pair intermediate ceases to exist as an ordinary species, and the reaction coordinate diagram would look something like Fig. 4.3c. The reaction could still proceed through the carbocation intermediate even if the barrier to collapse of the $[R^+Y^-]$ ion-pair back to reactant disappeared as indicated by the diagram of Fig. 4.3d. Unless there is some factor causing $[X^-R^+Y^-]$ to become more stable than the free carbocation, the reaction proceeds by pathways outside the solid lines of Fig. 4.2.

An unusual mechanism can exist in situations where the ion-triplet is of lower free energy than, or comparable free energy to, the carbocation, and the carbocation moiety is so reactive that the barriers for collapse of the ion-triplet to either reactants or to products become smaller than diffusional barriers. Since there is a diffusional barrier between the ion-triplet and the ion-pairs, the reaction must proceed through what is called a pre-association mechanism through formation of the encounter complex $[X^- \cdot R\text{-}Y]$ by the k_{as} path, on to the ion-triplet through the k_{2i} pathway, and to product through the k'_{-2i} and k'_{-as} path. A possible reaction coordinate diagram is shown in Fig. 4.4a where the barriers between the ion-triplet and the ion-pairs are diffusional barriers, as are the barriers between the reactant and the encounter complex and between the other encounter complex and the product. The necessity for the pre-association mechanism is most clearly seen by considering the paths from the ion-triplet to either reactant or product.

The pre-association mechanism of Fig. 4.4a will exhibit second-order kinetics, and, therefore, be classified as an SN2 mechanism, but structural effects on rates will be very similar to those on SN1 reactions

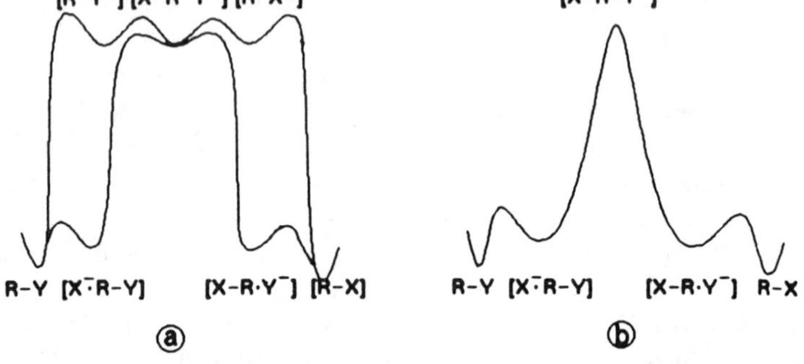

Figure 4.4: *Possible Reaction Coordinate Diagrams for SN2 Reactions. (a) The pre-association mechanism through an ion-triplet. (b) The classical SN2 mechanism.*

since the R^+ moiety of the ion-triplet has essentially all the attributes of a free carbocation.

The classical SN2 mechanism is shown by the reaction coordinate diagram of Fig. 4.4b. It differs from the pre-association mechanism in that the species $[XRY]^-$ is not an intermediate. The course of the reaction is through the encounter complex by the k_{as} path, to the other encounter complex by the k_2 path, and to product by the k'_{as} path. The transition state could lie anywhere within the solid-line-boxed region of Fig. 4.2, and have more or less carbocation character depending on its distance from the upper-right corner of that diagram. It would be extremely difficult to distinguish a classical SN2 mechanism with an 'exploded' transition state in the upper-right corner from a pre-association mechanism through an ion-triplet.

4.7 Reaction 'Clocks' and Lifetimes of Intermediates

At several points in the above general discussion, we have touched on the question of just how reactive some of the carbocations encountered might be. As we will discuss in detail in a later chapter, some carbocations are so stable that they are not even considered 'reactive' intermediates. For example, Crystal Violet, which is a triphenylmethyl cation having p-dimethylamino groups substituted on each of the three phenyls, has $pK_R = -\log K_R = 9.39$, where K_R is the equilibrium constant for the reaction:

$$R^+ + H_2O \xrightleftharpoons{K_R} ROH + H^+.$$

Even with approximately $10^{-3}M$ hydroxide ion present in water to force the reaction to the right, the half-life of this cation is approximately an hour at room temperature.

Studies of such stable cations [17] have established that a fairly constant scale of relative reactivities of nucleophiles exists for all nucleophiles which have rates slower than diffusion control. Some of the more reactive nucleophiles are azide ion, alkylthiolate ions, and peroxide ion, which have values for Log k_{nuc}/k_w of ~ 7.5, ~ 9, and ~ 8, respectively, where k_w is the pseudo-first-order rate constant for reactions of cations with water, and k_{nuc} is the second-order rate constant for reactions of cations with the given nucleophiles.

Special techniques have even allowed the direct study of rates of reactions of triphenylmethyl cation with nucleophiles in 2:1 water:acetonitrile solution [18]. This cation has $pK_R \approx -6.6$ and $k_w \approx 1.4 \times 10^5 sec^{-1}$. Its reaction with azide ion is almost certainly diffusion controlled with

$k_{azide} \approx 5 \times 10^9 M^{-1} sec^{-1}$. The triphenylmethyl cation is still much more stable than most carbocations encountered in studies of SN1 reactions, and the steric hindrance provided by the three propeller-like phenyl groups should make its reactions considerably slower than most other cations even of similar stability.

The key to getting at rates of reactions of some of the reactive intermediate carbocations is the fact that diffusion limited reactions are all expected to have nearly the same rate constant [19]. For carbocations reacting with anions in water, we may take this rate constant to be essentially that observed for the reaction of azide ion with triphenylmethyl cation mentioned above; $k_{dc} \approx 5 \times 10^9 M^{-1} sec^{-1}$. By carrying out reactions of alkyl-X compounds in the presence of known concentration of azide ion and some other nucleophile, then determining the amount of products formed from the competition of azide ion and other nucleophile with the carbocation, we obtain relative rate constants for the two nucleophiles, and, since the rate constant for azide ion reaction is known, we can evaluate the other rate constant. The reaction of azide ion is being used as a 'clock' to measure the rate of the other reaction. The idea of using reactions with known rate constants in competition with other reactions as 'clocks' to get at the rate constants for those other reactions has a rather long history of applications, particularly in the area of free radical chemistry [20].

A study [19] of reactions of a series of ring-substituted 1-phenylethyl derivatives in 1:1 water:trifluoroethanol, using azide ion or thiolate ion reactions as clocks, gave rate constants for reactions of the intermediate carbocations with solvent ranging from $2 \times 10^3 sec^{-1}$ for the 1-(p-dimethylaminophenyl)ethyl cation ($pK_R \approx -0.5$) to $4 \times 10^9 sec^{-1}$ for the 1-(p-methylphenyl)ethyl cation ($pK_R \approx -12.6$). A Hammett plot of the rate constants obtained for reactions of solvent with the derivatives having electron-donating substituents gave $\rho_+ = +4.5$. Use of this ρ_+ and the σ^+ values for electron-withdrawing substituents led to estimates of rate constants greater than $10^{11} sec^{-1}$ for the unsubstituted, the m-methoxy, and m-Br derivatives, and greater than $10^{12} sec^{-1}$ for the p-cyano and p-nitro derivatives, reacting with solvent.

In our discussion of the general mechanisms of SN1 and SN2 reactions in the previous section, the reactions with solvent as nucleophile were not considered. Solvent differs from other nucleophiles in that it is always in the immediate vicinity of any intermediate formed, and there is no diffusion step necessary to allow its reaction. The rates of reactions of intermediates with solvent will directly limit mechanisms in two ways. First, if the barrier for reaction of a carbocation with solvent disappears, the carbocation ceases to exist as an ordinary species. Second, if the barrier for reaction of solvent with the ion-pair $[R^+Y^-]$ becomes smaller than the diffusional barrier for separation of the ion-pair, the free carbocation cannot be an intermediate in the reaction.

When the carbocation becomes so reactive that it ceases to exist, the mechanism of the reaction must become SN2. The change from SN1 to SN2 mechanism in the reactions of the ring-substituted 1-phenylethyl derivatives has been shown [21] to occur as one goes from electron-donating to electron-withdrawing substituents. Those derivatives with substituents having $\sigma^+ <\sim -0.08$ give product ratios from reactions with azide ion and solvent which are proportional to azide ion concentrations, but show rates which are independent of azide ion concentration. Those derivatives having $\sigma^+ >\sim 0$ show rates of reactions linearly dependent on azide ion concentrations. The ρ_+ value for solvolyses of the derivatives with electron-donating substituents is ~ -4.8, while that for derivatives with electron-withdrawing substituents is ~ -2.9. This latter value indicates a transition state having appreciably less than the ca. 50–70% charge expected (see section 2.6) of an SN1 transition state, and, therefore an SN2 mechanism. The selectivities of these compounds between azide ion and solvent water depend on substituent exactly as expected from the above discussions.

Defining the selectivities,

$$S \equiv \frac{[RN_3]_\infty}{[N_3^-][ROH]_\infty},$$

in terms of product ratios obtained in the reactions, we should observe several different types of variations of S with substituents. For the most electron-donating substituents, where a carbocation is formed that reacts below the diffusion-controlled limit with azide ion and solvent, S should be $\sim 10^{7.5} M^{-1}$ which is the ratio of rate constants for reactions of azide ion and water with stable carbocations found in direct studies [17]. This value is approached by the estimated (from actual data for reaction of an alkylthiolate [19]) $S = 2.5 \times 10^6 M^{-1}$ for the p-dimethylamino substituted compound. For somewhat more reactive carbocations, there should be a range of reactivities where reactions with azide ion are diffusion-controlled, but reactions with solvent are normal activation-controlled. This is the range already discussed where the azide ion reaction serves as a clock for measuring the rates of reactions of the cations with solvent; values of S decrease as the rate of reaction with solvent increases. As the carbocation becomes still more reactive, the rate of reaction with solvent becomes greater than the rate of dissociation of the ion-pair. This situation is indicated for reaction of the p-methyl derivative; selectivities of this compound toward different solvent components depend somewhat on leaving group even though the reactions with solvent have most of the characteristics of SN1 reactions [21]. Values of S reach a minimum of $\sim 40M^{-1}$ for the p-methyl and p-fluoro substituted compounds, and then increase for more electron-withdrawing substituents, reaching a value of $\sim 8 \times 10^3 M^{-1}$ for the p-nitro derivative. All of those compounds having substituents with $\sigma^+ > 0$ show rates which are first-order with respect to azide ion concentrations, and, as already mentioned, the ρ_+ value indicates that the reactions of the latter compounds with solvent are also SN2, although still with considerable positive charge on carbon at the transition state. Values of log S are plotted vs. σ^+ in Fig. 4.5.

It is particularly interesting to note that the studies [7] of these 1-phenylethyl derivatives show that the SN1 and SN2 pathways can co-exist. Even though the carbocation is a normal species represented by a potential energy minimum in the potential energy surface for the p-fluoro derivative, the reaction of azide ion occurs by a concerted SN2

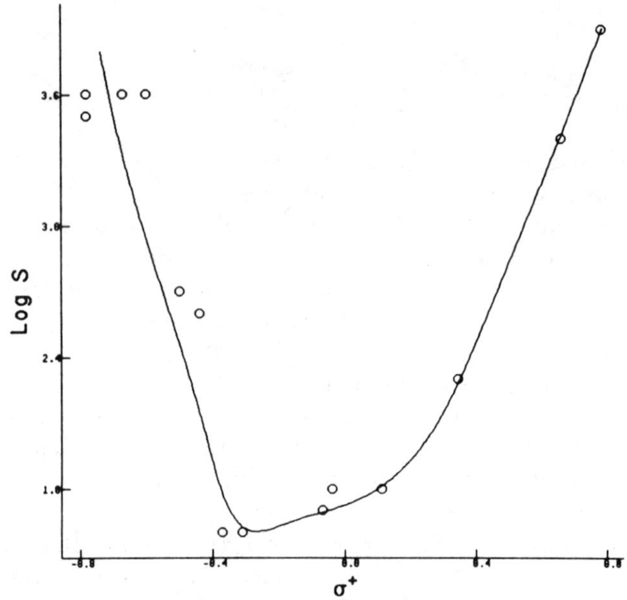

Figure 4.5: *Observed [7, 19, 21] Selectivities for Azide Ion/Water Reactions of m- and p-Substituted 1-Phenylethyl Derivatives in 1:1 Water:Trifluoroethanol Solvent Plotted Against* σ^+.

path. The reactions with solvent, a poorer nucleophile than azide ion, proceed by the SN1 mechanism as long as the carbocation can exist.

4.8 REFERENCES

[1]. R. A. More O'Ferrall, **J. Chem. Soc. (B)**, 274 (**1970**).

[2]. W. P. Jencks, **Chem. Rev., 72**, 705 (**1972**).

[3]. W. P. Jencks, **Accts. Chem. Res., 13**, 161 (**1980**).

[4]. E. F. Caldin, **Fast Reactions in Solution**, Blackwell Scientific Publ., Oxford, England, **1964**, pgs. 10–12 and 279–285.

[5]. E. Grunwald, **Prog. Phys. Org. Chem., 3**, 317 (**1965**); see particularly pg. 322.

[6]. J. Hine, **J. Amer. Chem. Soc., 93**, 3701 (**1971**).

[7]. J. P. Richard and W. P. Jencks, **J. Amer. Chem. Soc., 106**, 1373–1383 (**1984**); see particularly pgs. 1381–1382.

[8]. W. G. Young, S. Winstein, and H. L. Goering, **J. Amer. Chem. Soc., 73**, 1958 (**1951**).

[9]. S. Winstein and G. C. Robinson, **J. Amer. Chem. Soc., 80**, 169 (**1958**).

[10]. H. C. Brown, C. J. Kim, C. J. Lancelot, and P. v. R. Schleyer, **J. Amer. Chem. Soc., 92**, 5244 (**1970**).

[11]. S. Winstein, P. Klinedienst, and G. C. Robinson, **J. Amer. Chem. Soc., 83**, 885 (**1961**).

[12]. S. Winstein, P. Klinedienst, and E. Clippinger, **J. Amer. Chem. Soc., 83**, 4986 (**1961**).

[13]. H. L. Goering and J. L. Levy, **J. Amer. Chem. Soc., 86**, 120 (**1964**).

[14]. H. L. Goering, R. G. Briody, and G. Sandrock, **J. Amer. Chem. Soc., 92**, 7401 (**1970**).

[15]. H. L. Goering and H. Hopf, **J. Amer. Chem. Soc., 93**, 1224 (**1971**).

[16]. H. L. Goering and R. W. Thiess, **J. Amer. Chem. Soc., 90**, 2968 (**1968**).

[17]. C. D. Ritchie, **Can. J. Chem., 64**, 2239 (**1986**).

[18]. R. A. McClelland, N. Banait, and S. Steenken, **J. Amer. Chem. Soc., 108**, 7023 (**1986**); **110**, 6913 (**1988**); **111**, 2929 (**1989**).

[19]. J. P. Richard, M. E. Rothenberg, and W. P. Jencks, **J. Amer. Chem. Soc., 106**, 1361 (**1984**).

[20]. D. Griller and K. U. Ingold, **Accts. Chem. Res., 13**, 317 (**1980**).

[21]. J. P. Richard and W. P. Jencks, **J. Amer. Chem. Soc., 106**, 1383 (**1984**).

4.9 PROBLEMS

Problem 4.1

Sketch a reasonable potential energy contour diagram for the reaction:

$$D\cdot + H\text{–}H \rightleftharpoons D\text{–}H + H\cdot$$

Note that potential energy is independent of nuclear mass, so that D and H are completely equivalent for all present purposes. Assume that a linear configuration of the three atoms is always of lowest energy, and that there are no ordinary triatomic chemical species for this system.
Hint: Choose the coordinates for the diagram as the distance from the central H to each of the other two atoms. Be sure your diagram shows the equivalence of H and D; the diagram will be symmetrical about a diagonal line through the diagram.

Problem 4.2

Sketch possible potential energy contour diagrams for the Diels-Alder reaction of ethylene with 1,3-butadiene to form cyclohexene. Consider the possibility that there may be finite, but small, barriers to closure of di-radical intermediates. Again, pay attention to the required symmetry of the diagram.

Problem 4.3

Sketch possible potential energy diagrams, showing possible intermediates, for the reaction:

Problem 4.4

Nitrogen molecule has an unusually strong triple bond. Di-imide, H–N=N–H, reacts with alkenes to form alkanes plus nitrogen. Show, by a potential energy diagram, why one might expect this reaction to take place in a concerted fashion with both hydrogens being transferred simultaneously.

Problem 4.5

Why is the ion-pair trapping mechanism shown in the last paragraph of section 4.5 very unlikely to occur?
Hint: Consider why a transition state for the k_Y *step could be lower in energy than the free carbocation, then consider what this means for energies along the pre-association pathway shown in Fig. 4.4a relative to the energies along an ion-pair pathway.*

Problem 4.6

What can you say about the transition state geometry for an 'identity' reaction in which X^- reacts with CH_3X in an SN2 reaction if you assume that there is a single, unique, transition state? What possibilities exist if you do not assume a single transition state?
Hint: The potential energy diagram must show the symmetry of the system. If there is a single transition state, the symmetry is just like that in Prob. 4.1, above. If more than one transition state is allowed, consider the possibility of enantiomeric transition states along one pathway, and the possibility of 'enantiomeric pathways'. If you're still having difficulty, see: [R. G. Pearson and R. L. Burwell, **J. Phys. Chem.**, **70**, 300 (**1966**)].

Problem 4.7

In your earlier courses, and in the present chapter, you have been thoroughly indoctrinated into the concept of a reaction coordinate diagram. We have carefully and uniquely defined what we mean by real species and by transition states in terms of locations on potential energy surfaces. Can you define the other points along the reaction coordinate diagrams, such as those in Figs. 4.3 and 4.4, in a unique manner?
Hint: No. Only stationary points can be uniquely defined; any other points that you define will change location if you change (for example, rotate) coordinate system. The path that a system takes between stationary points is indeterminate.

Problem 4.8

The elimination reactions of several 2-arylethylquinuclidinium ions catalyzed by substituted alkoxide ions were studied [J. R. Gandler and W. P. Jencks, **J. Amer. Chem. Soc., 104**, 1937 **(1982)**.]:

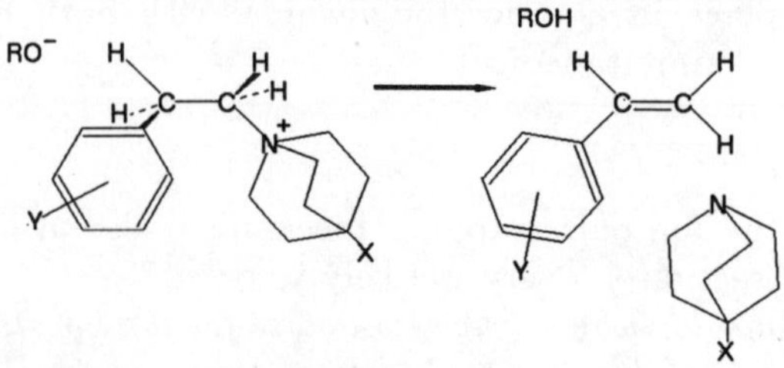

For variation of the alkoxide ions,

$$\frac{\Delta \log k}{\Delta \log K} \simeq 0.7$$

for variation of the substituents on the quinuclidine ring,

$$\frac{\Delta \log k}{\Delta \log K} \simeq 0.2$$

and, for variation of the substituents on the benzene ring, ρ=+3.5 (Equilibrium formation of a carbanion would be expected to give $\rho \simeq$ +5.0.) Sketch a potential energy diagram for the reaction and mark the approximate location of the transition states. How would you expect the location of the transition state to change if the leaving group were changed to a tosylate anion, or if the aryl group were changed to a methyl group?

Hint: Use the C-N bond length and another length related to the proton transfer as the two coordinates, similar to the type of diagram shown in Fig. 4.1b.

Problem 4.9

The effects of substituents in both aryl groups on the rates and equilibria of the SN2 reactions:

$$ArSO_3^- + CH_3SO_3Ar' \underset{k_r}{\overset{k_f}{\rightleftharpoons}} ArSO_3CH_3 + Ar'SO_3^-$$

in sulfolane solution have been determined [E. S. Lewis and D. D. Hu, **J. Amer. Chem. Soc., 106**, 3292 **(1984)**.].

For variation of the substituents in the Ar group, keeping Ar′ constant, the Hammett ρ for the equilibrium constants is -2.9, and the ρ for the forward rate constants is -1.1. Note that this requires that the ρ for the reverse rate constants is +1.8.
When the substituents in both Ar and Ar′ are varied, with Ar=Ar′, obviously the equilibrium constants are all equal to 1.0, and the ρ for the equilibrium constants must be zero. It is observed, however, that the rate constants do vary. A plot of the $\log k_f$ values against the ordinary Hammett σ values gives ρ=0.6.
What do these four ρ values imply about the transition states for these reactions? In particular, what does the last one, ρ=0.6, imply about the variability of the location of the transition state on the potential energy surface?
Hint: Draw the diagram for the potential energy surface just as suggested in Prob. 4.6. Assume only one transition state, and that ρ_{rate}/ρ_{eq} *is a measure of the extent of bond formation at the transition state. You may find it useful, and amusing, to sketch or to visualize a three- dimensional Hammett plot in which the* σ *values for the substituents in Ar are along one coordinate, the* σ *for the substituents in Ar′ are along the second coordinate, and the* $\log k$ *values are along the third coordinate.*

Problem 4.10

The acid-catalyzed hydrolyses of substituted acetophenone ketals are believed to follow the mechanism:

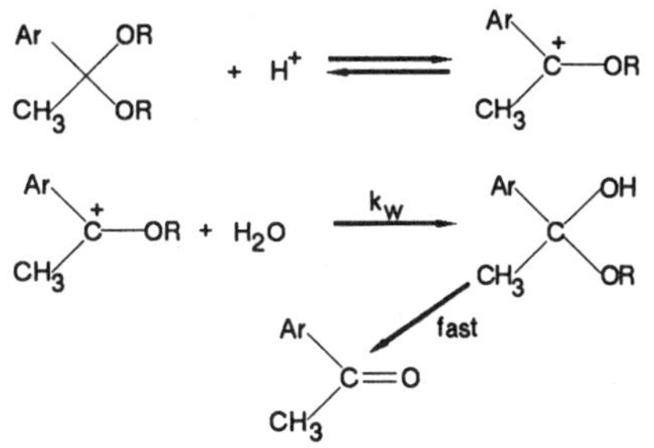

It was found possible to trap the intermediate oxy-carbocation, (I), with sulfite ion in much the same manner as azide ion was used in the SN1 reactions described in this chapter. The rates of reactions of sulfite ion with the oxy-carbocations are almost certainly diffusion-controlled with $k \simeq 5 \times 10^9 M^{-1} sec^{-1}$.

From the product ratios obtained in the presence of various concentrations of sulfite ion, the differently substituted acetophenone ketals were found to have rate constant ratios, k_w/k_S (k_w is the pseudo-first-order rate constant for reaction of (I) with water, and k_S is the second-order rate constant for reaction of sulfite ion with (I).) as follows: p-OCH_3, $1.4 \times 10^{-3} M$; p-CH_3, $1.0 \times 10^{-2} M$; H, $1.9 \times 10^{-2} M$; p-Cl, $4.1 \times 10^{-2} M$; p-Br, $5.9 \times 10^{-2} M$.

(a). Calculate the value for k_w for each derivative.

(b). Make a Hammett plot for k_w and determine the ρ_+ value for the water reaction with (I).

(c). Use the Hammett relationship to estimate k_w for the p-NO_2 substituted ketal.

(d). It has been estimated that replacement of a phenyl group by hydrogen would increase the rates of reaction of (I) by a factor of at least 10^4, and that similar replacement of the methyl group by hydrogen would increase rates by another factor of at least 10^2. Assuming that these estimates are roughly reliable, comment on the mechanisms of hydrolyses of the ketals of acetaldehyde and of formaldehyde.

Hint: You shouldn't need to, but, just to convince yourself that you are right, see: [P. R. Young and W. P. Jencks, **J. Amer. Chem. Soc., 99**, 8238 **(1977)**].

Problem 4.11

In buffered basic aqueous solutions, 1-phenylcyclopropanol undergoes a ring opening reaction to produce 1-phenyl-1-propanone. At one time, the reaction was thought to proceed through an extremely reactive carbanion intermediate, $^-CH_2CH_2COC_6H_5$. Studies of the rates of reactions of other carbanions with water, however, led to estimates of $> 10^{15} sec^{-1}$ for the rate constant for reaction of this carbanion with water. Experimentally, the rate of the ring opening reaction is found to depend on the concentrations of the buffer acids used to maintain constant pH. Propose a mechanism for this reaction.

Hint: Consider how a figure similar to Fig. 4.4a might be pertinent to this reaction. If all else fails, see: [A. Thibblin and W. P. Jencks, **J. Amer. Chem. Soc., 101**, 4963 **(1979)**].

Problem 4.12

A mechanism has been proposed for nucleophilic substitutions at phosphate monoesters which involves the following steps:

$$X\text{-}PO_3^- \underset{k_{-1}}{\overset{k_1}{\rightleftharpoons}} [X^- \cdots PO_3]$$

$$[X^- \cdots PO_3] + Y^- \underset{k_{-2}}{\overset{k_2}{\rightleftharpoons}} [X^- \cdots PO_3 \cdots Y^-]$$

$$[X^- \cdots PO_3 \cdots Y^-] \underset{k_{-3}}{\overset{k_3}{\rightleftharpoons}} Y\text{-}PO_3^- + X^-$$

What is wrong with this as a general mechanism for the substitution reaction?

Hint: First, consider the case where X and Y are the same. Second, look again at Fig. 4.4a, and be sure you understand it; if you don't, read ref. [3].

Problem 4.13

Rate constants and product yields for the acetolyses of eight substituted 3-aryl-2-butylbrosylates are shown in the Table below.

Acetolyses of 3-aryl-2-butylbrosylates at 75°C

Data from Ref. [10].

Aryl	$10^5 k_t$, sec^{-1}	% alkene formed	% erythro acetate formed	% threo acetate formed
p-$CH_3OC_6H_4$-	1060	0.3	0	99.7
p-$CH_3C_6H_4$-	81.4	12	0	88.0
m-$CH_3C_6H_4$-	28.2	31	1	68
C_6H_5-	18.0	38	3	59
p-ClC_6H_4-	4.53	53	6	39
m-ClC_6H_4-	2.05	76	11	12
m-$CF_3C_6H_4$-	1.26	75	14	11
p-$NO_2C_6H_4$-	0.495	68	12	1

Assume that the threo-acetate products are formed only from the phenonium ion intermediate, and that the other products of reactions are formed by competing SN2 solvolyses or concerted eliminations, calculate the Hammett ρ_+ value for the solvolysis through the phenonium ion, and the ρ for the SN2 solvolysis.

Hint: Use the reported % yields to dissect the rate constants into those for the individual paths. Make Hammett plots of these to obtain the ρ's.

Problem 4.14

The solvolysis of 1-p-tolyl-2-propyltosylate in 80% ethanol was carried out in the presence of various concentrations of sodium azide. The following data on rates and products were obtained under pseudo-first-order conditions:

Data from [Schleyer, et al, **J. Amer. Chem. Soc., 93**, 4829 (**1971**)].

$[NaN_3]$, M	$10^4 k_\psi$, sec^{-1}	% RN_3	% ROH	% ROEt
0.00	2.40	0	54	46
0.02	3.18	27	39	35
0.04	4.28	46	27	27
0.06	5.19	55	21	24

Some of the RN_3 product is formed by an SN2 pathway, while the remainder is formed by an SN1 path. Calculate the % RN_3 formed from each path.

Hint: Plot k_ψ against $[NaN_3]$ to obtain the rate constant for the SN2 reaction, then use this to calculate the amount of product produced by the second-order reaction.

The solvolysis of the isomeric 2-p-tolyl-1-propyltosylate was carried out under the same conditions as used above. Most of the products from this isomer are the same ones as formed from the 1-p-tolyl-2-propyltosylate, and are believed to be formed from the phenonium ion intermediate. The following amounts of the rearranged products were observed:

1-p-tolyl-2-propyl products from 2-p-tolyl-1-propyltosylate

$[NaN_3]$, M	% RN_3	% ROH	% ROEt
0.00	0	49	49
0.02	5	43	47
0.04	10	39	41
0.06	12	36	39

Is it likely that some of the ROH and ROEt from the 1-p-tolyl-2-propyltosylate was formed by an SN2 path?

Hint: From your answer above concerning the RN_3 product, and the trapping ratios obtained in the last table, how much ROH and ROEt were formed from the phenonium ion in the reaction of 1-p-tolyl-2-propyltosylate?

Assume that the reaction of azide ion with the phenonium ion is diffusion-controlled with $k \simeq 5 \times 10^9 M^{-1} sec^{-1}$. Calculate the pseudo-first-order rate constants for reactions of water and ethanol with the phenonium ion.

CHAPTER 5
Structure and Reactivity: Quantum Mechanics

5.1 Introduction

The empirical relationships between structure and reactivity which we discussed in Chapter 3 are extremely useful in many cases where basic theory presently offers no help. In fact, the empirical relationships are not anticipated by basic theory and it is still difficult even to rationalize the existence of some of them. There are also, however, many examples where the empirical relationships offer no help, but in which basic theory provides rational explanations of reactivities. For example, the Woodward-Hoffman rules were derived from molecular orbital theory after numerous efforts to understand reactivities and stereochemistries of electrocyclic reactions on empirical bases had failed.

In the present chapter, we will start with some concepts of atomic orbitals, molecular orbitals, and energies with which you probably already have at least a nodding acquaintance. The idea of molecular orbitals will then be developed into Hückel Π-MO theory, and some examples of the applications of this theory will be discussed. We'll then back up to the basic concepts of quantum mechanics and develop molecular orbital theory in a more rigorous manner.

5.2 Orbitals and Orbital Energies

Ever since your first course in chemistry, you have probably known that there are such things as atomic orbitals. Atomic orbitals were first developed in connection with attempts to understand the properties of hydrogen atoms. The electron of a hydrogen atom occupies one of the various orbitals and the atom has an electronic energy that is related to the particular orbital which is occupied. If we denote the orbitals as η's, the Hamiltonian operator for the hydrogen atom as $\mathcal{H}$, and the electronic energies of the atom as ϵ's, the Schrödinger equation can be written as:

$$\mathcal{H}\eta_i = \epsilon_i \eta_i.$$

The atomic orbitals, or wave functions, η_i describe the electron in the sense that the square of η_i gives the electron density at a particular location. Just as a concrete example, let's actually look at the 1s orbital for a hydrogen atom. Using the subscript zero to indicate the 1s orbital,

$$\eta_0 = 1.468e^{-1.892r}$$

where r is the distance from the nucleus expressed in Å units. If we wish to know the probability of finding the electron within some particular volume element of space, we would integrate the square of η over the particular volume element. The factor of 1.468 in the expression for η_0 is chosen such that the integral of the square of the function over all space is equal to unity. Functions having this property are called *normalized* functions. The factor 1.892 in the exponent is directly related to the nuclear charge of the atom. The 1s orbital for an electron on a carbon nucleus would simply have 6×1.892 as the factor in the exponent; this would require that the factor 1.468 be changed to $6^{3/2} \times 1.468$ for the function to remain normalized.

All atomic orbitals for a given atom contain the same exponential term. The 2s orbital will have a pre-exponential term of $(r - b)$, where b is simply a number; the $2p_x$ orbital will have a pre-exponential term of x; the $2p_y$ orbital will have a pre-exponential term of y; etc., each multiplied by a constant chosen to keep the functions normalized. The general form for a $2p_z$ orbital, for example, on any atom, then, can be written as

$$\eta_{2p_z} = Aze^{-\alpha r},$$

where A and α are constants whose values depend on the nuclear charge, z is distance along the z-Cartesian coordinate, and r is the distance from the nucleus.

We'll look in much more detail at the Schrödinger equations a little later, but for now you should realize that the equations cannot be solved to give analytical forms for the wave functions of any atom or molecule containing more than one electron. What we do is to try to approximate the wave functions for these many-electron systems in terms of the atomic orbitals obtained for one-electron systems. The general idea is somewhat similar to approximating a complicated sound wave as a sum of simple sine and cosine functions. In order to develop this idea at all, we need to understand the concept of an *expectation value* for a property of a system.

If any system is described by a normalized wave-function, ψ_i, not necessarily a solution of the Schrödinger equation for the system, the average value that would be observed for the energy of the system, called the *expectation value* for the energy and symbolized $\langle E \rangle$, is given by the equation:

$$\langle E \rangle = \int_{-\infty}^{+\infty} \int_{-\infty}^{+\infty} \int_{-\infty}^{+\infty} \cdots \int_{-\infty}^{+\infty} \psi^* \mathcal{H} \psi \; dx_1 dy_1 dz_1 \cdots dz_n,$$

where the integration is over all coordinates for each particle of the system, and ψ^* is the complex conjugate of ψ. Almost all of the wave functions that we will need in this book are real, and, for these, $\psi^* = \psi$. In order to save much writing, we will use *Dirac brackets* notation which is defined in the general equation:

$$\langle \psi | \mathcal{A} | \psi \rangle \equiv \int_{-\infty}^{+\infty} \int_{-\infty}^{+\infty} \int_{-\infty}^{+\infty} \cdots \int_{-\infty}^{+\infty} \psi^* \mathcal{A} \psi \; dx_1 dy_1 dz_1 \cdots dz_n = \langle A \rangle,$$

where $\mathcal{A}$ is any quantum mechanical operator, and A is the observable associated with the operator. Using this notation, we have, for normalized wave functions:

$$\langle E \rangle = \langle \psi | \mathcal{H} | \psi \rangle \quad \text{if} \quad \langle \psi | \psi \rangle = 1. \tag{5.1}$$

When the operator is omitted, as in $\langle \psi | \psi \rangle$, it is understood to be the *identity operator*, which leaves any function on which it operates unchanged (recall the identity matrices of matrix algebra, or the number 1 of ordinary algebra). It is not always convenient to use normalized wave functions. We then use the more general definitions for expectation values; for the energy, for example:

$$\langle E \rangle = \frac{\langle \psi | \mathcal{H} | \psi \rangle}{\langle \psi | \psi \rangle} \tag{5.2}$$

Note carefully that if ψ is a solution of the Schrödinger equation, $\mathcal{H}\psi = \epsilon\psi$, then the expectation value for the energy is just ϵ:

$$\langle \psi | \mathcal{H} | \psi \rangle = \langle \psi | \epsilon | \psi \rangle = \epsilon \langle \psi | \psi \rangle, \quad \text{therefore:} \quad \langle E \rangle = \epsilon$$

Now, suppose we have n atomic nuclei, each with a nuclear charge Z_i, and each associated with an atomic orbital η_i. We won't bother, for the moment, to even specify what types of atomic orbitals the η's are, nor to specify the locations of the nuclei. Let's find an approximate orbital for one electron associated with these nuclei. The Hamiltonian operator for the system depends on the nuclear positions and nuclear charges, but we will simply write it as $\mathcal{H}$. We're going to construct some electronic orbitals, ϕ_j, from the atomic orbitals, η_i, according to the equation:

$$\phi_j = \sum_{i=1}^{n} \eta_i c_{ij}$$

This is called a *linear combination of atomic orbitals*, and the technique we're using to construct the electronic orbitals is called the LCAO method.

According to Eq. (5.2), the expectation value for the energy of an electron in an orbital, ϕ_r, is:

$$\langle E_r \rangle = \frac{\langle \phi_r | \mathcal{H} | \phi_r \rangle}{\langle \phi_r | \phi_r \rangle} = \frac{\left\langle \sum_{k=1}^{n} \eta_k c_{kr} \middle| \mathcal{H} \middle| \sum_{\ell=1}^{n} \eta_\ell c_{\ell r} \right\rangle}{\left\langle \sum_{k=1}^{n} \eta_k c_{kr} \middle| \sum_{\ell=1}^{n} \eta_\ell c_{\ell r} \right\rangle}$$
$$= \frac{\sum_{k=1}^{n} \sum_{\ell=1}^{n} c_{kr} c_{\ell r} \langle \eta_k | \mathcal{H} | \eta_\ell \rangle}{\sum_{k=1}^{n} \sum_{\ell=1}^{n} c_{kr} c_{\ell r} \langle \eta_k | \eta_\ell \rangle},$$

where we have assumed that the η's and c's are all real.

The 'best' orbital will be the one with the lowest possible energy for the electron. We want, then, to choose the coefficients, c_{kr}, to give the lowest possible expectation value for E_r. This is a standard minimization problem... we take the partial derivatives of E_r with respect to the coefficients and set them equal to zero:

$$\frac{\partial \langle E_r \rangle}{\partial c_{kr}} = 0, \quad k = 1, n$$

Since the only things in the expression for $\langle E_r \rangle$ that depend on the c's are the c's, themselves, the differentiation is quite simple, requiring a little care only in handling the sums. Note that:

$$\frac{\partial \left(\sum_{k=1}^{n} \eta_k c_{kr} \right)}{\partial c_{mr}} = \eta_m.$$

Proceeding step-by-step, and assuming that the ϕ's are real, we obtain:

$$\frac{\partial \langle E_r \rangle}{\partial c_{kr}} = \frac{\partial \left(\langle \phi_r | \mathcal{H} | \phi_r \rangle / \langle \phi_r | \phi_r \rangle \right)}{\partial c_{mr}} = \frac{\partial \langle \phi_r | \mathcal{H} | \phi_r \rangle}{\langle \phi_r | \phi_r \rangle \, \partial c_{mr}} - \frac{\langle \phi_r | \mathcal{H} | \phi_r \rangle \, \partial \langle \phi_r | \phi_r \rangle}{\langle \phi_r | \phi_r \rangle^2 \, \partial c_{mr}}$$
$$= \frac{\langle \phi_r | \phi_r \rangle \, \partial \langle \phi_r | \mathcal{H} | \phi_r \rangle / \partial c_{mr} - \langle \phi_r | \mathcal{H} | \phi_r \rangle \, \partial \langle \phi_r | \phi_r \rangle / \partial c_{mr}}{\langle \phi_r | \phi_r \rangle^2} = 0.$$

Multiplying through the last line by $\langle\phi_r|\phi_r\rangle^2$, we get:

$$\begin{aligned}0 &= \langle\phi_r|\phi_r\rangle \frac{\partial\langle\phi_r|\mathcal{H}|\phi_r\rangle}{\partial c_{mr}} - \langle\phi_r|\mathcal{H}|\phi_r\rangle \frac{\partial\langle\phi_r|\phi_r\rangle}{\partial c_{mr}} \\ &= \langle\phi_r|\phi_r\rangle \left(\langle\eta_m|\mathcal{H}|\phi_r\rangle + \langle\phi_r|\mathcal{H}|\eta_m\rangle\right) - \langle\phi_r|\mathcal{H}|\phi_r\rangle \left(\langle\eta_m|\phi_r\rangle + \langle\phi_r|\eta_m\rangle\right)\end{aligned}$$

It is very easy to show that $\langle\phi_r|\eta_m\rangle = \langle\eta_m|\phi_r\rangle$ by simply writing the integrals with ϕ_r expanded as the sum of η's. It is not so easy to show, but it is true that $\langle\phi_r|\mathcal{H}|\eta_m\rangle = \langle\eta_m|\mathcal{H}|\phi_r\rangle$. Any operator for which this relationship is true is called a *Hermitian* operator (Sounds suspiciously like matrix terminology, doesn't it.). We'll prove later that the Hamiltonian operator for any system is Hermitian...just accept it for now. With these equalities, we have:

$$2\langle\phi_r|\phi_r\rangle\langle\phi_r|\mathcal{H}|\eta_m\rangle - 2\langle\phi_r|\mathcal{H}|\phi_r\rangle\langle\phi_r|\eta_m\rangle = 0,$$

which can be divided through by 2, and by $\langle\phi_r|\phi_r\rangle$, then expanded to give:

$$\sum_{k=1}^{n}\langle\eta_m|\mathcal{H}|\eta_k\rangle c_{kr} - \langle E_r\rangle \sum_{k=1}^{n}\langle\eta_m|\eta_k\rangle c_{kr} = 0, \quad \text{for all } m.$$

Now, let's simplify the appearance of the equation by defining some quantities:

$$H_{ij} \equiv \langle\eta_i|\mathcal{H}|\eta_j\rangle \quad \epsilon_i \equiv \langle E_i\rangle \quad S_{ij} \equiv \langle\eta_i|\eta_j\rangle,$$

so that

$$\sum_{k=1}^{n} H_{mk}c_{kr} = \epsilon_r \sum_{k=1}^{n} S_{mk}c_{kr}, \quad \text{for all } m.$$

It should be nearly obvious to you now that this is a matrix equation:

$$\mathcal{H}\vec{c}_r = \mathcal{S}\vec{c}_r\epsilon_r,$$

which actually has n different solutions for $\vec{c}_r$ and ϵ_r. We can write a single matrix equation for all of the solutions:

$$\mathcal{HC} = \mathcal{SCE},$$

where the $\vec{c}_r$ are the columns of the square matrix $\mathcal{C}$, and the ϵ_r are the elements of the diagonal matrix $\mathcal{E}$.

If we chose our original atomic orbitals, η, such that the two conditions:

$$\langle\eta_i|\eta_j\rangle = \begin{cases} 1, & \text{if } i = j; \\ 0, & \text{if } i \neq j, \end{cases}$$

are satisfied, the matrix $\mathcal{S}$ would be the identity matrix, and our matrix equation problem would become the common one of matrix diagonalization:

$$\mathcal{HC} = \mathcal{CE}, \quad \text{so} \quad \mathcal{C}^\dagger \mathcal{HC} = \mathcal{E},$$

in which $\mathcal{C}$ is a unitary matrix. If the η's satisfy the two conditions stated, they are said to form an *orthonormal* set; the functions are normalized, and they are *orthogonal* one to another. In most of the simplest molecular orbital treatments, the atomic orbitals used are assumed to form an orthonormal set *even when they obviously don't.* If the η's are orthonormal, then the ϕ's formed from the columns of the unitary $\mathcal{C}$ matrix will also be orthonormal. The column of $\mathcal{C}$ corresponding to the lowest ϵ will be the 'best' possible orbital that can be formed from a linear combination of the chosen η's. The column of $\mathcal{C}$ corresponding to the next lowest ϵ will be the next best possible orbital that is orthogonal to the best one, and so on up to the column corresponding to the highest ϵ.

In a real problem, we choose our set of η's, evaluate the H_{ij} elements of the $\mathcal{H}$ matrix by hook or crook, and then diagonalize the matrix to obtain the c's and the ϵ's. As a particularly simple example, let's consider two hydrogen nuclei, each associated with a 1s atomic orbital centered on it. Let's call these two atomic orbitals η_1 and η_2. The first task is to evaluate the elements of the $\mathcal{H}$ matrix. Since the two hydrogen nuclei are the same, and the two atomic orbitals are identical except that one is on one nucleus and the other on the other nucleus, we must have $\langle\eta_1|\mathcal{H}|\eta_1\rangle = \langle\eta_2|\mathcal{H}|\eta_2\rangle$, and $\langle\eta_1|\mathcal{H}|\eta_2\rangle = \langle\eta_2|\mathcal{H}|\eta_1\rangle$. Let's simply call the values of these elements α and β, respectively, for now. The $\mathcal{H}$ matrix, then, is:

$$\mathcal{H} = \begin{pmatrix} \alpha & \beta \\ \beta & \alpha \end{pmatrix},$$

which can be diagonalized by the unitary matrix:

$$\mathcal{C} = \begin{pmatrix} 1/\sqrt{2} & 1/\sqrt{2} \\ 1/\sqrt{2} & -1/\sqrt{2} \end{pmatrix}$$

to obtain:

$$\begin{pmatrix} 1/\sqrt{2} & 1/\sqrt{2} \\ 1/\sqrt{2} & -1/\sqrt{2} \end{pmatrix} \begin{pmatrix} \alpha & \beta \\ \beta & \alpha \end{pmatrix} \begin{pmatrix} 1/\sqrt{2} & 1/\sqrt{2} \\ 1/\sqrt{2} & -1/\sqrt{2} \end{pmatrix} = \begin{pmatrix} \alpha+\beta & 0 \\ 0 & \alpha-\beta \end{pmatrix}$$

so that:

$$\phi_1 = \frac{1}{\sqrt{2}}(\eta_1 + \eta_2), \quad \epsilon_1 = \alpha + \beta$$

$$\phi_2 = \frac{1}{\sqrt{2}}(\eta_1 - \eta_2), \quad \epsilon_2 = \alpha - \beta.$$

There are several things to notice here that will recur again and again in applications of the LCAO method. First, the quantity α would be just the energy of an electron on a hydrogen atom in the 1s state if the other nucleus were not present. Since energies are relative to electron and nuclei separated by infinite distance, α is a negative number. It should actually be somewhat less than the energy of an electron on a hydrogen atom, since the presence of the other nucleus would be expected to lower the energy. Second, the quantity β is not so simple to interpret, but it can be thought of as the energy of an electron in the region where both η_1 and η_2 have large values; it is certainly a negative quantity, although not so negative as α. This means that ϕ_1 is the orbital having the lowest energy. Third, ϕ_1 is positive in all regions of space since both η_1 and η_2 are positive in all regions of space, but ϕ_2 is negative for all regions of space closer to the second nucleus than to the first. At any point equidistant from the two nuclei, $\phi_2 = 0$. We say that ϕ_2 has a *node* at all points equidistant from the nuclei, or that is has a *nodal surface* which is the plane of points equidistant from the two nuclei. The molecular orbital, ϕ_1, places greater electron density between the nuclei than the average of the densities of the two atomic orbitals; the molecular orbital, ϕ_2, because of the node, places less electron density between the nuclei than the average of the densities of the atomic orbitals. In the LCAO treatments, in general, the lowest energy orbital will have no nodes that are not present in the atomic orbitals; the next highest energy orbital will have one nodal surface not present in the atomic orbitals, and higher energy orbitals will have greater numbers of nodal surfaces. You should be able to see that this increasing number of nodes is a direct consequence of the orthogonality of the ϕ's. Finally, since all of the matrices in the problem are square, if we start with n atomic orbitals, we will get n molecular orbitals ϕ. In the present case, we only use one of these because we only have one electron.

Suppose that we add another electron to the above system to get the hydrogen molecule. The simplest treatments recognize that two electrons can occupy each molecular orbital if they have different spins, and use the molecular orbitals found from the collection of nuclei and atomic orbitals, which are independent of the number of electrons in the system. In the case of the lowest energy state for the hydrogen molecule, we would place both electrons in ϕ_1, and each electron would have the energy of ϵ_1. The lowest electronic energy of the hydrogen molecule would then be $2\alpha + 2\beta$. We could even go on to the H_2^- radical anion by placing a third electron in ϕ_2 with an energy of ϵ_2; the total energy of the species would be $2\epsilon_1 + \epsilon_2 = 3\alpha + \beta$.

5.3 Hückel II Molecular Orbitals

The development given above can be carried over directly to planar π-conjugated molecules. Instead of starting with bare nuclei, we start with the nuclei and all σ-electrons in the xy-plane. Instead of atomic 1s orbitals, we use atomic $2p_z$ orbitals which are orthogonal to all of the orbitals involved in the σ-framework. The π-electron problem for ethylene looks exactly the same as the two hydrogen nuclei problem discussed above, although the quantities α and β will have different values from those above, and the ϕ's for ethylene will have a nodal plane, the nodal xy-plane associated with the $2p_z$ atomic orbitals, that those for hydrogen did not have. A sketch of the atomic orbitals and of the two molecular orbitals for the π-system of ethylene is shown:

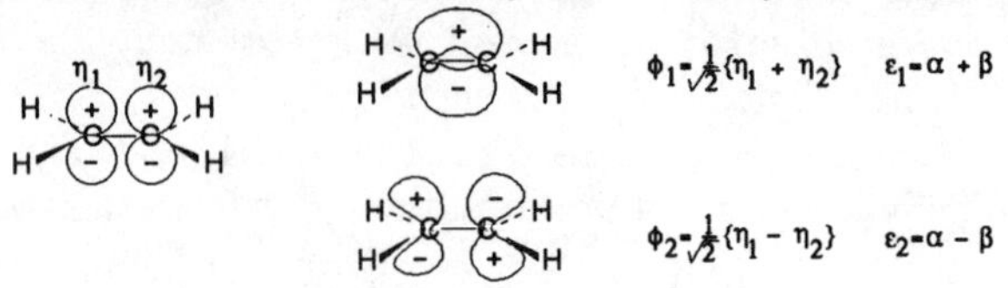

For ethylene, with only two atomic orbitals for the π-system, we still have only two parameters, α and β, for the two Hamiltonian matrix elements $\langle\eta_1|\mathcal{H}|\eta_1\rangle$ and $\langle\eta_1|\mathcal{H}|\eta_2\rangle$, respectively, which we didn't need to evaluate in order to find the molecular orbitals. If we consider the allyl system, H_2C-CH-CH_2, something new comes in. We start with a p_z orbital on each of the carbons, η_1, η_2, and η_3, which are orthogonal to the σ-framework:

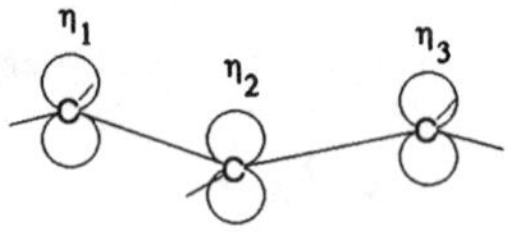

The H_{11}, H_{22}, and H_{33} matrix elements may all still be taken as essentially equal to the energy of an electron in the p_z orbital of a carbon atom, and set equal to α, just as for ethylene. Similarly, the H_{12}, H_{21}, H_{23}, and H_{32} are essentially the same as for ethylene, and may be set equal to β. The H_{13} and H_{31} elements, however, are clearly different from the others. In the Hückel molecular orbital treatment, all Hamiltonian matrix elements between atomic orbitals centered on atoms that are not bonded by the σ-framework system are set equal to zero. In any π-system, the Hückel treatment completely specifies the $\mathcal{H}$ matrix as:

$$\langle\eta_i|\mathcal{H}|\eta_j\rangle = \begin{cases} \alpha, & \text{if } i = j; \\ \beta, & \text{if atom } i \text{ is } \sigma\text{-bonded to atom } j; \\ 0, & \text{otherwise,} \end{cases}$$

and the $\mathcal{H}$ matrix will always consist of α's at all positions along the principal diagonal, and β's or zero's at all off-diagonal positions. The $\mathcal{H}$ matrix can be constructed immediately on looking at a diagram of the σ-framework of the conjugated system.

For the allyl system, we have:

$$\mathcal{H} = \begin{pmatrix} \alpha & \beta & 0 \\ \beta & \alpha & \beta \\ 0 & \beta & \alpha \end{pmatrix}.$$

It is never necessary to specify actual values for α or β in order to diagonalize the $\mathcal{H}$ matrices. The algebraic $\mathcal{H}$ matrix may be converted into a numerical matrix, $\mathcal{H}'$ by the following operations:

$$\mathcal{H}' \equiv \frac{1}{\beta}\mathcal{H} - \frac{\alpha}{\beta}\mathcal{I}$$

where $\mathcal{I}$ is the identity matrix. The multiplication by $1/\beta$ converts all of the diagonal elements to α/β, all elements that were β into 1's, and leaves the zero elements unchanged. The subtraction of $(\alpha/\beta)\mathcal{I}$ then converts all of the diagonal elements into zeroes without changing off-diagonal elements. The resulting $\mathcal{H}'$ matrix always has zeroes along the principal diagonal, ones at all positions corresponding to atoms bonded in the σ-framework, and zeroes at all other positions. For the allyl system, this gives:

$$\mathcal{H}' = \begin{pmatrix} 0 & 1 & 0 \\ 1 & 0 & 1 \\ 0 & 1 & 0 \end{pmatrix}.$$

Again, it is trivial to construct the $\mathcal{H}'$ matrix for any conjugated system. Draw the σ-framework and number the carbons of the conjugated system; place ones at all positions in the matrix corresponding to bonded carbon atoms, and zeroes everywhere else. The $\mathcal{H}'$ matrix is sometimes called a *connectivity* matrix. Notice that the connectivity matrix uniquely specifies the σ-framework.

The usefulness of the $\mathcal{H}'$ matrix lies in the fact that it may be diagonalized on a computer, that the unitary matrix which diagonalizes it also diagonalizes the original $\mathcal{H}$ matrix, and that the diagonalized matrices are simply related. The fact that the same unitary matrix diagonalizes both $\mathcal{H}'$ and $\mathcal{H}$ can be proved quite easily as follows. If

$$\mathcal{U}^{\dagger}\mathcal{H}'\mathcal{U} = \Lambda,$$

where Λ is a diagonal matrix, then substituting the definition of $\mathcal{H}'$:

$$\begin{aligned}\Lambda &= \mathcal{U}^\dagger [\frac{1}{\beta}\mathcal{H} - \frac{\alpha}{\beta}\mathcal{I}]\mathcal{U} \\ &= \mathcal{U}^\dagger [\frac{1}{\beta}\mathcal{H}]\mathcal{U} - \mathcal{U}^\dagger [\frac{\alpha}{\beta}\mathcal{I}]\mathcal{U} \\ &= \frac{1}{\beta}\mathcal{U}^\dagger \mathcal{H}\mathcal{U} - \frac{\alpha}{\beta}\mathcal{U}^\dagger\mathcal{U} \\ &= \frac{1}{\beta}\mathcal{U}^\dagger \mathcal{H}\mathcal{U} - \frac{\alpha}{\beta}\mathcal{I},\end{aligned}$$

where the second line follows from the distributive law of matrix multiplication, the third line from the fact that constants ($1/\beta$ and α/β) commute with any matrix, and the fourth line from the definition of unitary matrices ($\mathcal{U}^\dagger\mathcal{U} = \mathcal{I}$). Since both Λ and $\mathcal{I}$ are diagonal matrices, so must be $\mathcal{U}^\dagger\mathcal{H}\mathcal{U} = \mathcal{E}$. Multiplication by β, and addition of $\alpha\mathcal{I}$ to both sides of the final equation gives:

$$\mathcal{E} = \beta\Lambda + \alpha\mathcal{I} \quad \text{so that:} \quad \epsilon_{ii} = \alpha + \beta\lambda_{ii}$$

as the relationship between the diagonal matrices.

Continuing with the the allyl system, you can easily verify that:

$$\begin{pmatrix} 1/2 & 1/\sqrt{2} & 1/2 \\ 1/\sqrt{2} & 0 & -1/\sqrt{2} \\ 1/2 & -1/\sqrt{2} & 1/2 \end{pmatrix} \begin{pmatrix} 0 & 1 & 0 \\ 1 & 0 & 1 \\ 0 & 1 & 0 \end{pmatrix}$$
$$\times \begin{pmatrix} 1/2 & 1/\sqrt{2} & 1/2 \\ 1/\sqrt{2} & 0 & -1/\sqrt{2} \\ 1/2 & -1/\sqrt{2} & 1/2 \end{pmatrix} = \begin{pmatrix} \sqrt{2} & 0 & 0 \\ 0 & 0 & 0 \\ 0 & 0 & -\sqrt{2} \end{pmatrix},$$

so that the molecular orbitals and orbital energies are:

$$\begin{aligned}\phi_1 &= \frac{1}{2}\eta_1 + \frac{1}{\sqrt{2}}\eta_2 + \frac{1}{2}\eta_3, & \epsilon_1 &= \alpha + \sqrt{2}\beta \\ \phi_2 &= \frac{1}{\sqrt{2}}\eta_1 - \frac{1}{\sqrt{2}}\eta_3, & \epsilon_2 &= \alpha \\ \phi_3 &= \frac{1}{2}\eta_1 - \frac{1}{\sqrt{2}}\eta_2 + \frac{1}{2}\eta_3, & \epsilon_3 &= \alpha - \sqrt{\beta}.\end{aligned}$$

The lowest energy orbital, ϕ_1, has positive coefficients for each of the atomic orbitals, and, therefore, no nodes other than the always present xy-planar node of the p_z functions. The next orbital up in energy, ϕ_2

has a zero coefficient for η_2 and a difference in sign for the other two coefficients. It has, therefore, a nodal plane that is equidistant from atoms 1 and 3, passing through atom 2. The third orbital, ϕ_3, has two nodal surfaces, passing between atoms 1 and 2, and between atoms 2 and 3. A sketch of the molecular orbitals might help visualize them:

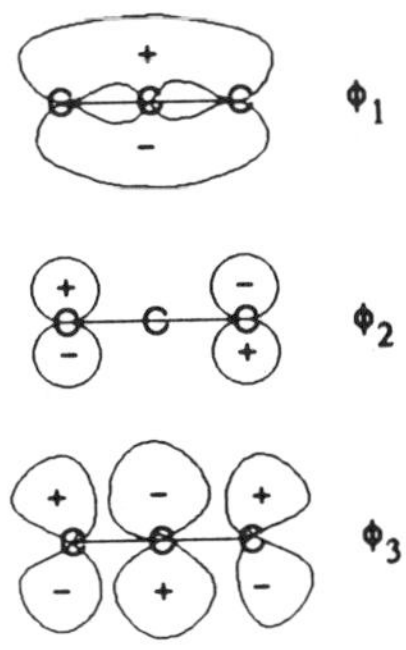

Note carefully in the allyl problem that the equivalence of carbons 1 and 3 is entered into the $\mathcal{H}$ matrix, and must appear in the electron densities of any of the allyl systems that we construct. If we place one electron in the π system, we would have the allyl di-cationic radical formed by placing the electron in ϕ_1. Adding another electron, we have the allyl cation with the electrons both in ϕ_1 with opposite spins. If we add a third electron, we have the allyl radical with the third electron going into ϕ_2. Addition of a fourth electron gives the allyl anion with the fourth electron going into ϕ_2 with the opposite spin of the third electron. The equivalence of electron densities on carbons 1 and 3 for all of these species requires that the squares of the coefficients of atomic orbitals 1 and 3 be equal, $c_{1i} = \pm c_{3i}$, for $i = 1, 2, 3$.

For all linear conjugated π-systems, the molecular orbitals will be similar to those we have just examined for the allyl system. The lowest energy orbital will have no nodes other than the nodal plane of the p_z orbitals, the next lowest will have one node, the next will have two nodes, the next will have three nodes, etc., in addition to the nodal plane of the p_z orbitals, up to the highest energy orbital which will have a node between each atom. Since the linear systems are symmetrical about the center, the nodal surfaces must be placed symmetrically about the center. One can, therefore, sketch the molecular orbitals of any linear system to show the sign of the orbital at each atom without even setting up the matrices. For example, the hexatrienyl system will have six molecular orbitals formed from the atomic orbitals with signs indicated by the sketch:

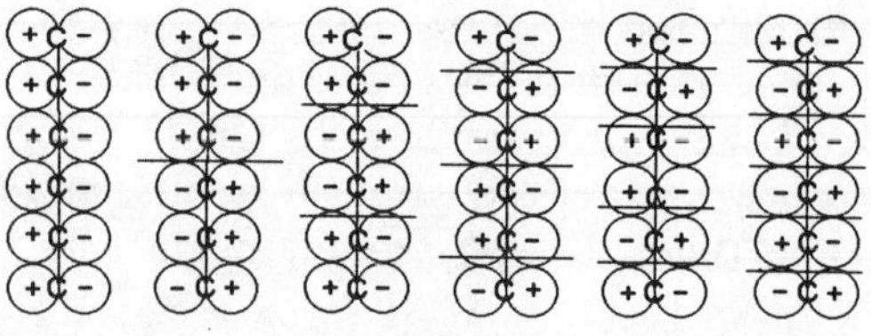

We also know from the symmetry of the system that $c_{1i} = \pm c_{6i}$, $c_{2i} = \pm c_{5i}$, and $c_{3i} = \pm c_{4i}$ for $i = 1, 2, 3, 4, 5, 6$.

When we start with the σ-framework of a conjugated π system, each of the carbon atoms has its two core, 1s, electrons and three electrons in the σ-bonds...each carbon, therefore, has a positive charge remaining. In the hexatrienyl system, for example, the σ-framework has six positive charges; adding six electrons to the system gives the neutral molecule, hexatriene. Since we can place two electrons, with opposite spins, in each of the molecular orbitals, the lowest three molecular orbitals are filled in hexatriene. Addition of another electron into ϕ_4 would form the anion-radical of hexatriene. Electrons are always added according to Hund's rules...electrons are placed into the lowest energy orbitals available; if there are two or more orbitals with the same energies, each of these is occupied singly, the electrons taking the same spins, before occupying them doubly with spin-paired electrons.

For the linear systems that we have discussed so far, the situation of two or more molecular orbitals with the same energies does not occur and occupation of the orbitals is always by paired electrons unless, of course, there are an odd number of electrons in the system, in which case, the unpaired electron is in the highest energy occupied orbital. We encounter the problem of orbital *degeneracy*, that is, two or more orbitals with the same energies, as soon as we leave the linear systems.

The cyclopropenyl σ-framework, $C_3H_3^{+3}$, has three identical carbons, each bonded to the other; the $\mathcal{H}'$ matrix has 1's in all off-diagonal positions, and you can verify that:

$$\begin{pmatrix} 1/\sqrt{3} & 1/\sqrt{3} & 1/\sqrt{3} \\ 1/\sqrt{6} & 1/\sqrt{6} & -2/\sqrt{6} \\ 1/\sqrt{2} & -1/\sqrt{2} & 0 \end{pmatrix} \begin{pmatrix} 0 & 1 & 1 \\ 1 & 0 & 1 \\ 1 & 1 & 0 \end{pmatrix}$$

$$\times \begin{pmatrix} 1/\sqrt{3} & 1/\sqrt{6} & 1/\sqrt{2} \\ 1/\sqrt{3} & 1/\sqrt{6} & -1/\sqrt{2} \\ 1/\sqrt{3} & -2/\sqrt{6} & 0 \end{pmatrix} = \begin{pmatrix} 2 & 0 & 0 \\ 0 & -1 & 0 \\ 0 & 0 & -1 \end{pmatrix}$$

We have, therefore:

$$\phi_1 = \frac{1}{\sqrt{3}}[\eta_1 + \eta_2 + \eta_3], \quad \epsilon_1 = \alpha + 2\beta,$$

$$\phi_2 = \frac{1}{\sqrt{6}}\left[\eta_1 + \eta_2 - 2\eta_3\right], \quad \epsilon_2 = \alpha - \beta,$$

$$\phi_3 = \frac{1}{\sqrt{2}}\left[\eta_1 - \eta_2\right], \quad \epsilon_3 = \alpha - \beta,$$

and ϕ_2 and ϕ_3 are *degenerate*, having the same energy, and the same number of nodal surfaces. If we add two electrons to the system to form the cyclopropenyl cation, the electrons occupy ϕ_1 with paired spins. The electronic energy of the cyclopropenyl cation, $2\alpha+4\beta$, is lower than that of the allyl cation, $2\alpha + 2\sqrt{2}\beta$. If we add two more electrons, to form the cyclopropenyl anion, one of these goes into ϕ_2 and the other into ϕ_3 with the same spin. The cyclopropenyl anion is therefore a di-radical species. The electronic energy of the cyclopropenyl anion, $4\alpha + 2\beta$, is higher than that of the allyl anion, $4\alpha + 2\sqrt{2}\beta$.

Notice that neither ϕ_2 nor ϕ_3 satisfy the symmetry condition on electron density imposed by the indistinguishability of the three carbons. If we sum the coefficients of each of the η's in ϕ_2 and ϕ_3, however, we find:

$$(c_{12}^2 + c_{13}^2) = \frac{1}{6} + \frac{1}{2} = (c_{22}^2 + c_{23}^2) = (c_{32}^2 + c_{33}^2) = \frac{2}{3},$$

showing that the three carbons equally share the two electrons in ϕ_2 and ϕ_3 of the cyclopropenyl anion di-radical. The problem of the symmetry of the electron density of the neutral cyclopropenyl radical is left for a problem at the end of the chapter.

The Hückel treatment for the cyclobutadienyl system is presented by:

$$\begin{pmatrix} 1/2 & 1/2 & 1/2 & 1/2 \\ 1/2 & 1/2 & -1/2 & -1/2 \\ 1/2 & -1/2 & -1/2 & 1/2 \\ 1/2 & -1/2 & 1/2 & -1/2 \end{pmatrix} \begin{pmatrix} 0 & 1 & 0 & 1 \\ 1 & 0 & 1 & 0 \\ 0 & 1 & 0 & 1 \\ 1 & 0 & 1 & 0 \end{pmatrix}$$
$$\times \begin{pmatrix} 1/2 & 1/2 & 1/2 & 1/2 \\ 1/2 & 1/2 & -1/2 & -1/2 \\ 1/2 & -1/2 & -1/2 & 1/2 \\ 1/2 & -1/2 & 1/2 & -1/2 \end{pmatrix} = \begin{pmatrix} 2 & 0 & 0 & 0 \\ 0 & 0 & 0 & 0 \\ 0 & 0 & 0 & 0 \\ 0 & 0 & 0 & -2 \end{pmatrix},$$

and, again, ϕ_2 and ϕ_3 have the same number of nodal surfaces and are degenerate. The neutral cyclobutadiene, formed by adding four electrons to the σ-framework, is predicted to be a di-radical species, and the electronic energy of $4\alpha+4\beta$ is the same as that of two isolated ethylenes.

The behaviors that we have just seen for the cyclopropenyl and cyclobutadienyl systems are general for all cyclic systems. All cyclic systems have one lowest energy molecular orbital with no nodal surfaces other than the xy-plane; higher energy orbitals occur in degenerate pairs... the first pair with one nodal surface in addition to the xy-plane, the second pair with two nodal surfaces in addition to the xy-plane, and so on up to the highest energy orbital. For cyclic systems having odd numbers of carbons, all orbitals above the lowest occur in degenerate pairs; for cyclic systems having even numbers of carbons, the lowest and the highest orbitals are non-degenerate and all orbitals in between occur in degenerate pairs. This is the origin of the Hückel (4n + 2)-π electron rule. Two electrons will always occupy the lowest energy orbital with paired spins. It then takes four more electrons before we fill the next pair of orbitals to obtain a species with all spins paired, then four more to fill the next pair, etc.

It is not necessary to have a cyclic system in order to have degenerate orbitals. The trimethylene methane system,

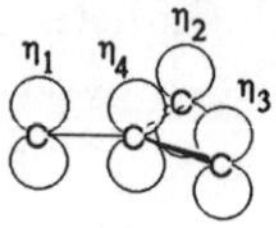

for example, is presented by:

$$\begin{pmatrix} 1/\sqrt{6} & 1/\sqrt{6} & 1/\sqrt{6} & 1/\sqrt{2} \\ 0 & 1/\sqrt{2} & -1/\sqrt{2} & 0 \\ -2/\sqrt{6} & 1/\sqrt{6} & 1/\sqrt{6} & 0 \\ -1/\sqrt{6} & -1/\sqrt{6} & -1/\sqrt{6} & 1/\sqrt{2} \end{pmatrix} \begin{pmatrix} 0 & 0 & 0 & 1 \\ 0 & 0 & 0 & 1 \\ 0 & 0 & 0 & 1 \\ 1 & 1 & 1 & 0 \end{pmatrix}$$
$$\times \begin{pmatrix} 1/\sqrt{6} & 0 & -2/\sqrt{6} & -1/\sqrt{6} \\ 1/\sqrt{6} & 1/\sqrt{2} & 1/\sqrt{6} & -1/\sqrt{6} \\ 1/\sqrt{6} & -1/\sqrt{2} & 1/\sqrt{6} & -1/\sqrt{6} \\ 1/\sqrt{2} & 0 & 0 & 1/\sqrt{2} \end{pmatrix} = \begin{pmatrix} \sqrt{3} & 0 & 0 & 0 \\ 0 & 0 & 0 & 0 \\ 0 & 0 & 0 & 0 \\ 0 & 0 & 0 & -\sqrt{3} \end{pmatrix},$$

which shows ϕ_2 and ϕ_3 to be degenerate.

5.4 Simple Orbital Symmetry Considerations

The Hückel π-MO method that we have been discussing is an extremely over-simplified treatment of many-electron systems. It's primary utility lies in the fact that it correctly handles the gross symmetry features and, usually, the energy ordering of the molecular orbitals. We will see shortly that it is the symmetry of the orbitals, or, more correctly, of the wave function, that is important in many problems of electrocyclic reactions [1].

In the above discussion of molecular orbitals, it was seen that any symmetry in a system immediately allowed us to say something about the coefficients in the linear combinations of the atomic orbitals. Specifically, for two atoms that are equivalent because of symmetry, the squares of the coefficients of the atomic orbitals centered on those atoms had to be equal; meaning that one coefficient had to be plus or minus the other. We can classify the molecular orbitals, in fact, by whether the plus or the minus sign is used. If the coefficients for equivalent centers are equal, we classify the orbitals as *symmetric*, and if one coefficient is minus the other, we classify the orbital as *anti-symmetric*. For the allyl system, for example, which has a plane of symmetry through carbon 2, ϕ_1 is symmetric, ϕ_2 is anti-symmetric, and ϕ_3 is symmetric. For the hexatrienyl system, with a plane of symmetry between carbons 3 and 4, ϕ_1 is symmetric, ϕ_2 is anti-symmetric, and so on.

Notice that the language used in the last paragraph changed a little... whereas we had talked previously about a system being symmetric about some atom or about some location, the last sentences specify that we are talking about a plane of symmetry for the systems. Even more precisely, we should recognize that we are talking about the system being symmetric *with respect to reflection in the plane passing through the center of the system.* The need for the precise description arises because many systems have more than one type of symmetry. The σ-framework of the allyl system has, for example, a two-fold axis of symmetry which lies in the xy-plane, passes through carbon 2, and is equidistant from carbons 1 and 3. If we rotate the system by 180° about this axis, carbons 1 and 3 change locations and the entire system is turned upside-down. Most importantly, this rotation converts ϕ_1 into $-\phi_1$, ϕ_2 into ϕ_2, and ϕ_3 into $-\phi_3$. We say that ϕ_1 and ϕ_3 are anti-symmetric, while ϕ_2 is symmetric with respect to the rotation. The xy-plane is also a plane of symmetry; all of the molecular orbitals are anti-symmetric with respect to reflection in this plane.

We talk about *symmetry elements* and *symmetry operations* for systems. The most common symmetry elements are planes and rotational axes, and the most common symmetry operations are reflections in the planes of symmetry and rotations about the rotational axes. Rotational axes are further distinguished by the number of rotational operations required to get back to where we started. For example, we had a two-fold axis for the allyl system; each operation involves rotation through 180°, and two operations is the same as no operation. The cyclopropenyl system has a three-fold axis of rotation which is perpendicular to the xy-plane and passes through the center of the system; rotation through 120° moves carbon 1 to the position originally occupied by either carbon

2 or 3, depending on how we number the system, etc. We can, in fact, state all of these things in terms of an *operator algebra.* The operators in this algebra operate on the coordinates of the system. For example if we have the yz-plane as a symmetry plane, then the reflection operator converts x to $-x$, y to y, and z to z in any function on which it operates. If we have the z-coordinate as a rotational axis, with a rotation operator rotating through θ degrees, then the operator converts x into $x\cos\theta + y\sin\theta$, y into $-x\sin\theta + y\cos\theta$, and leaves z unchanged.

If we denote a rotation about a symmetry axis as $\mathcal{R}$, then we can indicate a rotation of one of the molecular orbitals, for example, by $\mathcal{R}\phi_i$ and specify that $\mathcal{R}^n = \mathcal{I}$ means that the rotational axis is an n-fold one when $\mathcal{I}$ is the identity operator. It is not at all difficult then to write any of the operators, in general form, as 3×3 matrices which operate on the x, y, and z cartesian coordinates of the system. We'll leave this matrix formulation for a problem at the end of the chapter, but will use the operator notation in the rest of the discussion. A symmetry operator can operate on any function of the coordinates of the system, including the Hamiltonian operator, or other symmetry operators. Frequently, when a symmetry operator operates on a function, it converts it into another function. For example, the reflection operator for the allyl system can operate on η_1 to convert it to η_3. The product of two symmetry operators is always another symmetry operator, and all of the symmetry operators form a closed set in the sense that the product of any two operators in the set is another operator in the set [2] , [3].

If a system has elements of symmetry, all of the molecular orbitals will be either symmetric or anti-symmetric with respect to the associated symmetry operations. The classification is probably pretty obvious to you; if, for a symmetry operation $\mathcal{S}$, $\mathcal{S}\phi_i = \phi_i$, then ϕ_i is symmetric, and, if $\mathcal{S}\phi_i = -\phi_i$, then ϕ_i is anti-symmetric, with respect to the operation. It is important to note that the Hamiltonian operator for any symmetrical system is symmetric with respect to all symmetry operations.

Some of the interesting types of things that symmetry operators can operate on are integrals involving molecular orbitals. Suppose we have an integral $\langle\phi_i|\mathcal{H}|\phi_j\rangle$, and suppose that we have some symmetry operator, $\mathcal{S}$, such that $\mathcal{S}\phi_i = -\phi_i$ and $\mathcal{S}\phi_j = \phi_j$. Operating on the integral with $\mathcal{S}$ actually operates on all of the coordinates. Since the integral is over all coordinates from $-\infty$ to $+\infty$, it doesn't make any difference whether we integrate over the original coordinates or the ones into which the original ones are converted by the symmetry operator. Since the operator converts ϕ_i into $-\phi_i$, while leaving ϕ_j and $\mathcal{H}$ unchanged, we can write: $\mathcal{S}\langle\phi_i|\mathcal{H}|\phi_j\rangle = -\langle\phi_i|\mathcal{H}|\phi_j\rangle$. Remember, however, that the integral is a number, not a function of the coordinates, whose value is therefore unaffected by $\mathcal{S}$. We have, then, that $\mathcal{S}\langle\phi_i|\mathcal{H}|\phi_j\rangle = \langle\phi_i|\mathcal{H}|\phi_j\rangle$. The only way that both of these equations can be true is if the value of the integral is zero. The same is true whenever one of the ϕ's is anti-symmetric and the other is symmetric with respect to $\mathcal{S}$, and is true for any operator, not just the Hamiltonian, which is symmetric with respect to $\mathcal{S}$. We can further generalize to the conclusion that **in order for an integral over all space to be non-zero, the integrand must be symmetric with respect to all symmetry operations of the system.**

5.5 Wave Functions and Reactions

A chemical reaction can always be regarded as the conversion of a system in one state into the system in another state represented by different wave functions. The wave functions for the different states are eigenfunctions of Hamiltonians that differ because of different relative locations of the nuclei. Suppose we have a reactant whose lowest energy wave-function is Ψ_R, a product whose lowest energy wave function is Ψ_P, and a Hamiltonian $\mathcal{H}(x^\ddagger)$ which is a function of the 'reaction coordinate' $x^\ddagger$. For the reactant, the Hamiltonian is $\mathcal{H}(x_r^\ddagger)$, and, for the product, the Hamiltonian is $\mathcal{H}(x_p^\ddagger)$. The wave functions satisfy the equations:

$$\mathcal{H}(x_r^\ddagger)\Psi_R = E_R\Psi_R \quad \text{and} \quad \mathcal{H}(x_p^\ddagger)\Psi_P = E_P\Psi_P.$$

As the system moves along the coordinate $x^\ddagger$, the Hamiltonian changes so that these equations are satisfied only for the coordinate values for the reactant and the product, respectively. For any values of the coordinate other than $x_r^\ddagger$ or $x_p^\ddagger$, there will be some lowest energy wave function, $\Psi_0(x^\ddagger)$, for the system which we may approximate as a linear combination of Ψ_R and Ψ_P:

$$\Psi_0(x^\ddagger) \simeq a\Psi_R + b\Psi_P,$$

with an energy expectation value given by:

$$\left\langle\Psi_0(x^\ddagger)|\mathcal{H}(x^\ddagger)|\Psi_0(x^\ddagger)\right\rangle \simeq \left\langle E_0(x^\ddagger)\right\rangle$$
$$= a^2\left\langle\Psi_R|\mathcal{H}(x^\ddagger)|\Psi_R\right\rangle + b^2\left\langle\Psi_P|\mathcal{H}(x^\ddagger)|\Psi_P\right\rangle + 2ab\left\langle\Psi_R|\mathcal{H}(x^\ddagger)|\Psi_P\right\rangle.$$

The problem now is to choose the coefficients a and b to minimize the expectation value for the energy. This is precisely the same problem that we solved for the LCAO method... we diagonalize the $\mathcal{H}$ matrix:

$$\begin{pmatrix} \left\langle \Psi_R|\mathcal{H}(x^\ddagger)|\Psi_R \right\rangle & \left\langle \Psi_R|\mathcal{H}(x^\ddagger)|\Psi_P \right\rangle \\ \left\langle \Psi_P|\mathcal{H}(x^\ddagger)|\Psi_R \right\rangle & \left\langle \Psi_P|\mathcal{H}(x^\ddagger)|\Psi_P \right\rangle \end{pmatrix},$$

assuming that $\langle \Psi_R|\Psi_P \rangle = 0$ and $\langle \Psi_R|\Psi_R \rangle = \langle \Psi_P|\Psi_P \rangle = 1$. Let's shorten the notation a bit by writing H_{RR}, H_{PP}, and $H_{RP} = H_{PR}$ for the matrix elements, and $\epsilon = \langle E_0(x^\ddagger) \rangle$ for the energy expectation value. The lowest expectation value for the energy will be given by solution of the equation:

$$(H_{RR} - \epsilon)(H_{PP} - \epsilon) - H_{RP}^2 = 0,$$

$$\epsilon = \frac{H_{RR} + H_{PP} \pm \sqrt{(H_{RR} - H_{PP})^2 + 4H_{RP}^2}}{2},$$

and the coefficients a and b, can be obtained as $a = \cos\theta$, $b = \sin\theta$, where θ is determined by $\tan\theta = -(H_{RR} - \epsilon)/H_{RP}$.

Note carefully that the matrix element values change as $x^\ddagger$ changes. The situation will be something like that shown in Fig. 5.1; starting from the reactant configuration, the value of H_{RR} must increase as we move away from the equilibrium geometry of the system characterized by $x^\ddagger = x_r^\ddagger$, and, similarly, starting with the product, the value of H_{PP} must increase as we move away from $x^\ddagger = x_p^\ddagger$. At points between the reactant and product configurations, everything depends on the value of H_{RP} as seen in the above expression for ϵ. If $H_{RP} = 0$, the lowest energy of the system will simply be either H_{RR} or H_{PP}, and there will be no mixing of the wave functions; that is, either a or b will be zero. The system would follow the solid lines in Fig. 5.1. If $H_{RP} \neq 0$, then the lowest energy for the system will be lower than either H_{RR} or H_{PP}, following the lower dashed line of Fig. 5.1, and there will be a smooth mixing of the two wave functions which gradually transforms the wave function Ψ_0 from Ψ_R to Ψ_P.

If the wave functions Ψ_R and Ψ_P are of different symmetries with respect to any symmetry operator existing for a particular value of $x^\ddagger$, then, as we saw above, H_{RP} must be equal to zero. Our conclusions can be summarized in the *non-crossing rule*: **the potential energy surfaces for states of a system having the same symmetry cannot cross**. For these states, the system will follow a smooth curve from the minimum of one state to the minimum of the other. For states not having the same symmetries, the potential energy curves may cross, and a

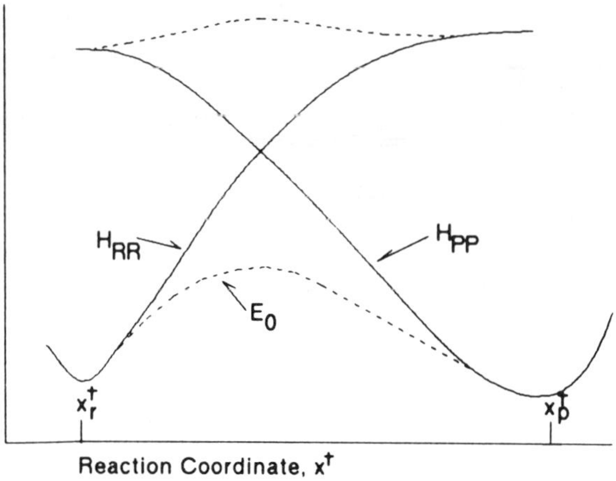

Figure 5.1: *Illustration of the Variation of Matrix Elements and Energy as Functions of the Reaction Coordinate for Symmetry Allowed (dashes) and Symmetry Forbidden (solid lines) Reactions.*

system cannot move smoothly from one state to the other; in fact, transitions between the two states cannot occur under usual circumstances [4], and reactions converting one state to the other are called *symmetry forbidden.*

One more idea is needed before we can begin to apply these ideas to some real many-electron systems. Thus far, in our discussion of molecular orbital theory, we have considered only the molecular orbitals, not the many-electron wave function for the molecule. For the present, we can regard the wave function for a system as a product of the occupied molecular orbitals. In particular, we can determine the symmetry of the total wave function as a product of the symmetries of the occupied orbitals. For example, for *s-cis*-butadiene, CH_2=CH-CH=CH_2, we have four electrons in the π-system which occupy the molecular orbitals, ϕ_1 and ϕ_2. With respect to the plane of symmetry perpendicular to the xy-plane and equidistant from the two end carbons, ϕ_1 is symmetric and ϕ_2 is antisymmetric. Since the product of a symmetric and an antisymmetric function is antisymmetric, the wave function for the π- system is antisymmetric.

Now, let's take a look at a conversion of butadiene into cyclobutene. The wavefunctions for the two molecules contain the same bonds, and, therefore, molecular orbitals, except for the π-bonds and the σ-bond between carbons 1 and 4 in cyclobutene. The π-bond in cyclobutene is like that in ethylene, and the molecular orbital is symmetric with respect to the plane passing between the two carbons. The σ-bond between carbons 1 and 4 involves positive overlap between sp^3 orbitals on carbons 1 and 4 and is symmetric with respect to the plane of symmetry. The pertinent molecular orbitals for the two molecules can be diagrammed as:

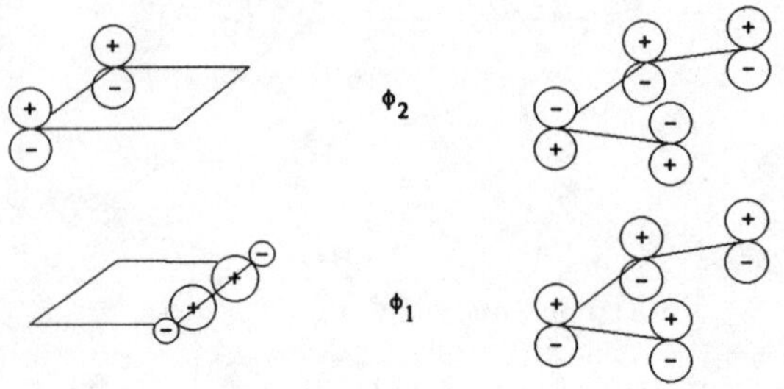

The part of the total wavefunction corresponding to the product of the two orbitals of cyclobutene, then, is symmetric. Since the two molecules have wave functions of different symmetries with respect to the plane of symmetry, reaction paths which preserve the plane of symmetry are symmetry forbidden. The two end carbons of butadiene must rotate along any reaction path so that the atomic orbitals on these two carbons can overlap to form the σ-bond of cyclobutene. There are two basically different reaction pathways by which the rotations could conceivably occur; either the two carbons could rotate in the same sense, or one could rotate clockwise while the other rotates counterclockwise. The different geometries along the two different paths would look something like:

(a) (b)

The geometry (a) results from what is called a *disrotary* motion in which the carbons rotate in opposite senses... this is a motion that preserves the plane of symmetry, and is symmetry forbidden because the wavefunctions of reactant and product are of different symmetries with respect to that plane. Geometry (b) does not preserve the plane of symmetry. It does, however, preserve the two-fold axis of rotation that lies in the xy-plane and bisects the system. Now, look back at the molecular orbitals of the two molecules... ϕ_1 of butadiene is antisymmetric while ϕ_1 of cyclobutene is symmetric, and ϕ_2 of butadiene is symmetric while ϕ_2 of cyclobutene is antisymmetric with respect to the rotation. The wave functions for the two molecules are both products of a symmetric and an antisymmetric function and therefore have the same symmetry with respect to the rotation. Reaction by the path giving geometries like (b) is symmetry allowed. The pathway employed, of course, has stereochemical consequences in the cases of 1,4-disubstituted butadienes giving cis- or trans-3,4-disubstituted cyclobutenes that are observed to be in accordance with these expectations.

The principles for ring opening or forming reactions should be clear. Applications to several more examples are given in problems at the end of the chapter.

Similar arguments can be applied to cycloaddition or retro-cycloaddition reactions. Suppose we have two π-systems, CH_2-$(CH)_n$-CH_2, and CH_2-$(CH)_m$-CH_2 which can react to form a cyclic system containing two new σ-bonds between the end carbons of the two molecules. Again, there are two basically different pathways that the cyclizations might follow. Each of the two open π-systems in the all cis- geometries have a plane of symmetry and a two-fold rotational axis of symmetry exactly like those we discussed for butadiene. The cyclic product can also have configurations that have these same elements of symmetry. There is one general type of pathway that will preserve the plane of symmetry, and another general type that preserves the two-fold rotational axis of symmetry; possible configurations along these two pathways are shown in Fig. 5.2.

You should be able to convince yourself quite quickly, by drawing a few orbital diagrams, that the symmetries of the molecular orbitals for *s-cis* acyclic π-systems alternate as the number of nodes increases, and therefore as the orbital energies increase. The lowest energy orbital will always be symmetric with respect to reflection in the plane of symmetry and antisymmetric with respect to rotation about the two-fold axis, the next orbital up in energy will always be antisymmetric with respect to the reflection and symmetric with respect to the rotation, etc. The symmetries of the π-wavefunctions, then depend only on the number of occupied orbitals. If we have two electrons in the π-system, occupying the first orbital, the wave function will have just the symmetry of the lowest molecular orbital; with four elecrons, the wave function will be antisymmetric with respect to both symmetry operations; with six electrons, the wave function will be antisymmetric with respect to reflection and symmetric with respect to rotation; with eight electrons, it will be symmetric with respect to both operations; and so on.

The cyclization product will contain two isolated π-systems and two isolated σ-bonds. We must form molecular orbitals from the σ-bonding orbitals that are either symmetric or antisymmetric with respect to all

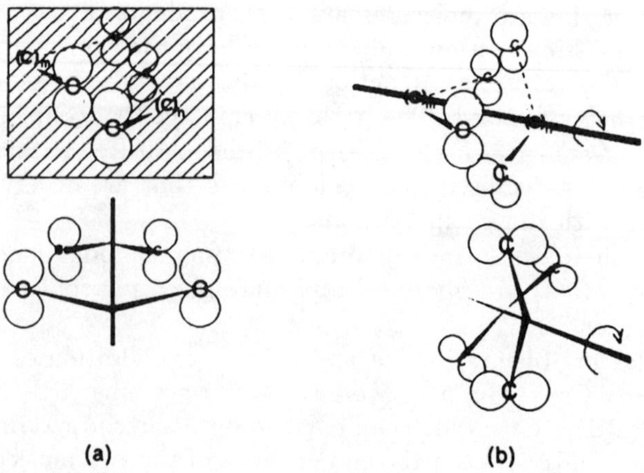

Figure 5.2: *Intermediate Configurations in Pathways Preserving a Plane of Symmetry (a) and a Two-fold Axis of Symmetry (b) for a Cyclization Reaction.*

elements of symmetry[1]. This can be done simply by taking one plus the other to form one molecular orbital, and one minus the other to form the second molecular orbital. Since the pertinent σ-system contains four electrons, the σ-wave function is a product of the two molecular orbitals, and is antisymmetric with respect to both symmetry operations. The π-orbitals of the two isolated systems follow the same pattern that we noted above, and the total wave function to be considered will have the symmetry that is a product of the σ- and the two π-wave functions.

A specific example should help make things clear. Let's consider the cycloaddition of butadiene and hexatriene:

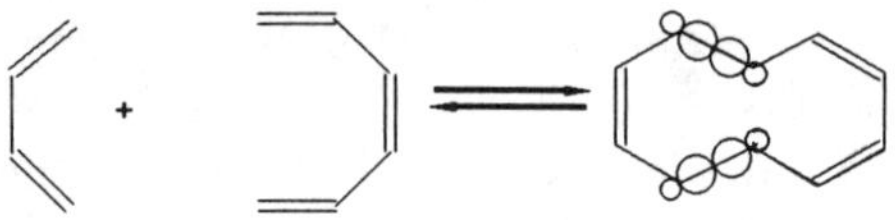

The π-wave function for butadiene is antisymmetric with respect to both the reflection and the rotation operations. The π-wave function for hexatriene is antisymmetric (product of SAS=symmetric, antisymmetric, and symmetric) with respect to reflection, and is symmetric (ASA) with

[1] Note that all of the isolated π-orbitals are already either symmetric or antisymmetric with respect to our symmetry elements.

respect to the rotation. The product of the two π-wave functions is AA=symmetric with respect to reflection, and AS=antisymmetric with respect to the rotation.

The pertinent wave function for the cyclodecatriene, is composed of the σ-part, discussed in the last paragraph, which is antisymmetric with respect to both operations, the π-ethylene part which is symmetric with respect to reflection and antisymmetric with respect to rotation, and the π-butadiene part which is antisymmetric with respect to both operations. The pertinent wave function, then, is ASA=symmetric with respect to the reflection, and AAA=antisymmetric with respect to the rotation.

In this case, then, the wavefunctions for reactants and product are both symmetric with respect to the reflection, and are both antisymmetric with respect to the rotation; **both pathways are symmetry allowed**. A few more examples are given in the problems at the end of the chapter.

Because of the regular patterns of symmetric and antisymmetric molecular orbitals which occur for the π-systems, it is a fairly simple matter to set up rules concerning symmetry allowed and forbidden processes based on the numbers of π-electrons in the reacting systems. This, and the recurring geometries, such as those in Fig. 5.1 and 5.2, which preserve individual symmetry elements, make it tempting to invent numerical rules and descriptive languages if one were so inclined [1].

The ideas developed here are applicable to reactions other than those of π-systems. An excellent presentation of the subject with applications to many types of reactions is given by Pearson [4].

5.6 Basic Quantum Mechanics

It should be clear from the examples presented above, if it wasn't before, that a familiarity with quantum mechanics is a virtual necessity for understanding much of chemistry. In order to keep up with new developments in the applications of quantum mechanics to problems of chemical reactivity, and in order to guard against foolish errors in using the developments, it is advisable to develop some knowledge of the basics.

The basic formulations of quantum mechanics are in terms of observables, operators, and state functions. We've already used some of each of these in our discussion of simple orbitals; for example, energy is an observable, the Hamiltonian is the operator associated with it, and the wave functions that we discussed are state functions for the electrons. It was also pointed out that the electronic wave functions could be interpreted in terms of probabilities in the sense that the integral of $\psi^\dagger\psi$ over any volume element of space represents the probability of finding the electrons in that volume element. Only a slight change in language is needed to make this last statement general: a state function, Ψ, may be interpreted in terms of probability in the sense that the integral of $\Psi^\dagger\Psi$ over any volume element of the system is proportional to the probability that the system is in that volume element.

This interpretation of state functions places some very stringent limitations on the types of functions that can serve as state functions. Since the integral of $\Psi^\dagger\Psi$ over all space is the probability that the system exists, it must be finite. In order to make any sense, the probability that a system is in some volume element must be single valued. If we wish to take the usual scale of probabilities from zero to one, with one representing certainty, then we will deal with *normalized* state functions such that $\langle\Psi|\Psi\rangle = 1$, as we did with atomic and molecular orbitals.

Just to give concreteness to the concepts, let's consider a specific function that could be a state function for some one-dimensional system:

$$\xi = \left(\frac{2\alpha}{\pi}\right)^{1/4} \exp(-\alpha x^2), \tag{5.3}$$

where α is some constant, and x is the position coordinate of a particle on a line. The definite integrals of the square of this function, by itself, multiplied by x, and multiplied by x^2, can be obtained from nearly any Handbook of integrals as:

$$\langle\xi|\xi\rangle = 1, \quad \langle\xi|x|\xi\rangle = 0, \quad \text{and} \quad \langle\xi|x^2|\xi\rangle = \frac{1}{4\alpha}. \tag{5.4}$$

This, in fact, is an extremely useful type of function as we'll see a little later. It is finite, single-valued, continuous, and can be differentiated any number of times.

Operators and observables are associated through equations of the form:

$$\mathcal{G}\Psi = \gamma\Psi, \tag{5.5}$$

where $\mathcal{G}$ is an operator, and Ψ is a state function for the system such that if the system is in that state, the precise value γ will be found for the observable corresponding to the operator. The state function Ψ is called an *eigenfunction* or *eigenstate* of the operator $\mathcal{G}$ and γ is the corresponding *eigenvalue*. When a system is in some state Θ that is not an eigenstate of the operator $\mathcal{G}$, then the observable has an expectation value of $\langle\gamma\rangle$, given by:

$$\langle\gamma\rangle = \frac{\langle\Theta|\mathcal{G}|\Theta\rangle}{\langle\Theta|\Theta\rangle}. \tag{5.6}$$

Such an expectation value is the average value that would be obtained by a large number of measurements of the observable.

All of the operators of quantum mechanics are linear operators; that is, they obey the same algebra that matrices do, and, in fact, matrix representations of the operators are frequently used. The common operators are those corresponding to the observables position, momentum, and total energy. The forms of these operators are shown in Table 5.1.

TABLE 5.1
Common Observables, Corresponding Operators, and Expectation Values for a State Θ
Note: $\hbar = h/2\pi$

Observable	Symbol	Operator	Expectation value
Position	q_i	q_i	$\langle\Theta\|q_i\|\Theta\rangle$
Momentum	$p_i = m\frac{\partial q_i}{\partial t}$	$-i\hbar\frac{\partial}{\partial q_i}$	$-i\hbar\langle\Theta\|\partial\Theta/\partial q_i\rangle$
Kinetic Energy	$T = \frac{p_i^2}{2m}$	$-\frac{\hbar^2}{2m}\frac{\partial^2}{\partial q_i^2}$	$-\frac{\hbar^2}{2m}\langle\Theta\|\partial^2\Theta/\partial q_i^2\rangle$

Classical expressions for observables in terms of positions and momenta can be used to construct other operators from those for position and momenta, as shown for kinetic energy in the table. For the total energy of a system having a potential energy, $V(q_1, q_2, \ldots, q_n)$, that can be expressed as a function of coordinates only, the Hamiltonian operator (the operator corresponding to total energy) of the system can be written as a sum of the kinetic energy operators for each coordinate of each particle of the system plus the potential energy function.

There is a further relationship, the time dependent Schrödinger equation, between the Hamiltonian of a system, and an operator involving derivatization with respect to time:

$$\mathcal{H}\Psi = i\hbar\frac{\partial\Psi}{\partial t}, \tag{5.7}$$

where Ψ is a function of coordinates and time. If $\mathcal{H}$ is not an explicit function of time, then this time dependent Schrödinger equation can always be satisfied by state functions of the form:

$$\Psi(q_1, q_2, \ldots, q_n, t) = \psi(q_1, q_2, \ldots, q_n)\exp[-E(i/\hbar)t],$$

which, when substituted into the time dependent Schrödinger equation gives:

$$\begin{aligned}\mathcal{H}\psi(q_1, q_2, \ldots, q_n)\exp[-E(i/\hbar)t] &= \exp[-E(i/\hbar)t]\mathcal{H}\psi(q_1, q_2, \ldots, q_n)\\ &= i\hbar\frac{\partial\Psi}{\partial t}\\ &= E\exp[-E(i/\hbar)t]\psi(q_1, q_2, \ldots, q_n),\end{aligned}$$

and, therefore:

$$\mathcal{H}\psi(q_1, q_2, \ldots, q_n) = E\psi(q_1, q_2, \ldots, q_n), \tag{5.8}$$

which is the time independent Schrödinger equation relating the Hamiltonian and the energy of a system.

In order to gain some familiarity with operators and expectation values, let's further consider the one-dimensional function ξ which we defined in Eq. (5.3). Suppose we have a particle, of mass m and confined to a line along the $x-axis$ of a coordinate system, whose state is given by ξ. The expectation values for the position, x, and for the momentum, p, of the particle are given by:

$$\langle x\rangle = \langle\xi|x|\xi\rangle = 0,$$

$$\langle p\rangle = \langle\xi| - i\hbar\frac{\partial}{\partial x}\xi\rangle = \langle\xi| - i\hbar(-2\alpha x)\xi\rangle = 0,$$

respectively, where we have used the results in Eqs. (5.4). The expectation value for the kinetic energy, T, is:

$$\begin{aligned}\langle T\rangle &= -\frac{\hbar^2}{2m}\langle\xi|\frac{\partial^2\xi}{\partial x^2}\rangle\\ &= -\frac{\hbar^2}{2m}\langle\xi|(-2\alpha + 4\alpha^2x^2)\xi\rangle\\ &= -\frac{\hbar^2}{2m}\left[-2\alpha\langle\xi|\xi\rangle + 4\alpha^2\langle\xi|x^2|\xi\rangle\right]\\ &= -\frac{\hbar^2}{2m}\left[-2\alpha + 4\alpha^2\frac{1}{4\alpha}\right]\\ &= \frac{\alpha\hbar^2}{2m},\end{aligned}$$

which shows, incidentally, that the expectation value of zero for the momentum does not mean that the particle is motionless; it means that the momentum is as often positive as it is negative. Another revealing expectation value is that for the root-mean-square displacement of the particle, $\langle x^2 \rangle^{1/2}$, which is:

$$\langle \xi | x^2 | \xi \rangle^{1/2} = \left(\frac{1}{4\alpha} \right)^{1/2},$$

which shows that the zero value for $\langle x \rangle$ does not mean that the particle is 'located' at $x = 0$, but that it is as often at negative x as at positive x.

Let's take the example where the particle under consideration is operated on by a Hooke's law force, given by force= kx. Then the potential energy of the particle is given by:

$$\int kx\, dx = \frac{1}{2} kx^2,$$

and the expectation value for the potential energy, $\langle V \rangle$, is:

$$\langle V \rangle = \frac{k}{2} \langle \xi | x^2 | \xi \rangle = \frac{k}{8\alpha}.$$

The expectation value for the total energy, $\langle E \rangle$, is the sum of the kinetic and potential energies:

$$\langle E \rangle = \langle T \rangle + \langle V \rangle = \frac{\alpha \hbar^2}{2m} + \frac{k}{8\alpha}.$$

As a final exercise with the function ξ, we may note that ξ could be an eigenfunction of the total energy operator for this one-dimensional particle operated on by a Hooke's law force if α is properly chosen. Step-by-step:

$$\begin{aligned} \mathcal{H}\xi &= -\frac{\hbar^2}{2m} \frac{\partial^2 \xi}{\partial x^2} + \frac{k}{2} x^2 \xi \\ &= \frac{-\hbar^2}{2m} \left[-2\alpha + 4\alpha^2 x^2 \right] \xi + \frac{kx^2}{2} \xi \\ &= \frac{\alpha \hbar^2}{m} \xi + \left[\frac{k}{2} - \frac{2\alpha^2 \hbar^2}{m} \right] x^2 \xi, \end{aligned}$$

and we see that if the term in $x^2 \xi$ were to vanish, then ξ would be an eigenfunction of $\mathcal{H}$ with an eigenvalue of $\alpha \hbar^2 / m$. This can be accomplished by making:

$$\frac{k}{2} - \frac{2\alpha^2 \hbar^2}{m} = 0 \quad \text{by setting} \quad \alpha = \frac{\sqrt{km}}{2\hbar},$$

and E is then equal to $(\hbar/2)\sqrt{k/m}$. The function ξ, in fact, is the state function for the lowest energy state of a harmonic oscillator when α is chosen in this way. We'll have further use for these results when we discuss bond vibrations and isotope effects in a later chapter.

The operator equations of quantum mechanics have certain important general characteristics that we will state without proof. A second-order differential equation, such as the time-independent Schrödinger equation, has an infinite number of solutions for the eigenfunctions and eigenvalues. One of the most important properties of the set of eigenfunctions is that they *form a complete set of functions in the system space.* This means that any arbitrary function of the coordinates of the system, which meets the usual requirements of being finite, continuous, and single-valued, can be expressed as a linear combination of the eigenfunctions. Thus, the time-independent Schrödinger equation, $\mathcal{H}\psi_i = E_i\psi_i$, for example, has an infinite number of solutions ψ_i, $i = 1, 2, \ldots, \infty$, each with an associated eigenvalue, E_i. Any arbitrary function of the coordinates of the system, say $\Upsilon(q_1, q_2, \ldots, q_n)$, can be expressed by the equation:

$$\Upsilon = \sum_{i=1}^{\infty} \psi_i c_i, \tag{5.9}$$

where the c's are numerical coefficients.

Another property of the eigenfunctions is that any two different eigenfunctions, ψ_i and ψ_j, are orthogonal if they have different eigenvalues:

$$\langle\psi_i|\psi_j\rangle = \langle\psi_j|\psi_i\rangle = 0 \quad \text{if} \quad E_i \neq E_j, \tag{5.10}$$

and can simply be made to be orthogonal if they have the same eigenvalue. If, for example, we have two eigenfunctions ψ_k and ψ_l with $E_k = E_l$, then, from the fact that any constant commutes with any operator, it follows that:

$$\mathcal{H}(a\psi_k + b\psi_l) = a\mathcal{H}\psi_k + b\mathcal{H}\psi_l = aE_k\psi_k + bE_l\psi_l = E_k(a\psi_k + b\psi_l),$$

showing that if ψ_k and ψ_l are eigenfunctions with the same eigenvalue, then any linear combination of the two functions is also an eigenfunction with the same eigenvalue. Suppose that $\langle\psi_k|\psi_l\rangle = \beta \neq 0$. Then we form two new functions, ψ_k' and ψ_l' as:

$$\psi_k' = \psi_k + \psi_l, \quad \text{and} \quad \psi_l' = \psi_k - \psi_l.$$

You may easily prove to yourself that $\langle\psi_k'|\psi_l'\rangle = 0$. We can also pay some attention to the business of normalization in this same vein. If ψ_j

is an eigenfunction of $\mathcal{H}$ with an eigenvalue of E_j, then any constant times ψ_j is also an eigenfunction of $\mathcal{H}$ with the same eigenvalue:

$$\mathcal{H}(a\psi_j) = a\mathcal{H}\psi_j = aE_j\psi_j = E_j(a\psi_j).$$

We can, whenever we wish, then, assume that the set of eigenfunctions of an operator is an orthonormal set.

In our earlier discussion of section 5.2, we minimized the expectation value of the energy for a linear combination of atomic orbitals. We should have worried there about the statement that the best molecular orbital would be the one with the lowest energy. Is it possible that we could obtain some ridiculously low energy expectation value and a correspondingly ridiculous orbital? We are now in a position to be able to show that the lowest possible expectation value for an energy is greater than or equal to the true lowest energy of the system. We've already seen that any arbitrary function of the coordinates of a system can be expressed as a linear combination of the eigenfunctions of the Hamiltonian operator. The expectation value of the energy for a system in the state described by a normalized function, Υ, is given by $\langle\Upsilon|\mathcal{H}|\Upsilon\rangle$, which may be expanded as follows:

$$\begin{aligned}
\langle\Upsilon|\mathcal{H}|\Upsilon\rangle &= \sum_{i=1}^{\infty}\sum_{j=1}^{\infty}\langle\psi_i|\mathcal{H}|\psi_j\rangle c_i^* c_j \\
&= \sum_{i=1}^{\infty}\sum_{j=1}^{\infty}\langle\psi_i|E_i|\psi_j\rangle c_i^* c_j \\
&= \sum_{i=1}^{\infty}\sum_{j=1}^{\infty} E_i\langle\psi_i|\psi_j\rangle c_i^* c_j \\
&= \sum_{i=1}^{\infty} E_i(c_i^* c_i),
\end{aligned}$$

where the last line uses the assumption that the eigenfunctions are an orthonormal set. Since Υ is normalized, we also have:

$$\begin{aligned}
\langle\Upsilon|\Upsilon\rangle &= \sum_{i=1}^{\infty}\sum_{j=1}^{\infty}\langle\psi_i|\psi_j\rangle c_i^* c_j \\
&= \sum_{i=1}^{\infty}(c_i^* c_i) = 1.
\end{aligned}$$

Now, we can subtract the lowest eigenvalue, E_0, from $\langle E \rangle$ that we obtained above:

$$\begin{aligned}\langle E \rangle - E_0 &= \sum_{i=1}^{\infty} E_i(c_i^* c_i) - E_0 = \sum_{i=1}^{\infty} E_i(c_i^* c_i) - \sum_{i=1}^{\infty} E_0(c_i^* c_i) \\ &= \sum_{i=1}^{\infty}(E_i - E_0)(c_i^* c_i) \geq 0,\end{aligned}$$

where the last line follows from the fact that both factors in each term of the sum are necessarily positive or zero.

Also, in the earlier discussion we stated that the Hamiltonian is a Hermitian operator; we can now show that to be true. Consider any two functions of the coordinates of the system:

$$\mathcal{A} = \sum_{i=1}^{\infty} \psi_i a_i, \quad \text{and} \quad \mathcal{B} = \sum_{i=1}^{\infty} \psi_i b_i,$$

and expand the two integrals $\langle \mathcal{A}|\mathcal{H}|\mathcal{B} \rangle$ and $\langle \mathcal{B}|\mathcal{H}|\mathcal{A} \rangle^*$:

$$\begin{aligned}\langle \mathcal{A}|\mathcal{H}|\mathcal{B} \rangle &= \sum_{i=1}^{\infty}\sum_{j=1}^{\infty}\langle \psi_i|\mathcal{H}|\psi_j \rangle a_i^* b_j \\ &= \sum_{i=1}^{\infty}\sum_{j=1}^{\infty} E_j \langle \psi_i|\psi_j \rangle a_i^* b_j \\ &= \sum_{i=1}^{\infty} E_i a_i^* b_i,\end{aligned}$$

and

$$\begin{aligned}\langle \mathcal{B}|\mathcal{H}|\mathcal{A} \rangle^* &= \sum_{i=1}^{\infty}\sum_{j=1}^{\infty}(\langle \psi_i|\mathcal{H}|\psi_j \rangle b_i^* a_j)^* \\ &= \sum_{i=1}^{\infty}\sum_{j=1}^{\infty} E_j (\langle \psi_i|\psi_j \rangle b_i^* a_j)^* \\ &= \sum_{i=1}^{\infty} E_i (b_i^* a_i)^* \\ &= \sum_{i=1}^{\infty} E_i (a_i^* b_i).\end{aligned}$$

The last line of each of these two expansions are equal, so that we have proved:

$$\langle\mathcal{A}|\mathcal{H}|\mathcal{B}\rangle = \langle\mathcal{B}|\mathcal{H}|\mathcal{A}\rangle^*,$$

for any two arbitrary functions $\mathcal{A}$ and $\mathcal{B}$. If both functions are real, the $*$ superscripts can simply be omitted.

5.7 Molecular Wave Functions

It is now simple enough to write the Hamiltonian operator for any system for which we know the potential function; it will be the sum of the kinetic energy operators for each of the three coordinates of each of the particles of the system, plus the expression for the potential energy as a function of coordinates of the particles. Throughout our discussions of molecules, we shall use the Born-Oppenheimer approximation that nuclear and electronic wave functions can be separated so that the electrons can be considered to exist in a system where the nuclei are stationary.

Suppose we have a system consisting of n electrons and N atomic nuclei. We can write the Hamiltonian as:

$$\mathcal{H} = -\frac{\hbar^2}{2m_e}\sum_{i=1}^{n}\left(\frac{\partial^2}{\partial x_i^2}+\frac{\partial^2}{\partial y_i^2}+\frac{\partial^2}{\partial z_i^2}\right) - \sum_{i=1}^{n}\sum_{j=1}^{N}\frac{Z_j e^2}{R_{ij}} + \sum_{i=1}^{n-1}\sum_{j=i+1}^{n}\frac{e^2}{r_{ij}},$$

where electron i has cartesian coordinates x_i, y_i, and z_i, nucleus j has a charge of $Z_j e$, the distance between nucleus j and electron i is R_{ij}, the distance between electron i and electron j is r_{ij}, and the electrons have a mass of m_e and a charge of e. The first summation over the electrons is, of course, the kinetic energy operator. The second summation is over electrons and nuclei, and is the coulombic potential energy of attraction between the nuclei and electrons. The third (double) summation is over pairs of electrons, and is the coulombic potential energy of repulsion between electrons. The appearance of the Hamiltonian can be simplified if we use a system of units in which the charge on an electron is taken as unity, the mass of an electron is taken as $\hbar^2$, and a unit of length is the Bohr radius, $5.292 \times 10^{-9} cm$. The unit of energy in this system of units is called an atomic unit, a.u., and $1\ a.u. = 627\ kcal/mole$. The Hamiltonian, then, is generally written as:

$$\mathcal{H} = -\frac{1}{2}\sum_{i=1}^{n}\left(\frac{\partial^2}{\partial x_i^2}+\frac{\partial^2}{\partial y_i^2}+\frac{\partial^2}{\partial z_i^2}\right) - \sum_{i=1}^{n}\sum_{j=1}^{N}\frac{Z_j}{R_{ij}} + \sum_{i=1}^{n-1}\sum_{j=i+1}^{n}\frac{1}{r_{ij}}, \quad (5.11)$$

It should be fairly obvious to you by this point that an exact solution of the Schrödinger equation with a Hamiltonian of the complexity for even the simplest organic molecules is not even thinkable at the present time. We will deal with approximate solutions in which we will use some form of wave function with adjustable parameters that we will proceed to optimize to obtain the lowest expectation value for the energy that is possible with that form of wave function. We have shown, in the last section, that the lowest expectation value obtainable for the energy is greater than or equal to the true energy, so that we can use minimization procedures to optimize the adjustable parameters.

The first problem that we need to solve, then, is that of the form of the wave function for a many-particle system. An n-particle system has $3n$ cartesian coordinates, three for each particle, so we need a wave function that is a function of all of these coordinates. In our earlier discussion of the symmetry of wave functions in the Hückel approximation, the idea of a wave function as a product of molecular orbitals was used. We might hope that a reasonable wave function for a many-electron system could be written as a product of one-electron functions, each a function of the coordinates of a single electron, such as:

$$\Psi_{guess} = \Phi_1(1)\Phi_2(2)\cdots\Phi_n(n),$$

where each of the Φ's contains only the coordinates of the electron, indicated in parentheses, which occupies it: that is, $\Phi_1(1)$ would be a function of x_1, y_1, and z_1. There are, however, several things wrong with this form for a wave function; the most important being that it makes the electrons distinguishable from one another. Electrons are indistinguishable one from another, and this is a type of symmetry which must be reflected in the wave function just as elements of symmetry of the entire molecule are required to be reflected in the molecular orbitals discussed earlier.

We can formulate the indistinguishability of electrons as requiring that $\Psi^\dagger\Psi$ be unchanged on changing the labels of the electrons. This condition could be satisfied in two quite distinct ways; either the interchange of any two electrons could leave Ψ unchanged, or the interchange of any two electrons could change the sign of Ψ. We can symbolize these possibilities by use of an operator, $\wp_{ij}$, which interchanges the labels of electrons i and j in a function on which it operates:

$$\wp_{ij}\Psi(1,2,\ldots,i,\ldots,j,\ldots,n) = \Psi(1,2,\ldots,j,\ldots,i,\ldots,n),$$

and we must have one of the two conditions:

$$\wp_{ij}\Psi = \begin{cases} \Psi, & \Psi \text{ is symmetyric, or;} \\ -\Psi, & \Psi \text{ is antisymmetric,} \end{cases}$$

with respect to interchange of two electrons. The two possibilities, in fact, lead to entirely different functions having entirely different expectation values. It is found, by trial and error, that electrons are described by antisymmetric wavefunctions. This antisymmetry requirement is synonymous with the Pauli exclusion principle. Electrons, and other particles requiring antisymmetric wave functions are called Fermions.There are other types of particles, called Bosons, which require symmetric wave functions.

For a two electron system, the antisymmetry requirement can be satisfied by writing:

$$\Psi = \Phi_1(1)\Phi_2(2) - \Phi_1(2)\Phi_2(1).$$

For a three electron system, there are three possible two-electron interchanges, $\wp_{12}$, $\wp_{13}$, and $\wp_{23}$, and the antisymmetry with respect to each can be satisfied by a function:

$$\begin{aligned}\Psi =&\Phi_1(1)\Phi_2(2)\Phi_3(3) + \Phi_1(2)\Phi_2(3)\Phi_3(1) + \Phi_1(3)\Phi_2(1)\Phi_3(2)\\ &- \Phi_1(2)\Phi_2(1)\Phi_3(3) - \Phi_1(3)\Phi_2(2)\Phi_3(1) - \Phi_1(1)\Phi_2(3)\Phi_3(2).\end{aligned}$$

The number of terms goes up by the factorial of the number of electrons. Fortunately, there already exists a symbolic form that is ready-made for this type of function: a determinantal form. We write the wave function as a determinant in which the row number is the electron label, and the column number is the subscript identifying the particular Φ:

$$\Delta = \begin{vmatrix} \Phi_1(1) & \Phi_2(1) & \dots & \Phi_n(1) \\ \Phi_1(2) & \Phi_2(2) & \dots & \Phi_n(2) \\ \vdots & \vdots & \ddots & \vdots \\ \Phi_1(n) & \Phi_2(n) & \dots & \Phi_n(n) \end{vmatrix}.$$

Since the interchange of any two rows of a determinant changes the sign of the determinant, this is a properly antisymmetric function. It expands into $n!$ terms, each containing a product of the n Φ's, with the proper sign for each term. Also, as stated above, the Pauli exclusion principle is incorporated since a determinant vanishes if any two columns are identical, or differ only by one being a multiple of the other; thus, we must have n different Φ's. The Φ's are almost always made an orthonormal set, so that $\langle\Delta|\Delta\rangle = n!$. If the wave function for the many electron system is to be normalized, we must have:

$$\Psi = \left(\frac{1}{n!}\right)^{1/2} \Delta.$$

Such a wave function is called a *Slater determinant*, and is the simplest possible wave function for a many electron system that satisfies the condition of antisymmetry with respect to interchange of electrons.

If we use a single Slater determinant as the form of the wave function, we introduce a serious approximation, called the *Hartree-Fock*, or simply HF, approximation, into the calculations. Since each of the Φ's is a function of the coordinates of one electron, there is no way that the wave function can accomodate the detailed correlated motions of the electrons, whereby movement of one electron influences the movement of another. Properly constructed Φ's will allow averaged electron positions to avoid one another, but the detailed correlated movements could only be accounted for if we had functions of inter-electron distances. The best possible single-Slater-determinant wave function would give an expectation value for the energy of the system called the Hartree-Fock energy limit. The true energy of the system is lower than the Hartree-Fock limit by an amount that is called the *correlation energy*.

Although the vast majority of calculations on organic molecules employ Hartree-Fock wave functions, as the capabilities of computers have improved over the years, there have been increasing numbers of calculations that go beyond the Hartree-Fock approximation. Nearly all of these employ wave functions that are linear combinations of Slater determinants:

$$\Psi = \sum_{i=1}^{m} \Delta_i a_i,$$

where the number of determinants used, m, may be as high as a few hundred. This type of calculation is called a *configuration interaction*, CI, scheme, wherein each of the determinants represents a *configuration* of the system. In principle, a CI wave function could give an exact representation of the true wave function of a system.

We will limit our further discussion to the Hartree-Fock approximation applied to closed-shell, that is, all paired electron, systems.

Before proceeding, it needs to be understood that electrons actually require a fourth coordinate, in addition to the three cartesian coordinates, for specification. This fourth coordinate is called the spin coordinate. A one electron wave function then consists of a 'space' part, which is a function of the three cartesian coordinates, and a 'spin' part, which is a function of the spin coordinate. All we need to know about the spin part is that it can be either one of two functions, $\varsigma_{\pm}(s)$, which are orthonormal, $\langle \varsigma_{+}(s) | \varsigma_{-}(s) \rangle = 0$, where the integration is over the spin coordinate, s, and have eigenvalues of the spin operator, S_z, which are $\pm\hbar/2$:

$$\mathcal{S}_z\varsigma_+(s) = \frac{\hbar}{2}\varsigma_+(s) \quad \text{or} \quad \mathcal{S}_z\varsigma_-(s) = -\frac{\hbar}{2}\varsigma_-(s).$$

We will take products of the space part and spin part to form *spin orbitals* which are proper one electron functions. That is:

$$\Phi_i(x, y, z, s) = \phi_j(x, y, z)\varsigma_k(s), \tag{5.12}$$

where ϕ_j is a *space orbital*, and Φ_i is a spin orbital for one electron. This means that we can form two orthogonal spin orbitals from a single space orbital, or, in the less precise language already familiar to you, we can place two electrons, with paired spins, into each molecular orbital.

Sooner or later in any application of the Hartree-Fock wave function, the expectation value for the energy, $\langle\Psi|\mathcal{H}|\Psi\rangle$, has to be expanded in terms of the spin orbitals, Φ. This is no trivial task, but the result is fairly simple, and, if done for the general Slater determinant form, $\Psi = (1/n!)^{1/2}\Delta$, needs to be done only once to handle all cases [5].

It will be convenient to write the Hamiltonian as a sum of one-electron, H_i, and two-electron, H_{ij}, operators:

$$H_i = -\frac{1}{2}\left(\frac{\partial^2}{\partial x_i^2} + \frac{\partial^2}{\partial y_i^2} + \frac{\partial^2}{\partial z_i^2}\right) - \sum_{k=1}^{N}\frac{Z_k}{R_{ik}} \quad \text{and} \quad H_{ij} = \frac{1}{r_{ij}},$$

so that the total Hamiltonian is:

$$\mathcal{H} = \sum_{i=1}^{n} H_i + \sum_{i=1}^{n-1}\sum_{j=i+1}^{n} H_{ij},$$

and, the expectation value for the energy is:

$$\langle E\rangle = \langle\Psi|\mathcal{H}|\Psi\rangle = \sum_{i=1}^{n}\langle\Psi|H_i|\Psi\rangle + \sum_{i=1}^{n-1}\sum_{j=i+1}^{n}\langle\Psi|H_{ij}|\Psi\rangle. \tag{5.13}$$

Now, when we expand Ψ, we will get several types of terms; for the one-electron H_1, for example, we will get:

$$\begin{aligned}
&\langle\Phi_1(i)\Phi_2(j)\cdots\Phi_r(1)\cdots\Phi_n(k)|H_1|\Phi_1(i)\Phi_2(j)\cdots\Phi_r(1)\cdots\Phi_n(k)\rangle \\
&\quad = \langle\Phi_r(1)|H_1|\Phi_r(1)\rangle\langle\Phi_1(i)|\Phi_1(i)\rangle\cdots\langle\Phi_n(k)|\Phi_n(k)\rangle \\
&\quad = \langle\Phi_r(1)|H_1|\Phi_r(1)\rangle,
\end{aligned}$$

where the product of Φ's on the left of H_1 is 'occupied' identically with the product of Φ's on the right. Note that the integral of products of Φ's over the coordinates of different electrons is equal to the product of integrals over the coordinates of each of the electrons. All other types of terms will give zero values for the one-electron integrals, since a permutation of electrons in the product of Φ's on the right of H_i from that on the left will cause a term in $\langle \Phi_p(j)|\Phi_q(j)\rangle = 0$ for each one of the two permuted electrons to appear in the product of integrals. Each of the n electrons will appear in each of the n Φ's for each of the n one-electron operators in one of the non-zero integrals. For a particular electron, say the j'th one, there will be $(n-1)!$ terms that give the identical value $\langle \Phi_k(i)|H_i|\Phi_k(i)\rangle$ for the k'th Φ. Since there are n H_i's, and since $\langle \Phi_k(i)|H_i|\Phi_k(i)\rangle = \langle \Phi_k(j)|H_j|\Phi_k(j)\rangle$, it follows that:

$$\sum_{i=1}^{n} \langle \Psi|H_i|\Psi\rangle = \sum_{i=1}^{n} \frac{1}{n!}\langle \Delta|H_i|\Delta\rangle = \sum_{k=1}^{n} \langle \Phi_k(1)|H_1|\Phi_k(1)\rangle .$$

If each of the spin orbitals, Φ, is a product of a space orbital, ϕ, and a spin function, as expressed in Eq. (5.12), and each of the ϕ's is 'doubly occupied' (that is, appears in two of the spin orbitals with different spin functions), then:

$$\sum_{i=1}^{n} \langle \Psi|H_i|\Psi\rangle = 2\sum_{k=1}^{n/2} \langle \phi_k(1)|H_1|\phi_k(1)\rangle . \tag{5.14}$$

The expansion of the two electron terms is slightly more difficult because permutations of the two electrons i and j between the product of Φ's on the left of H_{ij} and the one on the right do not always vanish. For example, we will have terms such as:

$$\begin{aligned} &\langle \Phi_1(p)\Phi_2(q)\cdots\Phi_k(i)\cdots\Phi_\ell(j)\cdots\Phi_n(m)| \\ &\qquad\qquad H_{ij}|\Phi_1(p)\cdots\Phi_k(j)\cdots\Phi_\ell(i)\cdots\Phi_n(m)\rangle \\ &\quad = \langle \Phi_k(i)\Phi_\ell(j)|H_{ij}|\Phi_k(j)\Phi_\ell(i)\rangle , \end{aligned}$$

which will be non-zero if the spin parts of Φ_k and Φ_ℓ are equal and all other terms on the right and left of the operator match. These terms will always have a negative sign since the two products have different numbers of pair interchanges. We will also have all of the terms where the products on the left and right of the operator are equal. Each type of term will appear $2[(n-2)!]$ times for each pair of spin orbitals with each pair of electrons. Since there are $n(n-1)/2$ distinct pairs of electrons, we get:

$$\sum_{i=1}^{n-1}\sum_{j=i+1}^{n}\langle\Psi|H_{ij}|\Psi\rangle = \sum_{k=1}^{n-1}\sum_{\ell=k+1}^{n}\Big(\langle\Phi_k(1)\Phi_\ell(2)|H_{12}|\Phi_k(1)\Phi_\ell(2)\rangle - \langle\Phi_k(1)\Phi_\ell(2)|H_{12}|\Phi_k(2)\Phi_\ell(1)\rangle\Big).$$

Taking account of the fact that each space orbital appears twice, we obtain:

$$\begin{aligned}\sum_{i=1}^{n-1}\sum_{j=i+1}^{n}\langle\Psi|H_{ij}|\Psi\rangle &= \sum_{k=1}^{n/2-1}\sum_{\ell=k+1}^{n/2}\Big(4\langle\phi_k(1)\phi_\ell(2)|\frac{1}{r_{12}}|\phi_k(1)\phi_\ell(2)\rangle \\ &\quad - 2\langle\phi_k(1)\phi_\ell(2)|\frac{1}{r_{12}}|\phi_k(2)\phi_\ell(1)\rangle\Big) \\ &\quad + \sum_{k=1}^{n/2}\langle\phi_k(1)\phi_k(2)|\frac{1}{r_{12}}|\phi_k(1)\phi_k(2)\rangle,\end{aligned}$$

which may be written a little more economically as unrestricted sums over the space orbitals:

$$\sum_{i=1}^{n-1}\sum_{j=i+1}^{n}\langle\Psi|H_{ij}|\Psi\rangle = \sum_{k=1}^{n/2}\sum_{\ell=1}^{n/2}\Big(2\langle\phi_k(1)\phi_\ell(2)|\frac{1}{r_{12}}|\phi_k(1)\phi_\ell(2)\rangle - \langle\phi_k(1)\phi_\ell(2)|\frac{1}{r_{12}}|\phi_\ell(1)\phi_k(2)\rangle\Big). \tag{5.15}$$

These expansions of the expectation values in terms of the space orbitals are correct for any form of space orbitals that we choose. In order to actually do a calculation, the next step is to decide on the form of the space orbitals to use.

5.8 The SCF-LCAO-MO Method

For reasons that will become clear as we proceed, the Hartree-Fock wave function calculations are often referred to as *self-consistent-field*,

SCF, calculations. When each of the space orbitals of the wave function is approximated as a linear combination of orbitals centered on the atomic nuclei of the system, we then have a *self-consistent-field linear-combination-of-atomic-orbitals, molecular orbital*, SCF-LCAO-MO, calculation. Virtually all calculations dealing with molecules consisting of more than two atoms use the LCAO method.

For the moment, we can leave the exact forms of the atomic orbitals unspecified. If we symbolize our atomic orbitals as η's, we will write the space orbitals as:

$$\phi_i = \sum_{j=1}^{m} \eta_j c_{ji}, \tag{5.16}$$

where m is the number of atomic orbitals to be used. It is not at all unusual for m to be on the order of one hundred. The coefficients, c_{ji}, are the adjustable parameters of a calculation. We wish to choose them to minimize the expectation value of the energy for a wave function which is a single Slater determinant composed of spin orbitals constructed according to Eqs. (5.12) and (5.16). In order to accomplish this, we need to write the expectation value of the energy in terms of the atomic orbitals and coefficients. With the results of Eqs. (5.13–5.15), it is not very difficult to obtain:

$$\begin{aligned}\langle E\rangle = &\sum_{i=1}^{n/2}\sum_{k=1}^{m}\sum_{\ell=1}^{m} 2\langle\eta_k(1)|H_1|\eta_\ell(1)\rangle c_{ki}c_{\ell i}\\ &+\sum_{i=1}^{n/2}\sum_{j=1}^{n/2}\sum_{k=1}^{m}\sum_{\ell=1}^{m}\sum_{p=1}^{m}\sum_{q=1}^{m}\Big(2\langle\eta_k(1)\eta_\ell(2)|H_{12}|\eta_p(1)\eta_q(2)\rangle\\ &\qquad - \langle\eta_k(1)\eta_\ell(2)|H_{12}|\eta_p(2)\eta_q(1)\rangle\Big)c_{ki}c_{pi}c_{\ell j}c_{qj},\end{aligned} \tag{5.17}$$

where the dummy indices i and j are over the space orbitals, and k, ℓ, p, and q are over the atomic orbitals.

The minimization of $\langle E\rangle$ with respect to the c's must be made subject to the constraint that the space orbitals remain orthonormalized. The constraint for each pair of orbitals can be introduced by means of a LaGrange undetermined multiplier [6], ϵ_{ij}, and the constrained minimization is given by setting:

$$\frac{\partial\langle E\rangle}{\partial c_{rs}} - 2\sum_{i=1}^{n/2}\sum_{j=1}^{n/2}\epsilon_{ij}\frac{\partial\langle\phi_i|\phi_j\rangle}{\partial c_{rs}} = 0, \tag{5.18}$$

for all r and s.

The evaluation of $\partial\langle E\rangle/\partial c_{rs}$ by differentiation of Eq. (5.17) is quite easy since the coefficients are the only variables. Watch out for the terms c_{ki}^2 and $c_{\ell j}^2$ which occur in the unrestricted sums when $k = p$ and $\ell = q$ and give a factor of two for these terms. The re-grouping of the two-electron terms requires a little care in re-naming the dummy variables of the sums, and the recognition that:

$$\begin{aligned}\langle\eta_k(1)\eta_\ell(2)|H_{12}|\eta_p(1)\eta_q(2)\rangle &= \langle\eta_p(1)\eta_\ell(2)|H_{12}|\eta_k(1)\eta_q(2)\rangle \\ &= \langle\eta_k(1)\eta_q(2)|H_{12}|\eta_p(1)\eta_\ell(2)\rangle = \langle\eta_p(1)\eta_q(2)|H_{12}|\eta_k(1)\eta_\ell(2)\rangle \\ &= \langle\eta_\ell(1)\eta_k(2)|H_{12}|\eta_q(1)\eta_p(2)\rangle = \langle\eta_q(1)\eta_k(2)|H_{12}|\eta_\ell(1)\eta_p(2)\rangle \\ &= \langle\eta_\ell(1)\eta_p(2)|H_{12}|\eta_q(1)\eta_k(2)\rangle = \langle\eta_q(1)\eta_p(2)|H_{12}|\eta_\ell(1)\eta_k(2)\rangle,\end{aligned}$$

in order to obtain:

$$\begin{aligned}\frac{\partial\langle E\rangle}{\partial c_{rs}} &= \sum_{k=1}^{m}\langle\eta_r(1)|H_1|\eta_k(1)\rangle c_{ks} \\ &\quad + \sum_{k=1}^{m}\sum_{\ell=1}^{m}\sum_{p=\ell}^{m}\left(\sum_{i=1}^{n/2} c_{\ell i}c_{pi}\right) 4c_{ks} \\ &\quad \times \Big(4\langle\eta_r(1)\eta_\ell(2)|H_{12}|\eta_k(1)\eta_p(2)\rangle \\ &\qquad - \langle\eta_r(1)\eta_k(2)|H_{12}|\eta_\ell(1)\eta_p(2)\rangle \\ &\qquad - \langle\eta_r(1)\eta_k(2)|H_{12}|\eta_p(1)\eta_\ell(2)\rangle\Big).\end{aligned} \tag{5.19}$$

Note the restricted sum over the index p to take care of the factor of two for $p = \ell$ as mentioned above. Also, note the grouping of the sum of coefficients over the index i of the occupied orbitals, which is convenient since none of the integrals involve this index. This sum is commonly defined as a *density matrix* element:

$$\sum_{i=1}^{n/2} c_{ki}c_{\ell i} \equiv D_{k\ell}. \tag{5.20}$$

The part of the differentiation containing the LaGrange multipliers is considerably easier, giving:

$$2\sum_{i=1}^{n/2}\sum_{j=1}^{n/2}\epsilon_{ij}\frac{\partial\langle\phi_i|\phi_j\rangle}{\partial c_{rs}} = 4\sum_{j=1}^{n/2}\epsilon_{js}\sum_{k=1}^{m}\langle\eta_r(1)|\eta_k(1)\rangle c_{ks}. \tag{5.21}$$

In order to simplify the writing of equations, a matrix, $\mathcal{F}$, called the Hartree-Fock matrix, is defined in terms of its elements, $F_{k\ell}$, as:

$$\begin{aligned} F_{k\ell} =& \langle \eta_k(1)|H_1|\eta_\ell(1)\rangle + \sum_{p=1}^{m}\sum_{q=p}^{m} D_{pq} \\ & \times \Big(4\langle \eta_k(1)\eta_p(2)|H_{12}|\eta_\ell(1)\eta_q(2)\rangle - \langle \eta_k(1)\eta_\ell(2)|H_{12}|\eta_p(1)\eta_q(2)\rangle \\ & - \langle \eta_k(1)\eta_\ell(2)|H_{12}|\eta_q(1)\eta_p(2)\rangle \Big), \end{aligned}$$

which can be interpreted as the energy of an electron shared by the atomic orbitals η_k and η_ℓ in the field due to the nuclei of the system and the other electrons of the system which are in the occupied orbitals as defined by the density matrix elements, D_{pq}. Another matrix, $\mathcal{C}$, is defined in terms of its elements, $c_{k\ell}$; an *overlap matrix*, $\mathcal{S}$, is defined to have elements $\langle \eta_k(1)|\eta_\ell(1)\rangle$; and, finally, a diagonal matrix, $\mathcal{E}$, is defined with the elements ϵ_s along the diagonal.

Using the results of Eqs. (5.19) and (5.21)) substituted into the constrained minimization requirement, Eq. (5.18), along with the matrix definitions, gives the final *Roothan equation*:

$$\mathcal{F}\mathcal{C} = \mathcal{S}\mathcal{C}\mathcal{E}, \tag{5.22}$$

as the equation which must be satisfied in order for the expectation value for the energy to be minimized with respect to the choice of coefficients in the LCAO's. The Roothan equation is the basic equation for all SCF-LCAO-MO calculations.

There are two difficulties that must be faced in solving the Roothan equation. The first, fairly trivial, matter is that the matrix $\mathcal{S}$ prevents the equation from being a matrix diagonalization problem. The second matter is that we can't evaluate the matrix elements $F_{k\ell}$ until we know the $\mathcal{C}$ matrix, which is obtained by solving the Roothan equation!

Let's take care of the first problem first. Since $\mathcal{S}$ is Hermitian, it can be diagonalized by a unitary transformation:

$$\mathcal{U}^\dagger \mathcal{S}\mathcal{U} = \Lambda.$$

This is equivalent to defining a new set of *basis functions*, ξ, from the atomic orbitals: $\xi_i = \sum_{j=1}^{m} \eta_j u_{ji}$, so that the new basis functions are orthogonal one to another, $\langle \xi_k|\xi_l\rangle = 0$ if $k \neq l$. Looked at in this way, it is obvious that the elements of Λ are all greater than, or equal to zero, since they are the integrals of squares, $\langle \xi_k | \xi_k \rangle$. This allows us to form a matrix, $\Lambda^{-1/2}$, whose elements are $\lambda_i^{-1/2} = \sqrt{1/\lambda_i}$. We can now write:

$$\Lambda^{-1/2}\mathcal{U}^\dagger \mathcal{S}\mathcal{U}\Lambda^{-1/2} = \Lambda^{-1/2}\Lambda\Lambda^{-1/2} = \mathcal{I},$$

where $\mathcal{I}$ is the identity matrix. Defining the matrix $\mathcal{R}$ as:

$$\mathcal{R} = \mathcal{U}\Lambda^{-1/2}, \quad \text{and} \quad \mathcal{R}^\dagger = \Lambda^{-1/2}\mathcal{U}^\dagger, \quad \text{and} \quad \mathcal{R}^{-1} = \Lambda^{+1/2}\mathcal{U}^\dagger,$$

$$\text{we have} \quad \mathcal{R}^\dagger \mathcal{S}\mathcal{R} = \mathcal{I},$$

We can now left-multiply both sides of Eq. (5.22) by $\mathcal{R}^\dagger$, insert the identity matrix, in the form $\mathcal{R}\mathcal{R}^{-1}$, preceding the $\mathcal{C}$ matrix on both sides, and obtain:

$$\mathcal{R}^\dagger \mathcal{F}\mathcal{R}\mathcal{R}^{-1}\mathcal{C} = \mathcal{R}^\dagger \mathcal{S}\mathcal{R}\mathcal{R}^{-1}\mathcal{C}\mathcal{E}$$
$$\left(\mathcal{R}^\dagger \mathcal{F}\mathcal{R}\right)\left(\mathcal{R}^{-1}\mathcal{C}\right) = \left(\mathcal{R}^{-1}\mathcal{C}\right)\mathcal{E},$$

and, defining $\mathcal{R}^\dagger \mathcal{F}\mathcal{R} \equiv \mathcal{F}'$ and $\mathcal{R}^{-1}\mathcal{C} \equiv \mathcal{C}'$, we get the matrix diagonalization problem:

$$\mathcal{F}'\mathcal{C}' = \mathcal{C}'\mathcal{E}, \quad \text{or} \quad \mathcal{C}'^\dagger \mathcal{F}'\mathcal{C}' = \mathcal{E}, \tag{5.23}$$

where we know that $\mathcal{C}'$ can be unitary since $\mathcal{F}'$ is Hermitian.

The difficulty that we cannot evaluate the Hartree-Fock matrix until we know the coefficients is handled by requiring self-consistency for the equations. We start the process by making a pure guess for the $\mathcal{C}$ matrix; we could, for example, use a 'bare nucleus' first-guess with $\mathcal{C} = 0$. The guess is used, along with the actual integral evaluations, to set up the $\mathcal{F}$ matrix, which is then transformed to $\mathcal{F}'$ and diagonalized to obtain a second-approximation for $\mathcal{C}$. The second-approximation $\mathcal{C}$ is used to re-evaluate $\mathcal{F}$, and the process repeated to obtain a third-approximation $\mathcal{C}$. This entire process is iterated until the $\mathcal{C}$ used to evaluate $\mathcal{F}$ is the same as the $\mathcal{C}$ that is obtained from the $\mathcal{C}'$ that diagonalizes $\mathcal{F}'$. This is where the SCF part of the name comes from.

In actual practice, there is another matter that must be settled: the choice of the atomic orbitals. In Section 5.1, it was pointed out that atomic orbitals obtained from the solution of the problem of a single electron in the field of a single nucleus are of the general form:

$$\eta_{Slater} = f(x, y, z)e^{-\alpha r},$$

where the $f(x, y, z)$ part determines whether the orbital is an s, p, d, of f-type, and α is proportional to the nuclear charge. As indicated by the subscript, these are called *Slater Type Orbitals*, or STO's. These STO's are seldom used in recent calculations because of the difficulty of evaluating the integrals, which cannot be expressed in analytical form and must be done by numerical integration techniques.

Another type of 'atomic orbitals' (which are not the solution of any real atomic problem) are used in virtually all recent calculations. These are called *Gaussian Type Orbitals*, GTO's, and have the general form:

$$\eta_{Gaussian} = f(x, y, z)e^{-\alpha r^2},$$

where the $f(x, y, z)$ is the same as for the STO's and determines whether the type is s, p, etc. All of the necessary integrals for GTO's can be expressed analytically and evaluated very rapidly. For a given accuracy of calculation, fewer STO's are required than GTO's (by a sizable factor of three to five), but the speed of integral evaluation is so much greater for the GTO's that the larger number of functions is still the best way to go. Many of the most recent calculations attempt a compromise wherein a single STO is approximated by a linear combination of several GTO's for purposes of integral evaluations. This reduces the number of integrals that need to be saved and re-formed into the Hartree-Fock matrix on each iteration.

The limitation on the molecules that can be treated by the SCF-LCAO-MO method comes from the number of basis functions (this is a more accurate term than 'atomic orbitals') required to get reasonable wave functions. A *minimal basis set* would consist of a $1s$-type orbital centered on each hydrogen, a $1s$, a $2s$, and three $2p$-type orbitals centered on each atom of the first full row of the periodic table, etc. Such minimal basis sets, even of STO's, are inadequate for most chemical purposes, and an equivalent of twice the minimal STO basis is a more reasonable starting point. The real problem comes from the number of two electron integrals, $\langle\eta_k(1)\eta_p(2)|H_{12}|\eta_l(1)\eta_q(2)\rangle$, that must be evaluated, stored, and formed into the Hartree-Fock matrices on each iteration. The problem has been called *the n^4 dilemma*, because the number of integrals increases with the fourth power of the number of basis functions used. Calculations on moderately large organic molecules are possible, but rare, even with the most modern computers.

The difficulty in obtaining chemically useful results can best be appreciated by realizing the accuracy necessary in the calculations. The total HF energy of methane is on the order of 20,000 kcal/mole. If we wish to obtain an accuracy of ca. 2 kcal/mole in our calculations, the calculation must be accurate to 0.01%. Correlation energies, in fact, are on the order of several electron volts even for very small molecules. In carefully chosen systems, errors due to insufficient basis sets and to correlation energies may be constant, so that relative values of calculated energies are useful. Unfortunately, it is not entirely clear when one may expect this cancellation of errors, and calculated results on all but the simplest molecules should be viewed with appreciable skepticism when distinctions of a few kcal/mole are important.

'Canned' programs, such as **GAUSSIAN 82**, are available [7] to carry out SCF-LCAO-MO calculations. All one needs to do is input the nuclear positions and charges and choose the basis functions to be used. The latter is not trivial; in order to obtain reasonable results, the α values and the types of functions must be carefully selected. There is a great deal of accumulated experience [8] to help the beginner.

5.9 First-Order Density Matrices

The elements of the density matrix, $\mathcal{D}$, were defined in Eq. (5.20) in order to simplify the writing of equations. The density matrix, along with the basis functions, in fact, contains all of the meaningful information obtainable from the Hartree-Fock wave function. For example, we can expand the space part of Ψ^2 to get electron density:

$$\begin{aligned} 2\sum_{i=1}^{n/2} \phi_i^2(1) &= \sum_{j=1}^{m}\sum_{k=1}^{m} 2\Big(\sum_{i=1}^{n/2} c_{ji}c_{ki}\Big)\eta_j(1)\eta_k(1) \\ &= \sum_{j=1}^{m}\sum_{k=1}^{m} 2D_{jk}\eta_j(1)\eta_k(1), \end{aligned} \tag{5.24}$$

or, we can collect the sums over occupied orbitals (both i and j indices) in Eq. (5.17), to get:

$$\begin{aligned} \langle E\rangle &= \sum_{k=1}^{m}\sum_{\ell=1}^{m} 2D_{k\ell}\langle\eta_k(1)|H_1|\eta_\ell(1)\rangle \\ &+ \sum_{k=1}^{m}\sum_{\ell=1}^{m}\sum_{p=1}^{m}\sum_{q=1}^{m} D_{kp}D_{\ell q}\Big(2\langle\eta_k(1)\eta_\ell(2)|H_{12}|\eta_p(1)\eta_q(2)\rangle \\ &- \langle\eta_k(1)\eta_\ell(2)|H_{12}|\eta_p(2)\eta_q(1)\rangle\Big). \end{aligned} \tag{5.25}$$

This last equation shows clearly that the energy is a function of the density matrix[2]. Eq. (5.24) shows that the electron density in any region of space is a function of the density matrix. This is important since modern concepts of bonds, lone-pairs, and even atoms, are most precisely given in terms of electron density, and, therefore, density matrices [9].

The difference between using the density matrix and using molecular orbitals can be appreciated if we look at the matrix formed from the columns of $\mathcal{C}$ corresponding to the occupied molecular orbitals. Let's call this matrix $\mathcal{C}_{omo}$; it will contain the $n/2$ columns of $\mathcal{C}$ corresponding to the lowest ϵ's of Eq. (5.22), and will have m rows where m is the number of basis functions used, so it is a rectangular matrix. The matrix $\mathcal{C}^\dagger_{omo}$ will have the columns of $\mathcal{C}_{omo}$ as its rows:

$$\mathcal{C}_{omo} = \begin{pmatrix} c_{1,1} & c_{1,2} & \cdots & c_{1,n/2} \\ c_{2,1} & c_{2,2} & \cdots & c_{2,n/2} \\ \vdots & \vdots & & \vdots \\ \vdots & \vdots & & \vdots \\ c_{m,1} & c_{m,2} & \cdots & c_{m,n/2} \end{pmatrix},$$

$$\mathcal{C}^\dagger_{omo} = \begin{pmatrix} c_{1,1} & c_{2,1} & \cdots & \cdots & c_{m,1} \\ c_{1,2} & c_{2,2} & \cdots & \cdots & c_{m,2} \\ \vdots & \vdots & & & \vdots \\ c_{1,n/2} & c_{2,n/2} & \cdots & \cdots & c_{m,n/2} \end{pmatrix}.$$

It should be fairly obvious that:

$$\mathcal{D} = \mathcal{C}_{omo}\mathcal{C}^\dagger_{omo}.$$

Now, suppose we right multiply $\mathcal{C}_{omo}$ by any unitary matrix, $\mathcal{U}$, to get a new matrix, $\mathcal{T}$; we then have:

$$\mathcal{T} = \mathcal{C}_{omo}\mathcal{U}, \quad \text{and} \quad \mathcal{T}^\dagger = \mathcal{U}^\dagger\mathcal{C}^\dagger_{omo},$$

and it follows that

$$\mathcal{T}\mathcal{T}^\dagger = \mathcal{C}_{omo}\mathcal{U}\mathcal{U}^\dagger\mathcal{C}^\dagger_{omo} = \mathcal{C}_{omo}\mathcal{I}\mathcal{C}^\dagger_{omo} = \mathcal{C}_{omo}\mathcal{C}^\dagger_{omo} = \mathcal{D},$$

showing that the occupation of any set of orbitals formed as linear combinations of the molecular orbitals which are eigenfunctions of the HF matrix gives the same density matrix, and, therefore, the same electron density, and same expectation value for any observable of the system.

---

[2] Note that the total energy is not the sum of the orbital energies. The evaluation of the difference between total energy and the sum of orbital energies is left for a problem at the end of this chapter.

5.10 REFERENCES

[1]. R. B. Woodward and R. Hoffmann, **Angew. Chem., International Ed., Engl., 8**, 781 (**1969**).

[2]. F. A. Cotton, **Chemical Applications of Group Theory**, John Wiley & Sons, Inc., New York, **1963**.

[3]. R. McWeeny, **Symmetry, An Introduction to Group Theory**, (Vol 3 in "The International Encyclopedia of Physical Chemistry and Chemical Physics"), McMillan Co., New York, **1963**.

[4]. R. G. Pearson, **Symmetry Rules for Chemical Reactions**, Wiley-Interscience, New York, **1976**.

[5]. R. G. Parr, **Quantum Theory of Molecular Electronic Structure**, Benjamin Co., New York, **1964**; The more general expansion of a CI wave function is given.

[6]. M. L. Boas, **Mathematical Methods in the Physical Sciences**, John Wiley & Sons, New York, **1966**, pg. 142ff.

[7]. J. A. Pople, et al, **GAUSSIAN 82**, Carnegie-Mellon University, Pittsburgh, Pa., **1982**.

[8]. S. Huzinaga, **Gaussian Basis Sets for Molecular Calculations**, Elsevier Co., Amsterdam, **1984**; W. J. Hehre, L. Radom, P. v. R. Schleyer, and J. A. Pople, **Ab Initio Molecular Orbital Theory**, John Wiley & Sons, New York, **1986**.

[9]. R. F. W. Bader, **Accts. Chem. Res., 18**, 9 (**1985**).

5.11 PROBLEMS

Problem 5.1

Show that $e^{-\alpha r}$ is an eigenfunction of the Hamiltonian for the hydrogen atom when α is properly chosen.
Hint: The Hamiltonian is given in Eq. (5.11) with a single term in the sums, and no electron-electron repulsion term. Write $r = \sqrt{x^2 + y^2 + z^2}$, *and proceed just as we did with* ξ *on pg. 177–178. Check your answer with the function given on pg. 152, but watch out for units.*

Problem 5.2

Write the Hückel $\mathcal{H}'$ matrix, containing ones and zeroes, for benzene, napthalene, the cyclopentadienyl system, and the indenyl system.

Problem 5.3

Without solving the matrix problem, using only the symmetries of the orbitals, write the Hückel molecular orbitals and calculate the orbital energies of the three lowest energy orbitals of benzene.
Hint: The lowest energy orbital must have all six coefficients equal. The next two orbitals are degenerate, having nodal planes at right angles to each other; let one have a nodal plane passing through carbons 1 and 4, and the other have a nodal plane passing midway between carbons 2 and 3 and between 5 and 6. The coefficients in the latter orbital can be calculated from the fact that occupation of the second and third orbitals must give equal electron densities on each of the six carbons. The orbital energies can be evaluated by expanding $\langle\phi_i|\mathcal{H}|\phi_i\rangle$.

Problem 5.4

The cyclopropenyl radical has three π-electrons. The electron density of the radical must be the same on each of the three carbons. What molecular orbital is occupied by the third electron?
Hint: Take a linear combination of the degenerate second and third orbitals using coefficients that may be complex $(a + ib)$. *Remember to take the complex conjugate,* $(a + ib)^\dagger = a - ib$, *when calculating* $\phi^\dagger\phi$.

Problem 5.5

Show that the Diels-Alder reaction of an alkene with a conjugated diene in the *s-cis* conformation is a symmetry allowed reaction when a plane of symmetry perpendicular to the σ-plane is preserved.

Problem 5.6

All attempts to carry out reactions in which cyclopropyl cation might be an intermediate lead to allyl products. Moreover, the acetolysis of trans,trans-2,3-dimethylcyclopropyl chloride occurs nearly 10^4 faster than the acetolysis of the cis,cis-isomer. Propose a mechanism for the acetolysis of cyclopropyl chloride to form allyl acetate.

Hint: Consider the possibility of the ring opening occurring simultaneously with the departure of the chloride ion, and look at symmetry reauirements. You should know that the **W***-form of 1,3-disubstituted allylic cations is much more stable than the* **U***-form.*

Problem 5.7

The *trace* of a matrix is defined as the sum of the diagonal elements of the matrix. The trace of a matrix is invariant under a unitary transformation, that is, if $\mathcal{U}^{\dagger}\mathcal{M}\mathcal{U} = \mathcal{N}$, then the trace of $\mathcal{N}$ is equal to the trace of $\mathcal{M}$ for any $\mathcal{U}$ (see Problem 1.11). What does this mean when applied to orbital energies in the Hückel approximation?

Problem 5.8

Note the similarity of the carbonate system to the trimethylenemethane system (pg. 164). Set up and solve the Hückel problem for the carbonate system, with the modification that the diagonal elements of the Hückel matrix corresponding to the oxygens should be set equal to $\alpha + \beta$ to account for the greater nuclear charge of oxygen than carbon.

Hint: Number the carbonate system such that 4 is the carbon atom. Use the first three rows and columns of the $\mathcal{U}$ matrix for the trimethylenemethane system to construct a unitary matrix, with $u_{i4} = u_{4i} = 0$, except $u_{44} = 1$; be careful that you re-normalize the columns so that you have a unitary matrix. Use this matrix to transform the Hückel matrix. This will not be a diagonal matrix, but note that:

$$\begin{pmatrix} 0.6 & -0.8 \\ 0.8 & 0.6 \end{pmatrix} \begin{pmatrix} 1 & \sqrt{3} \\ \sqrt{3} & 0 \end{pmatrix} \begin{pmatrix} 0.6 & 0.8 \\ -0.8 & 0.6 \end{pmatrix}$$

is a diagonal matrix. See the Hint to Problem 1.14, also.

Problem 5.9

Use the expression $\langle E \rangle = (\alpha\hbar^2/2m) + (k/8\alpha)$, pg. 177, to minimize the expectation value for the energy with respect to α, and show that this is the same result as obtained on pg. 178 by making ξ an eigenfunction of the Hamiltonian for the harmonic oscillator.

Problem 5.10

Show that $\langle \Delta | \Delta \rangle = n!$ for a determinant, Δ, containing n orthonormal spin orbitals.

Hint: If you don't see how to proceed, actually expand the expression for the cases of two and three electrons.

Problem 5.11

Fill in all of the steps between Eq. (5.21) and Eq. (5.22). Watch out for the order of multiplication in the product $\mathcal{SCE}$.

Problem 5.12

From the definition of the density matrix, Eq. (5.20), and the Roothan equation, Eq. (5.22), it is easy to obtain the expression:

$$\epsilon_i = \sum_{k=1}^{m} \sum_{\ell=1}^{m} F_{k\ell} c_{ki} c_{\ell i}, \quad \text{and} \quad \sum_{i=1}^{n/2} \epsilon_i = \sum_{k=1}^{m} \sum_{\ell=1}^{m} F_{k\ell} D_{k\ell}.$$

Using this result and Eq. (5.25), show that:

$$\langle E \rangle = \sum_{i=1}^{n/2} 2\epsilon_i - \sum_{k=1}^{m} \sum_{\ell=1}^{m} \sum_{p=1}^{m} 2D_{k\ell} D_{pp} \Big(2\langle \eta_k(1)\eta_p(2)|H_{12}|\eta_\ell(1)\eta_p(2)\rangle - \langle \eta_k(1)\eta_\ell(2)|H_{12}|\eta_p(1)\eta_p(2)\rangle \Big)$$

Problem 5.13

The ammonia molecule, which has a trigonal pyramidal geometry, has a three-fold axis of rotation passing through the nitrogen apex of the pyramid and the center of the base; in addition, there are three planes of symmetry, each containing the nitrogen and one hydrogen and bisecting the other H-N-H angle. Assign a symbol $\mathcal{R}_1$ to the symmetry operation which rotates the molecule by 120 degrees, a symbol $\mathcal{R}_2$ to the operation which rotates the molecule through 240 degrees, a symbol $\mathcal{I}$ to the operation which rotates through 360 degrees, and symbols of $\mathcal{S}_1$, $\mathcal{S}_2$, and $\mathcal{S}_3$ to the three reflection operations. Work out the multiplication table showing the result of any two successive operations, such as $\mathcal{R}_1\mathcal{R}_1 = \mathcal{R}_2$, and $\mathcal{R}_1\mathcal{R}_2 = \mathcal{I}$. Do these elements form a closed group; that is, does the successive operation by any two elements give one of the elements already defined?

Problem 5.14

Suppose we have a system for which $\mathcal{H}\psi_1 = \epsilon_1\psi_1$, and $\mathcal{H}\psi_2 = \epsilon_2\psi_2$, and in which $\mathcal{H}$ is not explicitly time dependent. At some time, t_0, the system is in a state represented by $\Psi(t_0) = (1/\sqrt{2})(\psi_1 + \psi_2)$. What will the state function be at some very short time later, $t_0 + \delta t$? *Hint: Assume that* $\Psi(t_0 + \delta t) = \Psi(t_0) + (\partial\Psi/\partial t)\delta t$, *and use the time dependent Schrödinger equation, Eq. (5.7).*

There is an operator called the 'time evolution operator' and defined as:

$$\mathcal{T} \equiv \exp\left(-\frac{i}{\hbar}(t - t_0)\mathcal{H}\right)$$

which is used in time dependent quantum mechanical problems. Can you figure out what is meant by this expression which involves the curious phenomenon of an operator in the exponent and see how it would be used?

Hint: The power series equivalent of e^x *is* $1 + x + x^2/2 + \cdots + x^n/n!$. *Write* $\Psi(t)$ *as a Taylor series about* t_0.

CHAPTER 6
Acid-Base Equilibria

6.1 Introduction

It should be clear by this point that many organic reactions involve reactive intermediates, and that the rates and courses of reactions are frequently influenced by the chemistry of the intermediates. Many of these intermediates, such as carbocations and carbanions, are involved in acid-base reactions that may, or may not, be part of the overall reaction. Carbocations in water may react with solvent to form alcohol and a proton:

$$R^+ + H_2O \xrightleftharpoons{K_R} ROH + H^+$$

so that the equilibrium concentration ratio of R^+ to ROH is governed by pH in much the same way that equilibrium concentration ratios of Brønsted acids and conjugate bases are. Carbanions are, of course, the Brønsted conjugate bases of hydrocarbons. Knowledge of the equilibrium constants governing the acid-base reactions of the intermediates can help greatly in understanding the overall reactions in which the intermediates are involved.

Traditionally, because of the ubiquity of water as a solvent, most measurements of acid-base reactions have been in aqueous solutions. This is unfortunate for many purposes; because of its nearly unique ability to strongly solvate small charge-localized anions, water masks many of the inherent differences in acidities and basicities of solute species, and because water is itself appreciably acidic and basic, it limits the range of acid-base equilibrium constants that can be determined in aqueous solution.

In the present chapter, we will start by considering acids and bases which may be studied in aqueous solution by traditional methods, and the limitations on these studies. We will then consider attempts to measure greater ranges of equilibrium constants than can be obtained in aqueous solution while still retaining an aqueous standard state. Finally, recent developments in which the aqueous standard state has been abandoned will be presented, and *intrinsic* acidities and basicities, obtained in the gas phase at low pressures, will be discussed and compared with the reactions in solutions.

6.2 Acidities in Aqueous Solution

The three most commonly encountered types of acid-base reactions may be generalized as

$$\mathrm{AH} \quad \underset{}{\overset{K_a}{\rightleftharpoons}} \quad \mathrm{A^-} \; + \; \mathrm{H^+} \quad \text{, with}$$

$$K_a = \left(\frac{[A^-]}{[HA]}\right)\left(\frac{\gamma_{A^-}}{\gamma_{HA}}\right) a_{H^+}, \tag{6.1}$$

$$\mathrm{BH^+} \quad \overset{K_a}{\rightleftharpoons} \quad \mathrm{B} \; + \; \mathrm{H^+} \quad \text{, with}$$

$$K_a = \left(\frac{[B]}{[BH^+]}\right)\left(\frac{\gamma_B}{\gamma_{BH^+}}\right) a_{H^+}, \tag{6.2}$$

and,

$$\mathrm{R^+} \; + \; \mathrm{H_2O} \quad \overset{K_R}{\rightleftharpoons} \quad \mathrm{ROH} \; + \; \mathrm{H^+} \quad \text{, with}$$

$$K_R = \left(\frac{[ROH]}{[R^+]}\right)\left(\frac{\gamma_{ROH}}{\gamma_{R^+}}\right)\left(\frac{a_{H^+}}{a_{H_2O}}\right), \tag{6.3}$$

where the distinction between (6.1) and (6.2) is in the charge type of the acid and conjugate base, and the distinction between (6.3) and the others is in the formal involvement of a water molecule as a reactant.

From an experimental standpoint, the study of these reactions always involves a measurement of the concentration ratio, written as the first term on the right hand side of each of the above equations for the equilibrium constants, and an independent measurement of a_{H^+}. In infinitely dilute aqueous solutions, the other terms on the right hand sides of the equations are all defined to be unity. In actual practice, the equilibrium constant values are determined by a series of measurements extrapolated to infinite dilution, or, in all but the most rigorous work, by measurements in moderately dilute solution, say $< 10^{-2}M$, and application of the Debye-Hückel limiting law to evaluation of the activity coefficients of the ionic species.

The determination of $pH = -\log a_{H^+}$ is an essential part of these measurements and is anything but trivial. The tremenduous amount of work that has gone into the methods of determination of pH in aqueous solution, however, has resulted in such convenient techniques that we hardly think about the difficulty [1]. The dozens of pH indicators available for use are themselves Brønsted acids or bases whose pK_a's have been precisely determined in aqueous solution. The use of potentiometric methods requires calibrations which depend on the availability of buffers consisting of acids and bases with known pK_a's in aqueous solution. One of the major practical reasons for studying acid-base reactions in aqueous solution is the existence of this data base which allows easy determination of pH.

Unfortunately, the study of many organic acids and bases cannot be carried out in aqueous solution. There is a fairly minor problem that many organic acids and bases are essentially insoluble in water, but the major problem arises from the magnitudes of the acid-base equilibrium constants. Relatively acidic hydrocarbons, such as cyclopentadiene, have $K_a \leq 10^{-18}M$, and relatively stable carbocations, such as triphenylmethyl cation, have $K_R \geq 10^6 M$. The problem comes in the measurement of the concentration ratios of conjugate acid to conjugate base in reasonably dilute aqueous solution.

The concentration ratios in Eqs. (6.1)–(6.3) are usually measurable only within the fairly narrow range of values of approximately $0.01 - 100$, or even narrower when high accuracy is desired. This is because of the difficulty of accurately measuring one of the two species in the presence of a large excess of the other. A common determination of the ratio in Eq. (6.1), for example, is to weigh a sample of HA into a measured volume of solvent, then to determine the concentration of A^- by spectrophotometry. The remaining concentration of HA is determined by difference. If the total concentration of $[HA]+[A^-]$, from the weight and volume is known to $\pm 0.1\%$, and the spectrophotometric measurement of $[A^-]$ is accurate to $\pm 0.1\%$, then a measured ratio of 99 would have an error of approximately $\pm 20\%$ (that is, a ratio of 99.1/0.8 to 98.9/1.2). Considerations such as these lead to the conclusion that the equilibrium constants can usually be determined only when the pH of the solution is within approximately two units, or less, of the pK ($= -\log K_a$ or $-\log K_R$). If we consider ionic strengths of greater than $10^{-2}M$ to be outside the "reasonably dilute" range where the Debye-Hückel equation can be used, then the range of pK's that can be determined in aqueous solution is approximately 2–12.

Some organic species with pK's in the range accessable in aqueous solution are shown in Table 6.1. A little more data is available [2], [3], [4], [5], [6], and [7], but this just about exhausts the list of carbon acids that can be measured in aqueous solution. For carbocations, either *aromatic* resonance such as that for the 6π-electron tropylium or 2π-electron cyclopropenium ions, or highly electron-donating groups such as the dimethylaminophenyl groups of tris-(p-dimethylaminophenyl)methyl cation, are required to make the pK_R's high enough to be accessable in water. For carbanions, either a nitro group or two other highly electron-withdrawing groups are necessary to make the species

TABLE 6.1
Carbon Acids in Aqueous Solution

Acid	pK	Ref.
Tris-(p-dimethylaminophenyl)methyl cation	9.39	[2]
Bis-(p-dimethylaminophenyl), phenylmethyl cation	6.94	[2]
p-Dimethylaminophenyl, diphenylmethyl cation	3.88	[3]
Tris-(p-methoxyphenyl)methyl cation	0.82	[3]
p-Dimethylaminophenyltropyl cation	7.35	[4]
p-Methoxyphenyltropyl cation	5.75	[4]
Phenyltropyl cation	4.84	[4]
p-Chlorophenyltropyl cation	4.55	[4]
Tropyl cation	4.76	[4]
1,2,3-Tri-i-propyl, cyclopropenyl cation	7.2	[5]
Dinitromethane	3.6	[6]
Nitromethane	10.2	[6]
Nitroethane	8.6	[6]
Acetylacetone	9.0	[6]
Malononitrile	11.2	[6]
Hydrocyanic acid	9.15	[7]

observable in water.

There are many interesting intermediates other than the carbon centered acids and bases that also fall outside the range measurable in aqueous solution. The Brønsted conjugate acids of ketones, esters, amides, alcohols, and alkenes are important reaction intermediates which are usually too acidic to measure in water. The Brønsted conjugate bases of alcohols, amines, and amides are usually too basic to measure in aqueous solution.

6.3 Acidity Functions

As soon as we leave dilute aqueous solutions, the second and third terms on the right-hand-sides of Eqs. (6.1)–(6.3) are no longer equal to 10^{-pH}, and as we discussed in Section 2.6, the activity coefficients become dependent on the precise natures of the solutes and solvents. The activity coefficient of a solute, S, in some solvent is directly related to the free energy of transfer of the solute from the solvent to the standard state: $\Delta G_t^\circ = -RT \ln \gamma_S$. If the free energies of transfers of solutes were due to additive contributions from the various parts of the solutes, then the difference in free energy of transfer of a base, B, and its conjugate acid, BH^+, would be due to the difference in contributions of the basic group and the protonated basic group, and would be independent of the structure of the rest of the molecule. Since group additive contributions to free energies of transfer would produce activity coefficients which are products of group contributions, this would mean that the activity coefficient ratio, γ_B/γ_{BH^+}, would be the same for all bases of a given type. For example, the activity coefficient ratio would be the same for aniline/anilinium ion as for 3,4-dinitroaniline/3,4-dinitroanilinium ion.

The most optimistic hope in initial studies was that group additive contributions to the transfer free energies would be such that only the difference of the proton 'group' between base and conjugate acid would contribute to the activity coefficient ratios. It was anticipated, however, that different charge types, such as Eq. (6.1) and (6.2), would have different activity coefficient ratios, and it was recognized that the equilibria in Eq. (6.3) were influenced by the activity of water as well as by the activity coefficient ratios. If these anticipations were realized, we could develop three very useful general functions.

The equilibrium constant expressions are most conveniently discussed in logarithmic form. For the three acid-base types, we have

$$pK_a = \log\left(\frac{[HA]}{[A^-]}\right) + \log\left(\frac{\gamma_{HA}}{\gamma_{A^-}}\right) + pH,$$

$$pK_a = \log\left(\frac{[BH^+]}{[B]}\right) + \log\left(\frac{\gamma_{BH^+}}{\gamma_B}\right) + pH,$$

and

$$pK_R = \log\left(\frac{[R^+]}{[ROH]}\right) + \log\left(\frac{\gamma_{R^+}}{\gamma_{ROH}}\right) + pH + \log a_{H_2O},$$

We can now define three *acidity functions*

$$H_- \equiv \log\left(\frac{\gamma_{HA}}{\gamma_{A^-}}\right) + pH$$
$$\text{so that} \quad pK_a = \log\left(\frac{[HA]}{[A^-]}\right) + H_-, \qquad (6.4)$$

$$H_0 \equiv \log\left(\frac{\gamma_{BH^+}}{\gamma_B}\right) + pH$$
$$\text{so that} \quad pK_a = \log\left(\frac{[BH^+]}{[B]}\right) + H_0, \qquad (6.5)$$

and

$$\begin{aligned} H_R &\equiv \log\left(\frac{\gamma_{R+}}{\gamma_{ROH}}\right) + pH + \log a_{H_2O} \\ \text{so that} \quad pK_R &= \log\left(\frac{[R^+]}{[ROH]}\right) + H_R. \end{aligned} \tag{6.6}$$

In dilute aqueous solution, all three of the H's are equal to pH. In any solution, for an acid with a given pK, the value of the H function determines the concentration ratio of acid and conjugate base. If the activity coefficient ratios were dependent only on charge type, the functions would apply to all acid-base equilibria of the same type, and we would have $H_R = H_0 + \log a_{H_2O}$. We might as well recognize right away that these hopes on which the acidity functions were founded are not realized.

The first acidity function to be established experimentally was an H_0 function based on a series of substituted anilinium ions [8]. Both m-nitroanilinium ($pK_a = 2.45$) and p-nitroanilinium ($pK_a = 1.00$) ions have pK_a's that are measurable in reasonably dilute aqueous solution. Anilinium ions having another electron-withdrawing group in addition to a nitro group are too acidic to be studied in water, but are protonated in mixtures of sulfuric acid and water. The pK_a of p-nitroanilinium ion is low enough that we can begin to move outside the range of dilute aqueous solutions. For example, in 5% H_2SO_4, the concentration ratio, $[BH^+]/[B]$, is found to be 10.0. With the known $pK_a = 1.00$, substitution into Eq. (6.5) allows us to evaluate $H_0 = 0.0$ for the 5% H_2SO_4-water solvent. If we can now find an anilinium ion whose concentration ratio can be measured in this 5% solvent, we can evaluate its pK_a. Then, we can use this anilinium ion to evaluate H_0 for a sulfuric acid-water mixture containing more sulfuric acid. The process can be continued in this manner as long as we can continue to find more acidic anilinium ions whose concentration ratios can be measured in a solvent mixture whose H_0 value has been determined by use of the slightly less acidic anilinium ion. This is called the *overlapping indicator method.* Some actual experimental data are given in a problem at the end of this chapter.

For any two of the overlapping indicators, B_1 and B_2, the assumption that the activity coefficient ratios are the same can be checked over the range where both concentration ratios can be measured. From the above equations:

$$pK_a^1 - pK_a^2 = \log\left(\frac{[BH_1^+]}{[B_1]}\right) - \log\left(\frac{[BH_2^+]}{[B_2]}\right) + \log\left(\frac{\gamma_{BH_1^+}}{\gamma_{B_1}}\right) - \log\left(\frac{\gamma_{BH_2^+}}{\gamma_{B_2}}\right),$$

and, if the activity coefficient ratios are equal,

$$pK_a^1 - pK_a^2 = \log\left(\frac{[BH_1^+]}{[B_1]}\right) - \log\left(\frac{[BH_2^+]}{[B_2]}\right).$$

Since the left-hand-side of this equation is a constant, the right-hand-side must also be a constant for all solutions in which both concentration ratios can be measured. Adherance to this equation over the observable range, however, does not guarantee that the activity coefficient ratios remain equal over the non-observable ranges.

Completely analogous studies, utilizing substituted triarylmethyl cations, have established the H_R acidity functions for aqueous acid mixtures [9]. Acidity functions have been evaluated for mixtures of water with nearly all strong acids and ranging up to, or nearly up to, the pure concentrated acids [10]. As was suspected before, the relationship, $H_R = H_0 + \log a_{H_2O}$, which would result from activity coefficient ratios depending only on charge type, does not hold; the two quantities differ by more than four units in solutions containing more than approximately 60% sulfuric acid.

Some years after the initial determinations of the H_0 and H_R scales, the overlapping indicator technique, but this time utilizing only primary anilinium ions, was applied to a re-determination of H_0 for sulfuric acid-water mixtures [11]. Appreciable discrepancies were found with the original study which had utilized primary, secondary, and tertiary anilinium ions without distinction. This was the first demonstration of just how stringent the limitations are on the range of structure allowable for a single acidity function. Appreciably different acidity functions apply to primary anilinium ions (H_0') and tertiary anilinium ions (H_0'''), and even more different functions apply to the protonation of azulenium ions (H_R') or to the protonation of carboxamides. Some of the functions for sulfuric acid-water mixtures are shown in Table 6.2.

The multitude of different acidity functions [12] calls into question the validity of any of them.

The H_- acidity functions [12] are usually defined for strong bases, and are established by the use of the overlapping indicator method in

TABLE 6.2
Acidity Functions in H_2O-H_2SO_4 Solutions [12]

Wt. % H_2SO_4	$-H_R$	$-H'_R$	$-H'''_0$	$-H'_0$	$-H_R + \log a_{H_2O}$
5	0.07	0.27	-0.02	-	0.10
10	0.72	0.51	0.53	0.31	0.70
20	1.92	1.50	1.47	1.01	1.86
30	3.22	2.55	2.44	1.72	3.10
40	4.80	3.85	3.46	2.41	4.55
50	6.60	5.53	4.54	3.38	6.15
60	8.92	7.34	5.91	4.46	8.13
70	11.52	9.40	7.65	5.80	10.18
80	14.12	-	9.44	7.34	11.84
90	16.72	-	11.14	8.92	13.23
95	18.08	-	11.89	-	-
98	19.64	-	-	10.41	-

a manner largely analogous to that applied to strong acids. Acidity functions have been established for aqueous sodium hydroxide, sodium hydroxide in water-alcohol mixtures, and for tetramethylammonium hydroxide in water-dimethyl sulfoxide mixtures by the use of carbon acids and by the use of amines ionizing to form amide ions.

A few of the pK_a's determined by the use of some of the acidity functions are shown in Table 6.3. In view of the uncertainties in the acidity functions, these pK's should be regarded with some suspicion. They are probably at least qualitatively correct in ranking various functional groups as to acidities or basicities.

Detailed examinations of individually determined activity coefficients [13] have now made it uncomfortably clear that acidity functions, at best, are of very limited validity[1] and must be applied with extreme care [14]. The major utility of the information from the study of acidity functions is in the evaluation of relative activity coefficients for a number of solutes and documentation of just how specific the variation of

[1] The word *fail* is not too strong. However, acidity functions were originated in 1932 [8], and were proliferated, applied, and taught for the next 50 years. Any concept that has gained so much vested interest will disappear very slowly.

TABLE 6.3
pK's Determined by Various Acidity Functions [12]

Acid	pK	Acid	pK
AzuleneH$^+$	-1.7	Cyanoform	-4.8
Tri-p-tolylmethyl cation	-3.6	Tri-m-tolylmethyl cation	-6.3
Triphenylmethyl cation	-6.6	9-Methyl, 9-fluorenyl cation	-16.6
2,4-Dinitroanilinium ion	-4.5	BenzamideH$^+$	1.7
PropionamideH$^+$	0.8	AcetamideH$^+$	0.6
IndoleH$^+$	-3.5	$C_2H_5OH_2^+$	~-3

activity coefficients is to precise structure of the solutes.

The entire concept of acidity functions was based on an underlying hope that relative acidities would be independent of solvent, and that the extension of the aqueous standard state to accomodate a greater range of pK values would result in, at worst, a few scales for different acid-base types. The large number of different acidity functions actually found means that such useful scales of acidities do not exist. Given that fact, the reasons for clinging to an aqueous standard state disappear. Although vestiges of the longing for a universal scale of acidities remain, most recent studies completely drop reference to the aqueous standard state. Virtually all of these more recent studies have concentrated on weak Brønsted acids and very few of the strong acids studied by the acidity function techniques have received further attention.

6.4 Acidities in Nonaqueous Solutions

Since all Brønsted acid-base reactions involve ions, studies of acidities will be simplest when carried out in solutions that allow independent ions to exist. As a very rough rule, solutions free of ion-pairing or other ionic aggregates at ionic concentrations in the millimolar range require solvents with dielectric constants greater than 30. Methanol is the only one of the common alcohols having a dielectric constant this large, but it is not much better than water with regard to the range of pK's that can be measured since the ion-product, $K_s = [H^+][CH_3O^-] \simeq 10^{-17}M^2$, is not much less than that for water [15]. The solvent that has received the most attention over the past 20 years is DMSO (dimethyl sulfoxide), which has a dielectric constant of about 48. The conjugate base of DMSO, the dimsyl ion, $CH_3SOCH_2^-$, can be prepared cleanly in solution, an operational pH scale can be established, and the ion product, $K_s = [H^+][CH_3SOCH_2^-] \simeq 10^{-33}M^2$, allows a large range of pK's to be determined [16]. At the present writing, over 1000 pK's have been determined for acids in DMSO solution [17]. A few of these are shown in Table 6.4.

Perhaps the most important point illustrated by the data is that large changes, even in relative acidities, occur on changing solvent from water to DMSO. For example, the reaction:

$$C_6H_5COOH + (C_2H_5)_3N \xrightleftharpoons{K_{ex}} C_6H_5COO^- + (C_2H_5)_3NH^+$$

is complete to the right in aqueous solution, with $K_{ex} = 10^{+6.5}$, while in DMSO, it is nearly complete to the left, with $K_{ex} = 10^{-2.1}$.

The generalizations regarding activity coefficients that were discussed in Section 2.6 are applicable here. Hydrogen bonding from water to anions stabilize the anions, with the smaller, more charge-localized, anions receiving the most stabilization. This factor is primarily responsible for the increases in the pK_a's of neutral acids on going from water to DMSO. Alcohols, in particular, show large effects; trifluoroethanol has a pK of 12.4 in water and of 23.5 in DMSO. Carboxylic acids and phenols, with some delocalization of charge in the anionic conjugate bases, are affected to a slightly lesser extent; phenol having a change of 9 units, p-nitrophenol a change of 3.7 units, and benzoic acid a change of 6.9 units.

Hydrogen bonding from the cationic acids to solvent, and solvation of the proton, are somewhat stronger in DMSO than in water. The better solvation of the proton in DMSO decreases pK's, while the better hydrogen bond acceptance from the cationic acids increases pK's relative to aqueous solution. These two factors nearly cancel for primary ammonium ions, and we see only small changes in pK on going from water to DMSO. For the tertiary ammonium ions, with only one hydrogen bond to donate, the solvation of the proton is the more important factor, and the pK's decrease on going from water to DMSO.

It is particularly enlightening to note that the pK differences of even closely related acids are not the same in DMSO and water. For example, in water the pK difference between benzoic acid and p-nitrobenzoic acid is 0.78 units, while in DMSO, the difference is 2.0 units. This is just another way of saying that the Hammett ρ value for ionization of benzoic acids is solvent dependent, as already noted in Section 3.3. Such changes in relative acidities mean that simple H acidity functions of the sort discussed in the last section cannot be valid for DMSO-water mixtures. This type of behavior for carboxylic acids had, in fact, been noted much

TABLE 6.4
pK's in DMSO Solution at 25° [17]

Acid	pK$_a$	*Acid*	pK$_a$
CF_3SO_3H	0.3	CH_3SO_3H	1.6
$C_6H_5N(CH_3)_2H^+$	2.5	CF_3COOH	3.5
$C_6H_5CH(CN)_2$	4.2	2,4-Dinitrophenol	5.1
$F_2CHCOOH$	6.4	$CF_3CH_2NH_3^+$	5.8
$(CH_3CO)_3CH$	8.6	$HONO$	7.5
9-Cyanofluorene	8.3	Morpholinium ion	8.7
$EtO_2CCH_2NH_3^+$	8.7	p-$NO_2C_6H_4COOH$	9.0
$(C_2H_5)_3NH^+$	9.0	C_6H_5SH	10.3
Fluoradene	10.5	9-CH_3O_2C-Fluorene	10.3
p-Nitrophenol	10.3	C_6H_5COOH	11.0
Piperidinium ion	10.6	*n*-Butylammonium ion	11.1
$CH_2(CN)_2$	11.1	CH_3COOH	12.3
$HO_2CCH_2NH_2$	12.8	$(CF_3)_3COH$	10.7
$C_6H_5CH_2NO_2$	12.3	$(CH_3CO)_2CH_2$	13.3
n-Butanethiol	15.4	$(CH_3)_2CHNO_2$	16.9
CH_3NO_2	17.2	Cyclopentadiene	18.0
Phenol	18.0	4-Cl, 2-Nitroaniline	18.9
p-Nitrotoluene	20.4	$C_6H_5CH_2CN$	21.9
Fluorene	22.6	CF_3CH_2OH	23.5
$CH_3COC_2H_5$	24.4	$C_6H_5COCH_3$	24.7
Acetone	26.5	Cyclohexanone	26.4
$C_6H_5C\equiv CH$	28.7	CH_3OH	29.0
$C_6H_5NH_2$	30.6	$(C_6H_5)_3CH$	30.6
CH_3CN	31.3	$(CH_3)_3COH$	32.2
$(C_6H_5)_2CH_2$	32.2	Xanthene	30.0

earlier in studies of water-alcohol mixtures and had led to a suggestion for a more complicated type of acidity function [18] which has not been much used (See problem 6.4 at the end of this chapter.).

In solvents having dielectric constants much lower than 30, ion-pairing and other ion aggregation become impossible to avoid at reasonable concentrations. The importance of low dielectric constant solvents for synthetic work makes studies of carbanion behaviors of considerable interest in spite of, or even because of, the complications. Primary attention has been given to the ether solvents tetrahydrofuran [19] and dimethoxyethane [20] which have dielectric constants of 7.6 and 7.2, respectively. In these solvents, delocalized carbanions exist almost exclusively as ion pairs. Relative **ion-pair acidities** are determined by equilibrating the hydrocarbon of interest with the lithium or cesium salt of a reference indicator. Fluorene has been used as the primary reference indicator, and ion-pair pK's assigned to other hydrocarbons relative to an assumed value of 22.6 for fluorene (This is the pK_a for fluorene in DMSO solution.). Ion-pair dissociation constants for some of the ion-pairs have been determined by conductivity studies, and these allow the calculation of free-ion relative pK_a's of the hydrocarbons in the ether solvents. For the highly delocalized carbanions studied, the relative free-ion pK_a's in the ether solvents are essentially identical with the relative values in DMSO solution.

It has been known for some time [21] that ion pairs in ether solvents can exist in two different forms, contact and solvent-separated, depending on the specific solvent, ion-pair, and even temperature. In tetrahydrofuran solutions near room temperature, the lithium salts of delocalized carbanions exist predominantly as solvent-separated ion-pairs, while the cesium salts exist as contact ion-pairs. This has the interesting effect of making the lithium ion-pair dissociation constants larger (by factors of 10^2–10^3) and less sensitive to the structure of the carbanion than the cesium ion-pair dissociation constants [19]. For this reason, the relative lithium ion-pair acidities are closer to the relative free-ion acidities than are the cesium ion-pair acidities.

Anions other than highly delocalized carbanions frequently form ionic aggregates in low dielectric constant solvents. For example, it has been estimated that the lithium enolate of acetophenone in tetrahydrofuran solution has an average aggregation of 3.5 ion-pairs per aggregate [22]. The equilibration of the conjugate acids of these anions with indicator carbanion salts leads to ***apparent pK_a's***. These apparent values are concentration dependent; for example, the apparent pK_a of acetophenone in tetrahydrofuran with cesium counterion varies from 24.4 at $2 \times 10^{-5} M$ to 22.8 at $5 \times 10^{-3} M$ of the enolate. One should obviously be extremely careful in using these ***apparent pK_a*** values.

6.5 Intrinsic Acidities

As data accumulated concerning the huge changes in relative acidities that are caused by change of solvent, it became more and more of interest to determine how much of the observed acidity order in any solvent is caused by the solvent and how much is due to the structure of the acid and conjugate base. Fortunately, negative ion mass spectrometry, particularly ion-cyclotron resonance techniques, became available at about the same time. This allowed the determinations of acidities at very low pressures in the gas phase where the acids and conjugate bases are completely free of intermolecular interactions.

It must be realized that the formation of ions from neutrals in the gas phase is an extremely unfavorable process. The ionization of hydrochloric acid, for example, in the gas phase has ΔG° of +328 kcal/mole; or, stated differently, the pK_a of HCl is approximately 240 in the gas phase. Nevertheless, measurements of relative acidities combined with indirect measurements of a few absolute values have produced a large body of data.

Some of the data for gas phase acidities of organic molecules are shown in Table 6.5.

It is sometimes convenient to think of these ionizations as occurring in two steps; the homolytic bond dissociation into a hydrogen atom and a radical and then the electron transfer from the hydrogen atom to the radical to form the anion and the proton. Since the removal of the electron from the hydrogen atom is common to all acid ionizations, the relative acidities of various HX are dependent on the HX bond dissociation energy and the electron affinity of X· or, equivalently, the ionization potential of X^-. As we go across any row of the periodic table, from CH_4 to HF for example, bond dissociation energies increase and so do the electron affinities of the radicals; the electron affinities increase more than do the bond energies, however, and acidities increase. The ΔG° values are 408, 396, 384, and 366 kcal/mole for ionizations of CH_4, NH_3, OH_2, and FH, respectively [23]. As we go up any column of the periodic table, again both bond energies to hydrogen and electron affinities of the radicals increase; in this case, the bond energies have the largest changes, and the acidities decrease. The ΔG° values are 318, 328, and 366 kcal/mole for HBr, HCl, and HF, respectively.

There are a number of surprises in relative acidities in the gas phase for those used to thinking about solution behaviors. Many of these surprises are due to the stabilization of charges by any polarizable matter in the vicinity of the charge. In solution, solvent molecules are always nearby, but in the gas phase, nothing is nearby but parts of the charged molecule. Thus, small non-complex ions are much less stable in the gas phase than in solution. Water, for example, is a much weaker acid ($\Delta G^\circ = 384$ *kcal/mole*) than toluene ($\Delta G^\circ = 374$ *kcal/mole*) in the gas phase. The addition of any matter around the charge gives large effects; the ΔG° values for ionizations of water, methanol, and ethanol

TABLE 6.5
Free Energies of Acid Ionizations in the Gas Phase at 25° [23] Kcal/mole

Acid	ΔG^o	*Acid*	ΔG^o
NH_4^+	198	$C_6H_5NH_3^+$	205
CF_3COOH	316	CH_3COOH	341
$C_6H_5CH(CN)_2$	315	*p*-Nitrophenol	321
$(CH_3CO)_2CH_2$	337	$(CF_3)_3COH$	324
p-$NO_2C_6H_4COOH$	321	Fluoradene	325
C_6H_5SH	334	Fluorene	344
C_6H_5COOH	333	Triphenylmethane	353
Piperidinium ion	222	*n*-Propylammonium ion	214
$CH_2(CN)_2$	329	Diphenylmethane	358
n-Butanethiol	347	CH_3NO_2	350
Cyclopentadiene	348	Acetophenone	354
Phenol	342	CH_3CN	365
p-Nitrotoluene	345	$C_6H_5CH_2CN$	346
CF_3CH_2OH	354	$C_6H_5C\equiv CH$	363
Acetone	362	9-Phenylfluorene	336
$CH_3SO_2CH_3$	358	CH_3SOCH_3	366
H_2S	345	HCN	344
C_2H_5OH	371	H_2O	384

are 384, 374, and 371 kcal/mole, respectively. This factor operates in stabilizing positive charges as well as negative ones, as seen in the ΔG^o's for acid dissociations of NH_4^+, $CH_3NH_3^+$, $(CH_3)_2NH_2^+$, and $(CH_3)_3NH^+$, which are 198, 210, 217, and 221 kcal/mole, respectively [24]. The magnitude of the polarizability effect in the gas phase is enough to counteract many of the resonance and polar effects seen in solutions; aniline is a much weaker base in solution than is ammonia, both because of the resonance stabilization from conjugation of the nitrogen lone pair of aniline with the benzene ring and the electron withdrawing effect of the benzene ring. In the gas phase, however, aniline is a stronger base than ammonia. Less naïve comparisons of compounds having similar masses in proximity to the charges reveal that the resonance and polar effects still operate in the gas phase. If we compare aniline with cyclohexylamine, aniline is the weaker base both in solution and in the gas phase.

The differences between acidities in the gas phase and those in solution are directly related to the free energies of transfer of the acids and conjugate bases from the gas phase into solution. Since the free energies of transfer of neutral species from gas to solution do not vary nearly as much as those of ionic species, comparisons of gas phase acidities with those in DMSO solution give a reasonable indication of the relative amounts of stabilization provided to the charged species by the DMSO solvent. As expected, small charge-localized ions are most stabilized by solvation. Even though the stabilization of these small ions by DMSO is less than that provided by hydrogen-bond donating solvents, as discussed above, it is still quite appreciable. Most strikingly, the acidities of the acids having small anionic conjugate bases relative to those of hydrocarbons, such as 9-phenylfluorene, having large delocalized anionic conjugate bases are greatly increased on going to DMSO from the gas phase. Water, relative to 9-phenylfluorene, shows an increased acidity amounting to ca. 30 kcal/mole in DMSO [23]. On the other hand, the acidities of the hydrocarbons having large delocalized conjugate bases, relative to one another, are essentially the same in the gas phase as in solution, indicating that these species all have nearly the same solvation energies.

Quantitative evaluations of free energies of transfer of ions from the gas phase to solution can be made from gas phase and solution acidities combined with solubility or vapor pressure measurements on the neutral species involved in the acid-base reactions [25]. The pertinent thermodynamic cycles are:

$$
\begin{array}{ccccc}
(HA)_g & \overset{-RT\ln K_a^g}{\rightleftharpoons} & (H^+)_g & + & (A^-)_g \\
\Delta G^\circ_{t(HA)}\downarrow\uparrow & & \Delta G^\circ_{t(H+)}\downarrow\uparrow & & \Delta G^\circ_{t(A-)}\downarrow\uparrow \\
(HA)_s & \overset{-RT\ln K_a^s}{\rightleftharpoons} & (H^+)_s & + & (A^-)_s
\end{array}
$$

for neutral acids, and

$$
\begin{array}{ccccc}
(BH^+)_g & \overset{-RT\ln K_a^g}{\rightleftharpoons} & (H^+)_g & + & (B)_g \\
\Delta G^\circ_{t(BH+)}\downarrow\uparrow & & \Delta G^\circ_{t(H+)}\downarrow\uparrow & & \Delta G^\circ_{t(B)}\downarrow\uparrow \\
(BH^+)_s & \overset{-RT\ln K_a^s}{\rightleftharpoons} & (H^+)_s & + & (B)_s
\end{array}
$$

for cationic acids, which lead to the equations:

$$\Delta G^{\circ}_{t(A^-)} = 2.303RT(pK_a^s - pK_a^g) + \Delta G^{\circ}_{t(HA)} - \Delta G^{\circ}_{t(H^+)}$$

and

$$\Delta G^{\circ}_{t(BH^+)} = 2.303RT(pK_a^g - pK_a^s) + \Delta G^{\circ}_{t(B)} + \Delta G^{\circ}_{t(H^+)}.$$

The free energy of transfer of the proton from the gas phase to water has been evaluated as -259.5 kcal/mole, and the use of this value along with measurements or estimates of the other quantities on the right-hand-sides of these equations gives the data shown in Table 6.6.

It is also possible, by use of similar thermodynamic cycles, to use bond dissociation energies obtained in the gas phase and solution acidity measurements, along with estimates of free energies of solutions of radicals, to obtain estimates of solution oxidation potentials of various Brønsted bases [26]. Some of these data are also reported in Table 6.6.

The major point of the data is that solvation energies of both cations and anions are very large, on the same order as bond energies, and differ appreciably even for such seemingly similar species as methoxide and trifluoroethoxide ions.

6.6 Comparisons of Lewis and Brønsted Acidities and Basicities

The Lewis acid-base concept of an acid as any species which can form a bond with a base, which is a species having a lone-pair of electrons, is far more generally applicable to chemical reactions than is the Brønsted concept. Unfortunately, the relative acidities of acids, or the relative basicities of bases, are dependent on the reaction partners, and any quantitative consideration of Lewis acidities or basicities quickly becomes complicated by all of the factors involved in chemical bonding. There are some interesting and useful relationships between various measures of acidities and basicities, and there are some glaring exceptions to such relationships [27], [28].

Any comparison of Brønsted and Lewis acids and bases should recognize at the outset that the real comparison involves an exchange reaction of the sort that we considered several times in Chapter 3. We can arrive at the pertinent exchange reaction by starting with the comparison of Lewis and Brønsted basicity of some particular base, B. Writing R^+ as the particular Lewis acid, we have the two reactions

$$RB \overset{K_R^B}{\rightleftharpoons} R^+ + B^- \quad \text{and} \quad HB \overset{K_H^B}{\rightleftharpoons} H^+ + B^-.$$

TABLE 6.6
Solvation Energies and Oxidation Potentials in Aqueous Solution, Standard states are the ideal gas at 1 atm. and the ideal solution at 1 M, both at 25°*C*

Ion	$-\Delta G^{o}_{soln}$ [26b][(b)]	$-\Delta G^{o}_{soln}$ [26a][(b)]	$\Delta G^{o(a)}_{ox}$ [26b][(b)]	$\Delta G^{o(a)}_{ox}$ [26a][(b)]
HOO^-	99	95	19.1	20.5
$C_6H_5O^-$	70	74	21.2	20.7
NO_2^-	70		20.5	21.5
n-$C_3H_7S^-$	74	78	17.1	21.7
$C_6H_5S^-$	65	68	17.5	22.3
CH_3O^-	93	93	25.4	27.3
$CF_3CH_2O^-$	81	84	29.5	32.9
N_3^-	72	75	29.3	34.0
CH_3COO^-	75	67	47.0	47.5
CN^-	75	74	58.1	59.1
OH^-	104	105	39.7	45.7
$NH_2NH_3^+$	80	72	3.6[c]	16.1
PiperidineH$^+$	55	56	21.3[c]	26.7
n-$C_3H_7NH_3^+$	61	63	30.0	36.5
H_3O^+	102	103	79.3	89.3

(a) For the reaction: $H^+ + X^- \xrightarrow{\Delta G^{o}_{ox}} \frac{1}{2}H_2 + X\cdot$ for anions, and for: $H^+ + B \xrightarrow{\Delta G^{o}_{ox}} \frac{1}{2}H_2 + B\cdot^+$ for BH^+ cations.

(b) Values from the two references differ because of slightly different assumptions in estimates of some of the quantities. The discrepancies are probably reasonable estimates of the uncertainties in values.

(c) The values reported in [26b] were based on gas phase ionization potentials which are quite different from the latest available data which were used in [26a]. The present values are based on assumptions given in [26b], but use the ionization potentials selected in [26a].

The ratio of the two equilibrium constants, K_R^B/K_H^B, is the equilibrium constant for the simple exchange reaction:

$$RB \;+\; H^+ \;\overset{K_{ex}^B}{\rightleftharpoons}\; R^+ \;+\; HB.$$

Obviously, the value of K_{ex}^{B} depends, among other things, on the relative stabilities of H^{+} and R^{+}. We could, in fact, write the standard free energy for this exchange as the differences in the standard free energies for bond dissociation of RB and HB into radicals minus the difference in oxidation potentials of $R\cdot$ and $H\cdot$ in the medium of interest. Since the difference in oxidation potentials will contribute the same amount for any base, the variations in Lewis as compared to Brønsted basicities for any two bases are dependent only on the relative bond strengths of the two bases with H and R. This may be made more apparent if we consider a second base, S, which we will take as some standard for comparison, in the simple exchange:

$$RS \quad + \quad H^{+} \quad \overset{K_{ex}^{S}}{\rightleftharpoons} \quad R^{+} \quad + \quad HS.$$

We can combine these two simple exchange reactions into the double exchange reaction:

$$RB \quad + \quad HS \quad \overset{K_{HB}^{RB}}{\rightleftharpoons} \quad RS \quad + \quad HB, \tag{6.7}$$

which is of exactly the form used in Chapter 3. Quite obviously, the standard free energy change for this exchange, $\Delta G_{ex}^{\circ} = -RT \ln K_{HB}^{RB}$, is the sum of the free energies of bond dissociations of RS and HB minus the sum of those for HS and RB. The thermodynamic quantities associated with this double exchange reaction are those pertinent to the discussion of the Lewis, as compared with the Brønsted, basicity of any base relative to the standard base toward the Lewis acid R^{+}. If Lewis basicities toward a given Lewis acid were proportional to Brønsted basicities, with K_{R}^{B}/K_{H}^{B} having a constant value, all K_{HB}^{RB} would be unity. Thus, values of K_{HB}^{RB} can be taken as direct measures of the deviations from such proportionality. Values larger than unity mean that B, relative to S, prefers to be bound to H in the double exchange reaction, and values smaller than unity mean that B, relative to S prefers bonding to R in the double exchange reaction of eq. (6.7).

The simplest organic Lewis acid is the methyl cation. A great deal of effort has gone into the determination of Lewis basicities toward the methyl cation, and the comparison of these with Brønsted basicities in the gas phase to avoid any complication from solvent effects. Since entropy changes for the Lewis and Brønsted reactions in the gas phase should be essentially identical, standard enthalpies of the reactions will show the same relationships as the standard free energies. A plot of *proton affinities* against *methyl cation affinities*, which are the standard

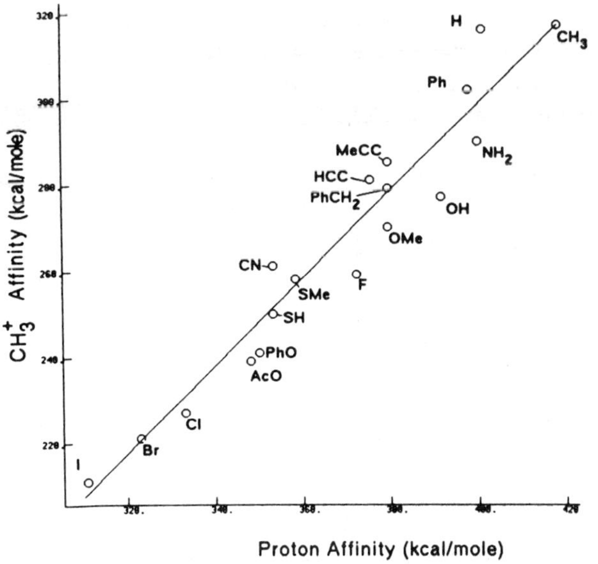

Figure 6.1: *Methyl cation affinities plotted against proton affinities for various bases. Data are taken from [28]. The line is of unit slope.*

enthalpies for combinations of the bases with proton and methyl cation, respectively, is shown in Fig. 6.1.

Although there is an obvious rough relationship between proton and methyl cation affinities, some of the deviations lead to huge values of $K_{HB}^{CH_3B}$. For F^- and H^- as bases, the equilibrium constant for the double exchange reaction, eq. (6.7), is greater than 10^{18}, favoring CH_3H plus HF. Cyanide ion similarly gives an equilibrium in which CH_3CN plus HF is favored over CH_3F plus HCN by about 10^{14}. In general, those bases having an electronegative first-row atom as the basic site most prefer the proton, while those with less electronegative atoms most prefer the methyl cation.

An interesting correlation has been reported between the heats of formation of various CH_3B and HB compounds [28] with

$$\Delta H_f^\circ(CH_3B) = 0.857\Delta H_f^\circ(HB) - 2.35\ kcal/mole. \qquad (6.8)$$

The deviations of K_{HB}^{RB} from unity were interpreted as arising primarily from the non-unit slope of this correlation. A deeper look, however, shows a fundamental error in reasoning. We can write the heat of formation of CH_3B as the heat of formation of $CH_3\cdot$ plus that of $B\cdot$ minus the heat of bond dissociation of CH_3-B, and, similarly for HB, as the heats of formation of H· and B· minus the heat of bond dissociation of H-B. If these are substituted into equation (6.8), the heat of formation of B· appears on the left hand side, but 0.857 times the heat of formation of B· appears on the right hand side. The fallacy in reasoning becomes apparent when it is realized that the heat of formation of B· is based on arbitrary definitions of standard states of the elements comprising B. Any change in the definitions of these standard states will lead to different changes in heats of formation for different B's. If the correlation involved a unit slope, the changes would cancel on the two sides of the equation, but, with non-unit slope the entire correlation is changed by the change in standard state. This is an excellent example of just how easy it is to fool oneself with statistical fits of data.

It is also interesting to note that values of K_{HB}^{RB} have been reported for the cases where R^+ is a ruthenium or a platinum cation, with ligands, in tetrahydrofuran or benzene solution [29]. Values close to unity, using hydroxide ion as the standard base, S, in the double exchange reaction, eq. (6.7), are found for phenylacetylide ion, anilide ion, and acetone enolate. Values significantly greater than unity are found for cyanide ion and attributed to back-donation of electrons. Since the situation is the same as for methyl cation where such back-bonding is not expected, the rationalization seems unlikely to be correct. Similarly, thiolate ions, relative to hydroxide ion, prefer bonding to both the metal cations and to methyl cation over bonding to the proton.

For reactions in aqueous solution, values of K_{HB}^{RB} for a fair variety of common bases, relative to hydroxide ion as the standard base, and R^+ equal either Pyronin-Y cation or Crystal Violet cation are shown in Table 6.7 [30].

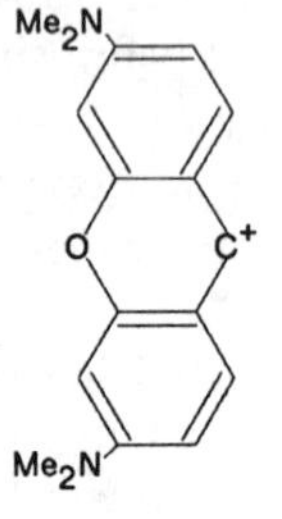

Pyronin-Y

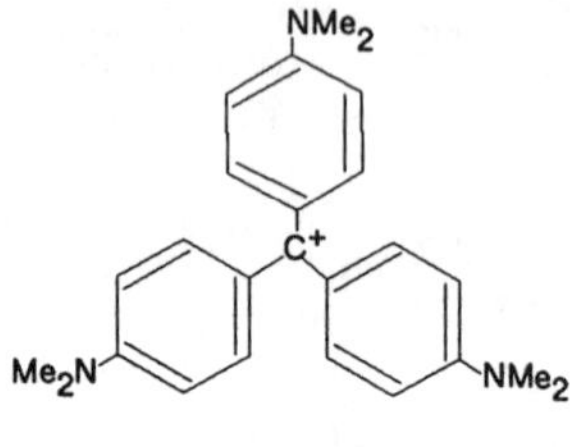

Crystal Violet

Most obviously, nearly all values of K are larger than unity, showing that nearly all bases have a greater preference than does hydroxide for bonding to carbon than to hydrogen just as seen for the methyl cation. Also, all of the equilibrium constants are greater for Pyronin-Y than for Crystal Violet. This is most likely a result of the smaller size of hydroxide than any other base shown, and the greater crowding around the triarylmethyl carbon than the xanthyl carbon. Other values shown are more difficult to rationalize, and a number of effects are probably operative.

TABLE 6.7
Values of $\log K_{HB}^{RB}$ for Eq. (6.7) in Aqueous Solution at 25°C. [30].

B	pK_a	$\log K_{HB}^{RB}$ Pyr-Y	C.V.
$CF_3CH_2O^-$	12.4	-0.3	
CO_3^{-2}	10.3	0.2	
HOO^-	11.6	4.1	
n-PrS^-	10.8	7.6	
$HOCH_2CH_2S^-$	9.5	7.2	4.6
$CH_3O_2CCH_2S^-$	8.0	7.6	4.9
$C_6H_5S^-$	6.5	7.7	
SO_3^{-2}	7.2	12.7	7.8
N_3^-	4.0		3.4
n-$BuNH_2$	10.5	3.0	0.2
$CH_3OCH_2CH_2NH_2$	9.7	2.4	
$CF_3CH_2NH_2$	5.6	1.9	-0.1
H_2NNH_2	8.1	4.7	1.2
CH_3ONH_2	4.6	4.0	3.0
Piperidine	11.4	3.2	
Morpholine	8.5	2.8	

The quantitative, or even semi-quantitative, understanding of Lewis acidities and basicities remains one of the major elusive goals of the physical organic chemist. The kinetics of Lewis acid-base reactions are no better understood than are the equilibria [30]. It seems likely that a better understanding of the entire concept of the chemical bond is a prerequisite to understanding these acid-base reactions.

6.7 REFERENCES

[1]. R. G. Bates, **Determination of pH; Theory and Practice**, Wiley and Sons, New York, **1964**.

[2]. C. D. Ritchie, D. J. Wright, D. Huang, and A. Kamego, **J. Amer. Chem. Soc., 97**, 1163 (**1975**).

[3]. R. A. Diffenbach, K. Sano, and R. W. Taft, **J. Amer. Chem. Soc., 88**, 3168 (**1966**).

[4]. C. D. Ritchie and H. Fleischhauer, **J. Amer. Chem. Soc., 94**, 3481 (**1972**).

[5]. R. Breslow, H. Höver, and H. W. Chang, **J. Amer. Chem. Soc., 84**, 3168 (**1962**).

[6]. R. G. Pearson and R. L. Dillon, **J. Amer. Chem. Soc., 75**, 2439 (**1953**).

[7]. R. P. Bell, **The Proton in Chemistry**, Cornell Univ. Press, Ithaca, N. Y., **1959**.

[8]. L. P. Hammett and A. J. Deyrup, **J. Amer. Chem. Soc., 54**, 2721 (**1932**).

[9]. N. C. Deno, J. J. Jaruzelski, and A. Schriesheim, **J. Amer. Chem. Soc., 77**, 3044 **1955**.

[10]. R. A. Cox and K. Yates, **Can. J. Chem., 61**, 2225 (**1983**).

[11]. M. J. Jorgenson and D. R. Hartter, **J. Amer. Chem. Soc., 85**, 878 **1963**.

[12]. R. H. Boyd in **Solute-Solvent Interactions**, Vol. 1, Ed. J. F. Coetzee and C. D. Ritchie, Dekker, New York, **1969**, Chap. 3.

[13]. K. Yates and R. A. McClelland, **Prog. Phys. Org. Chem., 11**, 323 (**1974**).

[14]. R. A. Cox and K. Yates, **Accts. Chem. Res., 20**, 27 (**1987**).

[15]. R. G. Bates in **Solute-Solvent Interactions**, Vol. 1, *loc. cit*, Chap. 2.

[16]. C. D. Ritchie in **Solute-Solvent Interactions**, Vol. 1, *loc. cit.*, Chap. 4.

[17]. F. G. Bordwell, **Accts. Chem. Res., 21**, 456 (**1988**).

[18]. B. Gutbezahl and E. Grunwald, **J. Amer. Chem. Soc., 75**, 565 (**1953**), and earlier references cited there.

[19]. M. J. Kaufman, S. Gronert, and A. Streitwieser, **J. Amer. Chem. Soc., 110**, 2829 (**1988**), and earlier references cited there.

[20]. A. A. Solov'yanov, I. P. Beletskaya, and O. A. Reutov, **Zh. Org. Khim., 19**, 1822 (Engl. Transl. pg. 1593) (**1983**), and earlier references cited there.

[21]. J. Smid, **Ions and Ion-Pairs in Organic Reactions**, Vol. 1, Ed. M. Szwarc, Wiley Interscience, New York, **1972**, Chap. 3.

[22]. M. J. Kaufman and A. Streitwieser, **J. Amer. Chem. Soc., 109**, 6092 (**1987**).

[23]. R. W. Taft and F. G. Bordwell, **Accts. Chem. Res., 21**, 463 (**1988**).

[24]. C. D. Ritchie in **Solute-Solvent Interactions**, Vol. 2, Ed. J. F. Coetzee and C. D. Ritchie, Dekker, New York, **1976**, Chap. 12.

[25]. E. M. Arnett, **Accts. Chem. Res., 12**, 404 (**1973**).

[26]. (a) C. D. Ritchie, **J. Amer. Chem. Soc., 105**, 7313 (**1983**); C. D. Ritchie in **Nucleophilicity**, (Proc. of ACS Symposium held Sept. 1985), Ed. M. Harris and S. McManus, ACS Advances in Chemistry Series, **1987**, pg. 169ff. (b) R. G. Pearson, **J. Amer. Chem. Soc., 108**, 6109 (**1986**).

[27]. J. Hine and R. D. Weimer, **J. Amer. Chem. Soc., 87**, 3387 **1965**).

[28]. J. Brauman and C-C. Han, **J. Amer. Chem. Soc., 110**, 5611 (**1988**); **111**, 3485 (**1989**).

[29]. H. E. Bryndza, L. K. Fong, R. O. Paciello, W. Tam, and J. E. Bercaw, **J. Amer. Chem. Soc., 109**, 1444 (**1987**).

[30]. C. D. Ritchie, **Can. J. Chem., 64**, 2239 (**1986**).

6.8 PROBLEMS

Problem 6.1

A chemistry student decided to try to convert triphenylmethanol to triphenylmethyl cyanide by reacting an aqueous solution of cyanide ion with triphenylmethanol. He hoped that there would be enough of the triphenylmethyl cation in equilibrium with the alcohol that it would react with the cyanide ion. In order to get cyanide ion, rather than HCN, he decided to use a solution with a pH of 10. He used $10^{-2}M$ cyanide ion and $10^{-4}M$ triphenylmethanol. Assuming that the triphenylmethyl cation reacts much more rapidly with hydroxide ion and water than with cyanide ion, that the pK_R of triphenylmethyl cation is -6.6, that the formation of the triphenylmethyl cyanide is essentially irreversible, and that the second-order rate constant for reaction of the cation with cyanide ion is $10^8 M^{-1} sec^{-1}$, what will be the half-life for the conversion?
Hint: Simply set up the steady-state expression for the reaction, and plug in values to obtain the pseudo-first-order rate constant. You probably don't want to wait around for the reaction half-life.

Problem 6.2

In the last problem, did the student help anything by choosing the pH of 10? Would it be possible to make the reaction a reasonable one by using a lower pH? What would be the pseudo-first-order rate constant for the reaction at pH of 2 with the total amount of cyanide, HCN plus CN^-, still $10^{-2}M$? Assume that HCN is in rapid equilibrium with cyanide ion.
Hint: Notice the product $[H^+][CN^-]$ in the steady-state expression; substitute something for it.

Problem 6.3

In aqueous solution, amino acids exist almost exclusively in the zwitterionic form, $^{-}O_2C\text{-R-}NH_3^+$. What form do you expect for DMSO solution?

Problem 6.4

The changes in pK's with changes in solvent for aqueous alcohol solutions of carboxylic acids were found [18] to be correlated by the equation:

$$pK_a^s - pK_a^w = \log f_s + m_{HA}Y_-^s,$$

where pK_a^s is the pK in the particular solvent, pK_a^w is the pK in water, f_s and Y_-^s depend only on the solvent, and m_{HA} depends only on the structure of the acid. The Y_-^s value for pure ethanol is 1.00, and the m_{HA} value for benzoic acid is 1.57. From the fact that the Hammett ρ for the ionizations of benzoic acids in ethanol is 1.957, derive an expression relating the m_{HA} values for substituted benzoic acids to the Hammett σ's of the substituents.

Hint: Write the above equation for both benzoic acid and an X-substituted benzoic acid, then write the Hammett equations for the pK's in water and for the pK's in ethanol.

Problem 6.5

A solution initially containing $10^{-4}M$ triphenylmethane and 1.6 × $10^{-3}M$ potassium dimsyl in DMSO solution was prepared. The red color of the triphenylmethide ion formed rapidly. The absorbance of the solution at 495 nm, in a 0.5 cm path cell, was 0.95. Calculate the molar extinction coefficient of the triphenylmethide ion at 495 nm.

Hint: Use the pK value of 30.6 for triphenylmethane, and the value of $10^{-33}M^2$ for $[H^+][CH_3SOCH_2^-]$ in DMSO to calculate the concentration of the triphenylmethide anion.

Problem 6.6

Explain how you might be able to use acidity functions to obtain a value for the equilibrium constant of the reaction in water:

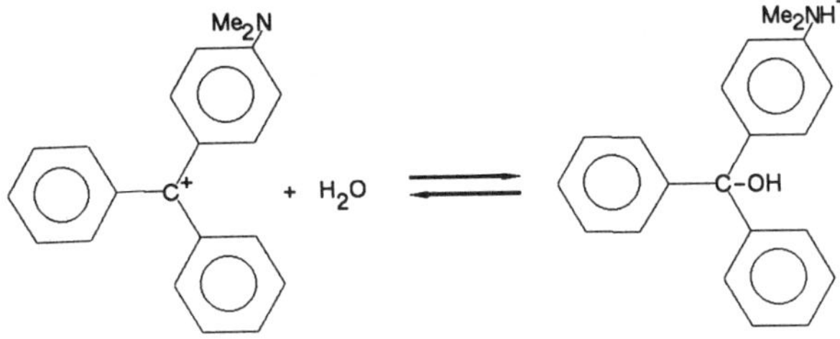

The problem here is that in dilute aqueous solution, the absorbance observed is for the equilibrium system; the true molar extinction coefficient for the carbocation cannot be determined in water. The pK_R for the carbocation shown is expected to be about 3 or 4. The pK_R for the carbocation with a protonated dimethylamino group should be less than -6.6.

Hint: Write some equilibria involving protonation of the dimethylamino group of the carbocation. Look at the differences in H_0 and H_R functions in Table 6.2. What effect would these differences have on the position of the equilibria?

Problem 6.7

The pK_a of 2,5-dichloro, 4-nitroanilinium ion is -1.78. The following table contains values of Log $[B]/[BH^+]$ found for various substituted anilinium ions in aqueous sulfuric acid solutions. Calculate the pK of each anilinium ion, and the H_0 function value for each of the solutions.

Log $[B]/[BH^+]$ Values

	Substituents on anilinium ion					
Wt. % H_2SO_4	2,5-Cl_2-4-NO_2	2-Cl-6-NO_2	2,6-Cl_2-4-NO_2	2,4-$(NO_2)_2$	2,6-$(NO_2)_2$	2,6-$(NO_2)_2$-4-Cl
19.38	0.80					
21.98	0.64					
29.35	0.13	0.74				
37.19	-0.44	0.17	1.10			
41.63	-0.77	-0.11	0.72			
43.95	-0.94	-0.31	0.49			
45.72		-0.47	0.36			
52.21		-0.98	-0.33	0.86		
57.38			-0.84	0.35		
59.77				0.10	1.21	
61.94				-0.15	0.88	
63.50				-0.50	0.61	1.21
68.65				-1.07	-0.05	0.58
74.82					-1.04	-0.49
77.50						-0.89

CHAPTER 7
Acid-Base Catalysis

7.1 Introduction

A very large number of Organic reactions are catalyzed by Brønsted acids and bases. The observation of catalysis, that the rates of reactions are increased by increasing concentrations of the acids or bases, means that the acids or bases are involved in stoichiometric formation of the transition states of the rate determining steps of the reactions. This is usually indicative of the involvement of a proton transfer step in the mechanism of the reaction. There are two broad classifications for these reactions. If the rate of a reaction increases with an increase in concentration of any acid, or base, present in solution, the reaction is classified as general acid, or base, catalyzed. If the rate of reaction increases only with an increase in the concentration of proton, or hydroxide ion, the reaction is classified as specific acid, or base, catalyzed. The distinction of these two broad types depends experimentally on the behavior of buffer systems, and we will begin this chapter by reviewing that behavior. Once we have seen how the distinction between general and specific catalysis can be accomplished, we will then discuss how acid or base structure influences the rates of the general acid-base catalyzed reactions. This will get us into the subject of rates of proton transfer reactions. Finally, the question of why some reactions involve catalysis, while other seemingly similar ones do not, will be discussed in terms of the properties of reaction intermediates.

7.2 Buffer Systems

A buffer system may be composed of any Brønsted acid-base pair in equilibrium:

$$BH \quad \overset{K_a}{\rightleftharpoons} \quad B \ + \ H^+, \tag{7.1}$$

where charges on BH and/or B are omitted for generality. The particular acid-base pair will have associated with it the equilibrium constant K_a°, defined by:

$$K_a^\circ = \frac{[B][H^+]}{[BH]}\frac{\gamma_B\gamma_{H^+}}{\gamma_{BH}}, \tag{7.2}$$

as discussed in the last chapter. In most studies where buffers are used, particularly in the study of acid-base catalysis, the ionic strengths of all solutions are kept constant to minimize variations in the activity coefficients. In the following discussion, we will assume that activity coefficients are indeed constant and work with the *practical* equilibrium constants defined as

$$K_a = \frac{[B][H^+]}{[BH]} = K_a^\circ \frac{\gamma_{BH}}{\gamma_B \gamma_{H^+}}. \tag{7.3}$$

Rearrangement of Eq. (7.3) gives us what might be called the *master equation* for buffer solutions:

$$[H^+] = K_a \frac{[BH]}{[B]}. \tag{7.4}$$

All of this is, of course, simply a review of freshman chemistry. The important point to be made here is that Eq. (7.4) shows that concentrations of BH and of B may be varied without varying the concentration of H^+, so long as the ratio of concentrations of B and BH remains constant, and that the concentration of H^+ may be varied without varying the concentration of BH if the concentration of B is varied, or without varying the concentration of B if the concentration of BH is varied.

From the practical standpoint, it is essential to realize that the values of [B] and [BH] in Eq. (7.4) are not necessarily equal to the values that we would calculate from the amounts of B and BH added to the solution. Suppose we weigh out a certain number of moles of B, W_B, and a certain number of moles of BH, W_{BH}, and dissolve these in a volume, V, of water. We then define:

$$C_B^\circ = \frac{W_B}{V} \quad \text{, and} \quad C_{BH}^\circ = \frac{W_{BH}}{V}.$$

In order to calculate the concentrations of B and BH actually present in the solution, we note that any conversion of BH into B will produce a proton, or that any conversion of B into BH will consume a proton. The only other source, or sink, for protons in the solution is the equilibrium

$$H_2O \overset{K_w}{\rightleftharpoons} H^+ \quad + \quad OH^- \quad \text{with} \quad K_w = [OH^-][H^+].$$

We will, for the moment, assume that we have a *practical* value for K_w just as for K_a. For every proton formed by water dissociation, there is also one hydroxide ion formed. Any inequality in the concentrations of proton and hydroxide ions must be due to conversion of B into BH or *vice versa*. We can, therefore, write the stoichiometric relationship:

$$[H^+] = [OH^-] + [B] - C_B^\circ, \tag{7.5}$$

or, on rearrangement,

$$[B] = C_B^\circ + [H^+] - [OH^-]. \tag{7.6}$$

Since the total concentration of B and BH must always be equal to $C_B^\circ + C_{BH}^\circ$, we also have the relationship

$$[BH] = C_{BH}^\circ - [H^+] + [OH^-]. \tag{7.7}$$

We may now substitute $K_w/[H^+]$ for $[OH^-]$ in Eqs. (7.6) and (7.7) and substitute the results into Eq. (7.4) to obtain

$$[H^+] = K_a \frac{C_{BH}^\circ - [H^+] + \frac{K_w}{[H^+]}}{C_B^\circ + [H^+] - \frac{K_w}{[H^+]}}. \tag{7.8}$$

Equation (7.8) is a **cubic** equation in $[H^+]$ which may be solved if K_a, C_B°, and C_{BH}° are known. Under almost all practical situations, Eq. (7.8) can be simplified to, at worst, a quadratic equation in $[H^+]$. For example, in kinetic studies where we wish to establish pseudo-first-order conditions, even if the reactant concentrations are kept very low, we would certainly want both C_B° and C_{BH}° to be above $10^{-3}M$. Then, in the pH range of 5 to 9, $[B] = C_B^\circ$ and $[BH] = C_{BH}^\circ$ to within 1% accuracy. At pH below 5, the concentration of hydroxide ion is less than 0.01% of the concentration of the proton, and the terms in $K_w/[H^+]$ can be dropped in both the numerator and denominator of Eq. (7.8). At pH above 9, the concentration of proton is less than 0.01% of that of hydroxide ion, and the $[H^+]$ terms may be dropped from numerator and denominator. These simplifications lead to the three limiting forms:

If C_B° and $C_{BH}^\circ \geq 10^{-3}M$, then:

$$[H^+] = \begin{cases} K_a \frac{C_{BH}^\circ}{C_B^\circ}, & \text{for } 5 \leq pH \leq 9; \\ \frac{-(C_B^\circ + K_a) + \sqrt{(C_B^\circ + K_a)^2 + 4K_a C_{BH}^\circ}}{2}, & \text{for } pH \leq 5; \\ \frac{(K_a C_{BH}^\circ + K_w) + \sqrt{(K_a C_{BH}^\circ + K_w)^2 + 4C_B^\circ K_a K_w}}{2C_B^\circ}, & \text{for } pH \geq 9. \end{cases}$$

If a buffer system is used to attain pseudo-first-order conditions for some reaction that consumes or produces protons, the proton transfers between solvent and buffer components must be much faster than the reaction under study. As we'll see shortly, it is not too difficult to find buffers for which the proton transfers to and from solvent occur with first-order rate constants greater than 10^8sec^{-1}. Attaining pseudo-first-order conditions, then, is usually a problem only when extremely fast reactions are studied.

7.3 Specific Acid or Base Catalysis

Specific acid catalysis is almost always indicative of a mechanism in which some initial reactant is rapidly and reversibly converted to its conjugate acid, which then undergoes slow conversion, by one or more steps, to product:

$$\mathrm{S} + \mathrm{H}^+ \underset{K^\circ_{SH}}{\overset{\text{fast}}{\rightleftharpoons}} \mathrm{SH}^+$$

$$\mathrm{SH}^+ \xrightarrow{\text{slow, } k_2} \text{Product}$$

$$\text{where} \quad K^\circ_{SH} = \frac{[S][H^+]}{[SH^+]} \frac{\gamma_S \gamma_{H^+}}{\gamma_{SH^+}},$$

$$\text{giving} \quad \frac{d[Product]}{dt} = k^\circ_2[SH^+]\frac{\gamma_{SH^+}}{\gamma_\ddagger},$$

where $\gamma_\ddagger$ is the activity coefficient for the transition state of the rate-determining step. We can obviously solve for $[SH^+]$ in terms of $[S]$ and $[H^+]$ using the expression for K°_{SH} and substitute this into the rate expression. There is a slight complication, and a source of possible confusion, in using this substitution for some systems; we must pay close attention to what is experimentally measured.

It is usually most convenient to work with the sum of $[S]$ and $[SH^+]$ when we put the rate expression into integrable form. Defining

$$[C] = [S] + [SH^+],$$

we will always be able to integrate the rate expression by using the stoichiometric equation

$$[Product] = [C]^\circ - [C],$$

where $[C]^\circ$ is the initial total concentration of reactant. We may substitute $[S] = [C] - [SH^+]$ into the expression for K°_{SH}, and solve for $[SH^+]\gamma_{SH^+}$,

$$[SH^+]\gamma_{SH^+} = \frac{[C][H^+]\gamma_S\gamma_{H^+}}{K^\circ_{SH} + [H^+]\gamma_S\gamma_{H^+}/\gamma_{SH^+}}, \tag{7.9}$$

which may be substituted into the rate equation. Eq. (7.9) is a fairly complex function of $[H^+]$, but it simplifies under the conditions where reactant is only slightly protonated, or where it is fully protonated. We have

$$[SH^+]\gamma_{SH^+} = \begin{cases} [C][H^+]\frac{\gamma_S\gamma_{H^+}}{K^\circ_{SH}}, & \text{if } K^\circ_{SH} \gg [H^+]\frac{\gamma_S\gamma_{H^+}}{\gamma_{SH^+}}; \\ [C]\gamma_{SH^+}, & \text{if } K^\circ_{SH} \ll [H^+]\frac{\gamma_S\gamma_{H^+}}{\gamma_{SH^+}}. \end{cases} \tag{7.10}$$

The first condition means that only a small fraction of the reactant is protonated at equilibrium, while the second condition means that the reactant is essentially all in the protonated form.

The most frequently encountered situation is that where the reactant is only slightly protonated, and the rate expression then integrates to give the pseudo-first-order rate constant

$$k_\psi = k^\circ_2[H^+]\frac{\gamma_S\gamma_{H^+}}{K^\circ_{SH}\gamma_\ddagger}, \tag{7.11}$$

which depends linearly on $[H^+]$ and is independent of the concentrations of buffer acids and bases. We may further note that if the ratio of activity coefficients, $\gamma_S/\gamma_\ddagger$, were equal to γ_B/γ_{BH^+} for some series of acids used to establish an acidity function, H, then $\log k_\psi$ would be a linear function of H with a slope of unity.

Experimentally, the situation in which neither of the limiting conditions in Eq. (7.10) are fulfilled is easily spotted; k_ψ *vs.* $[H^+]$ looks just like a titration curve, and spectrophotometry would almost certainly reveal the presence of appreciable amounts of both S and SH^+. The situation where reactant is completely protonated is usually spotted by spectrophotometric distinction of S and SH^+. The observation of specific acid catalysis is, however, the only way in which the situation involving extremely small extents of protonation can be spotted.

The typical mechanism leading to specific base catalysis is completely analogous to that leading to specific acid catalysis, but involves the de-protonation of a reactant to form the conjugate base of the reactant in a rapid equilibrium followed by rate-determining conversion of conjugate base into product:

$$\mathrm{SH + OH^-} \underset{K_b^\circ}{\overset{\text{fast}}{\rightleftharpoons}} \mathrm{S^- + H_2O}$$

$$\mathrm{S^-} \xrightarrow{\text{slow, } k_2} \text{Product}$$

$$\text{where} \quad K_b^\circ = \frac{[S^-]}{[SH][OH^-]}\frac{\gamma_{S^-}}{\gamma_{SH}\gamma_{OH^-}},$$

$$\text{giving} \quad \frac{d[Product]}{dt} = k_2^\circ[S^-]\frac{\gamma_{S^-}}{\gamma_\ddagger}.$$

The algebra to get to the pseudo-first-order rate constant is left as a problem at the end of this chapter.

7.4 General Acid and Base Catalysis

General acid catalysis or *general base catalysis* are almost always indicative of mechanisms involving a proton transfer in the rate determining step of the reaction. For general acid catalysis, any Brønsted acid may transfer the proton to the reactant, and since there are at least three acids, the proton, the buffer acid, and water, present in a buffered solution, there will be at least three elementary steps which contribute to the rate of reaction. A frequently encountered situation is one in which there is a slow proton transfer to the reactant to form its conjugate acid, followed by a fast conversion of the conjugate acid to product:

$$\mathrm{S + H^+} \xrightarrow{k_H} \mathrm{SH^+}$$

$$\mathrm{S + HA} \xrightarrow{k_{HA}} \mathrm{SH^+ + A^-}$$

$$\mathrm{S + H_2O} \xrightarrow{k_w} \mathrm{SH^+ + OH^-}$$

$$\mathrm{SH^+} \xrightarrow{\text{fast}} \text{Product}$$

$$\text{giving} \quad \frac{d[Product]}{dt} = \left(k_H[H^+]\frac{\gamma_{H^+}}{\gamma_\ddagger} + k_{HA}[HA]\frac{\gamma_{HA}}{\gamma'_\ddagger} + k_w\frac{\gamma_w}{\gamma''_\ddagger}\right)[S]\gamma_S,$$

so that the pseudo-first-order rate constant, ignoring activity coefficients, is

$$k_\psi = k_H[H^+] + k_{HA}[HA] + k_w. \tag{7.12}$$

There is a completely analogous situation in which a reactant transfers a proton to hydroxide ion, the buffer base A^-, and to water, in rate determining steps, followed by rapid conversion of the conjugate base of the reactant into product. Again ignoring activity coefficients, the pseudo-first-order rate for this general base catalysis will be

$$k_\psi = k_{OH}[OH^-] + k_A[A^-] + k_w. \qquad (7.13)$$

The individual rate constants in these expressions for k_ψ can, in principle, be obtained by proper use of buffer systems. For example, in Eq.(7.13), we could measure k_ψ in a series of buffers in which both $[HA]$ and $[A^-]$ are varied while keeping the ratio $[HA]/[A^-]$ constant. A plot of k_ψ *vs.* $[A^-]$ should give a straight line with a slope equal to k_A. Another set of measurements in a series of buffers with a constant $[A^-]$ and varying $[HA]$ should produce a linear plot of k_ψ *vs.* $[OH^-]$ with a slope equal to k_{OH}. Once we have k_A and k_{OH}, then k_w can be obtained from any one of the measured k_ψ's by substitution into Eq. (7.13).

7.5 Specific Plus General Catalysis

There is a fairly frequently encountered situation in which a reactant is converted into its conjugate acid in a rapid equilibrium step, specific acid catalysis, followed by rate-determining conversion of the conjugate acid into product by a proton transfer reaction, general base catalysis. Consider the case where HS′ is a reactive isomer of the reactant HS and is formed by a slow proton transfer from the conjugate acid of HS. We could have a reaction with the mechanism

$$\mathrm{HS} + \mathrm{H}^+ \underset{K_{SH_2}}{\overset{\text{fast}}{\rightleftharpoons}} \mathrm{H_2S^+}$$

$$\mathrm{H_2S^+} + \mathrm{OH^-} \xrightarrow{k_{OH}} \mathrm{HS'} + \mathrm{H_2O}$$

$$\mathrm{H_2S^+} + \mathrm{A^-} \xrightarrow{k_A} \mathrm{HS'} + \mathrm{HA}$$

$$\mathrm{H_2S^+} + \mathrm{H_2O} \xrightarrow{k_w} \mathrm{HS'} + \mathrm{H_3O^+}$$

$$\mathrm{HS'} \xrightarrow{\text{fast}} \text{Product},$$

which gives

$$\frac{d[Product]}{dt} = \left(k_{OH}[OH^-] + k_A[A^-] + k_w\right)[H_2S^+], \qquad (7.14)$$

if activity coefficients are ignored. We need to substitute for $[H_2S^+]$ in terms of total concentration of $[HS] + [H_2S^+] = [C]$, just as we did in the case of specific acid catalysis discussed above. Under the condition that $K_{SH_2} \gg [H^+]$, we have $[H_2S^+] = [C][H^+]/K_{SH_2}$ from the equation analogous to Eq. (7.10). Making the substitution into Eq. (7.14), and carrying $[H^+]$ inside the bracket, we obtain

$$\frac{d[Product]}{dt} = \left(k_{OH}[OH^-][H^+] + k_A[A^-][H^+] + k_w[H^+]\right)[C]. \quad (7.15)$$

Since $K_w = [OH^-][H^+]$ and $K_a[HA] = [A^-][H^+]$, this can be written as

$$\begin{aligned} \frac{d[Product]}{dt} &= \left(k_{OH}K_w + k_A K_a[HA] + k_w[H^+]\right)[C] \\ &= \left(k'_w + k'_{HA}[HA] + k'_H[H^+]\right)[C] \\ k_\psi &= k'_w + k'_{HA}[HA] + k'_H[H^+], \end{aligned} \quad (7.16)$$

where the second line defines the k''s and the last line shows that the situation is kinetically indistinguishable from that of general acid catalysis with k_ψ given by Eq. (7.12). Thus, specific acid catalysis followed by general base catalysis appears as general acid catalysis.

In fact, the mechanism that we have just considered is really just the reverse of the mechanism that was written in Section 7.4, above, for general acid catalysis. If there is a slow proton transfer from acid to reactant followed by rapid loss of the proton from product in the forward direction, then, in the reverse direction, there is a rapid protonation of reactant followed by a slow proton transfer to a base to form product.

There is a completely analogous situation in which specific base catalysis is followed by general acid catalysis and which is kinetically indistinguishable from general base catalysis:

$$\mathrm{HS} + \mathrm{OH^-} \underset{K^b_{SH}}{\overset{\text{fast}}{\rightleftharpoons}} \mathrm{S^-} + \mathrm{H_2O}$$

$$\mathrm{S^-} + \mathrm{H^+} \xrightarrow{k_H} \mathrm{HS'}$$

$$\mathrm{S^-} + \mathrm{HA} \xrightarrow{k_{HA}} \mathrm{HS'} + \mathrm{A^-}$$

$$\mathrm{S^-} + \mathrm{H_2O} \xrightarrow{k_w} \mathrm{HS'} + \mathrm{OH^-}$$

$$\mathrm{HS'} \xrightarrow{\text{fast}} \text{Product},$$

which gives

$$\frac{d[Product]}{dt} = \left(k_H[H^+] + k_{HA}[HA] + k_w\right)[S^-], \quad (7.17)$$

if activity coefficients are again ignored. The algebra involved in showing that this leads to an equation for the pseudo-first-order rate constants which is indistinguishable from Eq. (7.13) is left as a problem at the end of this chapter.

7.6 Bell-Shaped pH-Rate Profiles

The most convenient way to set up experiments to study acid-base catalysis is to start with master solutions of reactants which are then diluted into the appropriate buffer solutions. This can sometimes lead to more complex behaviors than those indicated in the discussions above.

Suppose we wish to study the reaction of a ketone with hydroxylamine to form an oxime [1]. Let's start by preparing separate master solutions of the ketone at $1.0 \times 10^{-2}M$ and of hydroxylamine at $1.7M$. In each experiment, we will add 1.00 ml of each master solution to a 100 ml portion of the particular buffer solution to be studied. Thus, each solution will contain $1.0 \times 10^{-4}M$ ketone and $1.7 \times 10^{-2}M$ of hydroxylamine. Since the concentration of hydroxylamine is much greater than that of the ketone, we should observe pseudo-first-order kinetics. If we use a series of buffers in which the pH is varied without varying the concentration of the buffer acid, we would find that the pseudo-first-order rate constants vary with pH as shown in Figure 7.1.

The bell-shaped curve results from the fact that hydroxylammonium ion, with $pK_a = 6.0$ is in rapid equilibrium with hydroxylamine,

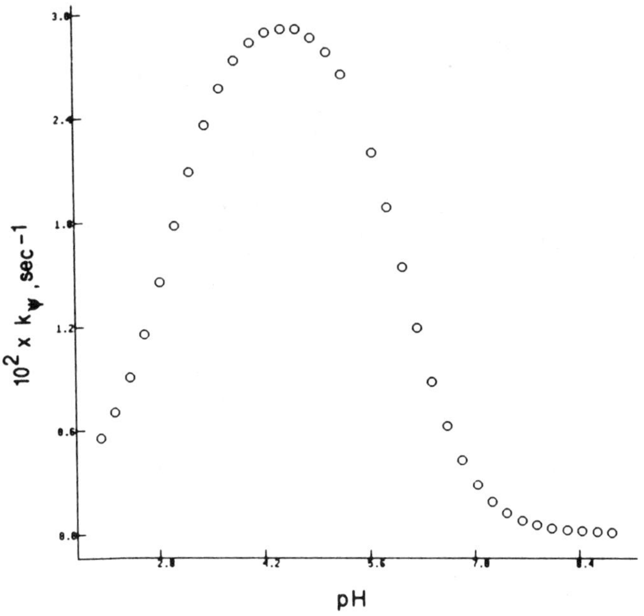

Figure 7.1: *The bell-shaped pH-rate profile for reaction of acetone with hydroxylamine [1].*

so that, at low pH, there is very little 'free' hydroxylamine present. Since the free hydroxylamine is the reagent that reacts with the ketone, decreases in its concentration lead to decreases in the observed rate constants. This decrease at pH between approximately 4 and 7 is more than compensated by the acid catalysis term for the reaction of hydroxylamine with the ketone. The numerical details are left as an exercise at the end of this chapter.

The lesson to be learned here is one that we have encountered before: one must be careful that the concentrations used in treating kinetic behaviors are the concentrations actually present in the solutions studied.

7.7 The Brønsted Relationship

As we saw in the above discussion of general catalysis, each of the three rate constants in Eqs. (7.12), (7.13), or their equivalents in the special cases of the last section, can, in principle, be evaluated by the proper use of buffer systems. In practice, difficulties are frequently encountered. These arise in cases where either the water catalysis or proton catalysis (hydroxide ion catalysis) terms are so large relative to the general acid (general base) catalysis term that the latter makes an insignificantly small contribution to the measured pseudo-first-order rate constants. If k_w has a numerical magnitude roughly equal to k_{HA} and k_H in Eq. (7.12), for example, then variations in the concentrations of HA and H^+ up to $10^{-2}M$ would produce only up to 2% variations in k_ψ. Unless extreme precision in rate measurements were attained, we would completely miss the existence of acid catalysis in this case. Similarly, if $k_H[H^+]$ were much larger than k_w and $k_{HA}[HA]$ under all conditions studied, we would think that we had only specific catalysis. This raises the question of what determines the magnitudes of the individual rate constants.

Empirically, it is often found that there is a logarithmic relationship between the rate constants, k_{HA}, for a series of acids, or k_A for a series of bases, and the equilibrium constants for acid ionizations, K_a^{HA} of the acids or conjugate acids of the bases:

$$\log k_{HA} = \alpha \log K_a^{HA} + const, \tag{7.18}$$

or

$$\log k_A = -\beta \log K_a^{HA} + const. \tag{7.19}$$

Both of these equations are called the Brønsted relationship, or, less correctly, the Brønsted catalysis law. The relationship, like the Hammett

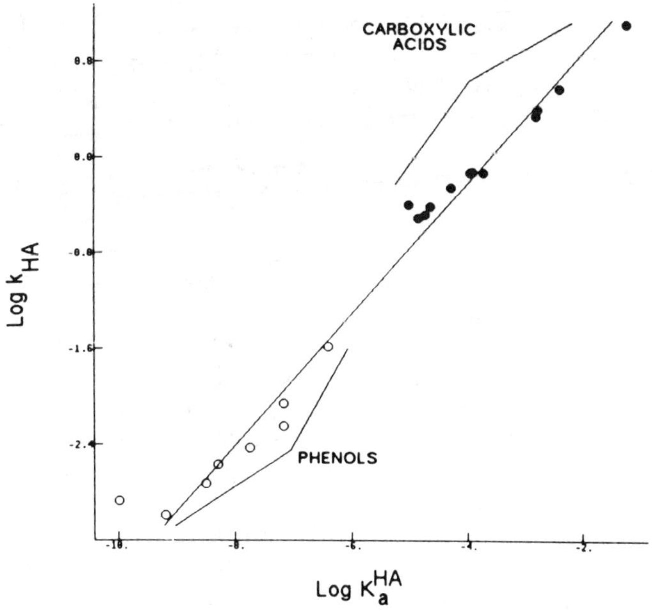

Figure 7.2: *Brønsted plot of the rate constants for the dehydration of acetaldehyde hydrate in water at 25°C. Data are from [2].*

and Taft relationships discussed in Chapter 3, is a linear free energy relationship; here, between the free energies of activations for the proton transfer reactions and the free energies of acid ionizations.

A typical example of the relationship is shown in Figure 7.2 for the general acid catalyzed dehydration of acetaldehyde hydrate by a series of phenols and a series of carboxylic acids [2]. The mechanism of the reaction is probably an example of a rapid protonation of the hydrate, followed by rate-determining general-base catalyzed loss of water [3]:

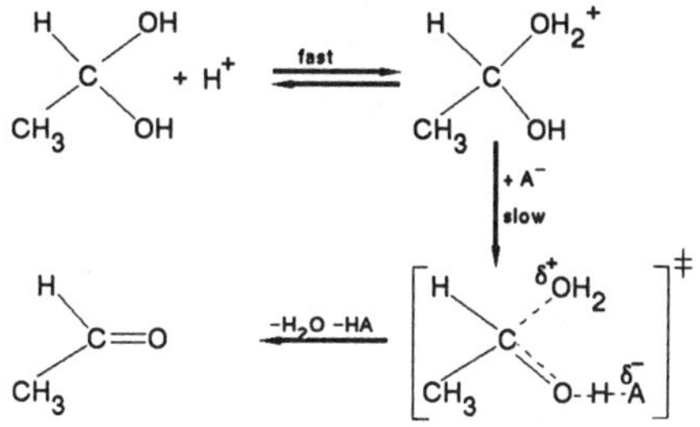

The correlation is reasonably good even for *ortho*-substituted phenols and for formic acid where the Hammett and Taft relationships often fail. There are, however, large deviations from a single line when the data for ammonium ions, carbon acids, and water are added to the plot [4].

There are some fundamental difficulties in applying the Brønsted relationship to k_w's and to k_H's due to the problem of defining what we mean by K_a for water and for the proton in water [5]. There is the related problem that k_w is a first-order constant while all others are second-order constants. The naïve resolution of the latter problem is to change our standard state definition for water; instead of defining it to be $a_w = 1$ for pure water as usual, we define it to be $a_w = 55.5M$, which is the molarity of water in pure water. We then convert k_w to a second-order rate constant by dividing the measured first-order rate constant by 55.5 M. A great deal of care is necessary when reading the literature to determine whether or not this 'correction' has been made. With this new standard state for water, the pK_a for H_3O^+ is usually assigned the value of -1.74 ($-\log 55.5$), and the pK_a for H_2O is assigned the value of 15.74, both at 25°C. In actual practice, the points for water and the proton seldom conform very well to the Brønsted relationship regardless of how we define these pK_a's, but the values given are sufficient for order-of-magnitude estimates in most cases.

In those cases where water obeys the Brønsted relationship for either acid or base catalysis, it would be very difficult to detect general catalysis if the slope of the relationship were close to zero. Since water is present at 55.5 M, if its second-order rate constant were equal to that of solute acids or bases, present at less than 1 M, the water catalysis would contribute at least 98% of the observed rates. At the other extreme of a Brønsted relationship with a slope near unity, either the proton or hydroxide ion, with their extreme pK_a's, are likely to dominate the kinetics so that general catalysis is hard to detect. It should come as no surprise to learn, then, that most observed cases of general catalysis give Brønsted slopes between 0.3 and 0.7. There are a few observed cases with slopes outside this range and for which water, hydroxide ion, or the proton, fall badly enough off the correlation line that the general catalysis can be detected.

The Brønsted slopes, α or β, are frequently interpreted as measures of the extent of proton transfer at the transition state for the reaction studied. The reasoning behind this interpretation is fairly obvious in terms of how much A, at the transition state, 'looks like' HA or A^-. Essentially, the interpretation assumes that all characteristics of A that change on going from HA to A^-, or *vice versa*, are linear functions of the extent of proton transfer. This, of course, is exactly the same type of reasoning that we discussed in Section 3.3 in connection with Hammett ρ values, and in Section 4.3 in connection with structural effects on the shapes of potential energy surfaces. The reasoning leads to the expectation, among others, that α or β must be between zero and unity just as it led to the expectation that the ρ value for the rate constants of a series of reactions would be smaller in magnitude than the ρ for the corresponding equilibrium constants.

As in the cases discussed earlier, the reasoning is somewhat naïve. For the Brønsted relationship, there are several examples where the reasoning is clearly incorrect because slopes outside the expected range have been observed. For the simple proton transfer reactions from a series of *meta*- and *para*-substituted phenylnitromethanes to hydroxide ion in water, the Brønsted α is 1.54 [6]. It is believed that the developing negative charge of the nitronates is still localized on the benzyl carbon at the transition state, while the nitronate ions have the charge predominantly localized on the oxygens of the nitro group. The proximity of the partial negative charge to the substituents at the transition state more than compensates the development of the full, but more distant, charge of the product. A β value of -0.10 has been reported [7] for base catalyzed hydrolyses of 2,4-dinitrophenylphosphate by a series of amines. Here, the loss of a water molecule hydrogen bonded to the lone pair of the amines is believed to be the most important change in going from reactant amine to a transition state involving very slight bonding to the amines.

It is also worth mentioning that the Brønsted relationship is frequently applied to reactions in which the bases act as nucleophiles rather than as proton acceptors. The hydrolysis reactions of carboxylic acid derivatives, for example, are often subject to catalysis by nucleophiles in a mechanism such as:

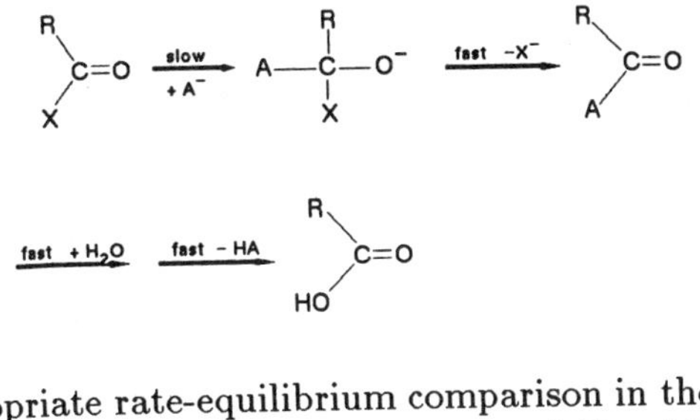

The more appropriate rate-equilibrium comparison in these cases would utilize the equilibrium constants for the nucleophilic addition steps, but since these are seldom known, it is assumed that they would be directly proportional to the Brønsted basicity constants, at least for a closely related series of nucleophiles. The base-catalyzed hydrolyses of the phosphate esters mentioned in the preceding paragraph are, in fact, believed to actually involve the bases as nucleophiles. In view of our rather poor understanding of the relationship between Lewis and Brønsted basicities, discussed in Chapter 6, one should be exceedingly cautious in the interpretations of the Brønsted slopes for these reactions.

7.8 Rates of Proton Transfer Reactions

The Brønsted relationship must fail for any reaction if a wide enough range of acids or bases are studied simply because the rates of the proton transfer steps will eventually reach the diffusion controlled limit. In order to understand the situation, it is simplest to consider a one-step proton transfer reaction:

$$\mathrm{HA} \quad + \quad \mathrm{B}^- \quad \underset{k_r}{\overset{k_f}{\rightleftharpoons}} \quad \mathrm{A}^- \quad + \quad \mathrm{HB}$$

where we can write the exact equations:

$$K = \frac{k_f}{k_r} = \frac{K_a^{HA}}{K_a^{HB}}. \tag{7.20}$$

The overall reaction can be written as involving three steps, the so-called *Eigen mechanism* [8]:

$$\mathrm{HA} \quad + \quad \mathrm{B}^- \quad \underset{k_{-1}}{\overset{k_1}{\rightleftharpoons}} \quad [\mathrm{B}\cdots\mathrm{HA}]^-$$

$$[\mathrm{B}\cdots\mathrm{HA}]^- \quad \underset{k_{-2}}{\overset{k_2}{\rightleftharpoons}} \quad [\mathrm{BH}\cdots\mathrm{A}]^-$$

$$[\mathrm{BH}\cdots\mathrm{A}]^- \quad \underset{k_{-3}}{\overset{k_3}{\rightleftharpoons}} \quad \mathrm{BH} \quad + \quad \mathrm{A}^-,$$

where the first and third steps are the diffusive encounters and separations of the reactant and product 'encounter complexes', respectively, as we discussed for other reaction types in Sections 4.4 and 4.6. If the encounter complexes are present only at negligible concentrations, we can apply the Bodenstein approximation to the reaction scheme to obtain

$$\begin{aligned} k_f &= \frac{k_1 k_2 k_3}{k_2 k_3 + k_{-1} k_{-2} + k_{-1} k_3}, \quad \text{and} \\ k_r &= \frac{k_{-1} k_{-2} k_{-3}}{k_2 k_3 + k_{-1} k_{-2} + k_{-1} k_3} \end{aligned} \tag{7.21}$$

for the measurable forward and reverse rate constants. There are three limiting situations corresponding to each of the three steps being rate-determining. If HA is a much weaker acid than HB, we can expect that

k_2 will be very small and that k_{-2} will be very large; for a limiting case we will have

$$k_f = K_1 K_2 k_3, \quad \text{and} \quad k_r = k_{-3} \quad \text{when} \quad k_2 \ll k_{-1} \quad \text{and} \quad k_{-2} \gg k_3. \tag{7.22}$$

As the relative strengths of the acids become more nearly equal, k_{-2} should decrease until we have

$$k_f = K_1 k_2, \quad \text{and} \quad k_r = k_{-2} K_3 \quad \text{when} \quad k_2 \ll k_{-1} \quad \text{and} \quad k_{-2} \ll k_3. \tag{7.23}$$

At the extreme of HA being a much stronger acid than HB, we will reach the limiting case

$$k_f = k_1, \quad \text{and} \quad k_r = k_{-1} K_2 K_3 \quad \text{when} \quad k_2 \gg k_{-1} \quad \text{and} \quad k_{-2} \ll k_3. \tag{7.24}$$

The rate constants k_1 and k_{-3} are the diffusion rate constants approximately equal to $5 \times 10^9 M^{-1} sec{-}1$, which are not expected to be very dependent on the structures of A or B. We have the same problem that we discussed in Sections 4.4 and 4.6 as to precisely what we mean by the encounter complexes, but we can assume here that they are simply the two species in contact, with $K_1 \simeq 10^{-1} M^{-1} \simeq 1/K_3$. Then, $k_{-1} \simeq k_3 \simeq 5 \times 10^8 sec^{-1}$.

Now, suppose we have a series of reactions in which we will keep HA constant, but study a series of B^-. If we start with weak bases whose conjugate acids have low pK_a's (giving small K's in Eq. (7.20)), we will have Eq. (7.22) applicable. Since the overall equilibrium constant $K_1 K_2 K_3$ is equal to K_a^{HA}/K_a^{HB}, and since both k_{-3} and K_a^{HA} are essentially invariant with change in HB, we must find that $k_f = k_{-3} K_a^{HA}/K_a^{HB} = constant/K_a^{HB}$. Taking the logarithms of both sides of the equation, we find that the Brønsted slope, β, relating k_f to K_a^{HB} must be unity. It should also be obvious that the Brønsted slope, α, relating k_r to K_a^{HB} must be zero for this set of reactions.

At the opposite extreme of a series of B^- whose conjugate acids have very large pK_a's, Eq. (7.24) will apply, and by the same reasoning as just used, we must find that the Brønsted slope, β, relating k_f and K_a^{HB} must be zero, while the α relating k_r and K_a^{HB} must be unity. It is only in the case where step 2 of the mechanism, the actual proton transfer, is rate-determining that the Brønsted slopes can be different from zero or unity. In this range, we must find that $\alpha + \beta = 1$. The situation is diagrammed in Figure 7.3. The interesting question, of course, is just where and how the slope of zero evolves into the slope of unity, or *vice versa*. There will be some range of bases for which the Brønsted

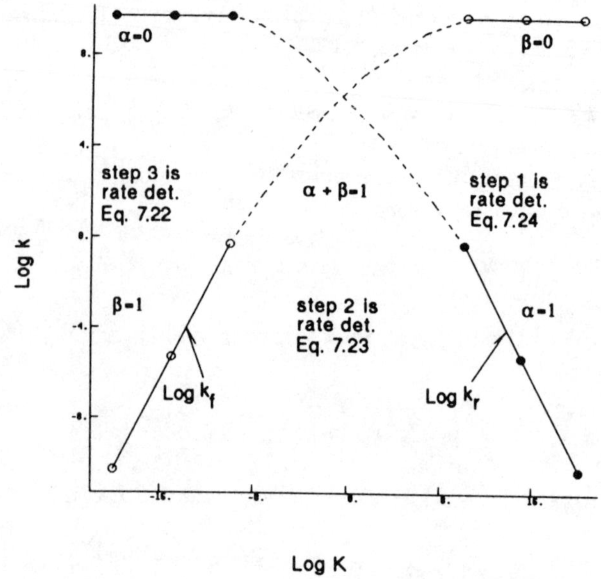

Figure 7.3: *Behavior of rate constants as a function of equilibrium constants for the Eigen mechanism of proton transfers.*

slopes will be 0.5. Will this slope remain constant as long as step 2 is rate-determining, or will the actual correlation be a smooth curve with the slope smoothly changing from zero to unity as we go from from one extreme to the other?

Experimental studies [8] have demonstrated that almost all proton transfers between acids and bases having an oxygen or a nitrogen atom as the basic site occur very near diffusion controlled limits in water or methanol solutions. When the equilibrium constant for the proton transfer is near unity, the rates of these reactions are only a factor of two below the diffusion limit, as expected for cases where step two is much faster than steps one and three. Most proton transfers which involve intramolecularly hydrogen bonded protons, or which involve substantial changes in geometries as the acid is converted to its conjugate base, or *vice versa*, have substantial free energies of activation for the second step of the mechanism unless the equilibrium constant for the transfer is extremely large or small. There are also a fair number of slow proton transfers between amines and ammonium ions when large steric effects are present [9]. It should be fairly clear that if we wish to go beyond these broad generalities concerning rates of proton transfer reactions, we must have some way of separating the effects of the magnitude of the equilibrium constant from purely kinetic effects.

7.9 Marcus Theory of Proton Transfers

If we limit our attention to pure proton transfer reactions, the potential energy surfaces should be fairly simple. In cases where detailed *ab initio* calculations have been carried out, the two basic site atoms and the transferring proton are found to be linear at the transition state [10]. This allows us to restrict our considerations to linear geometries. For a general proton transfer reaction between HA and B^-, we may choose the distances between A and H, d_{AH}, and between B and H, d_{BH}, as the two coordinates to be considered in the simplified potential energy diagrams of the sort discussed in Section 4.3 and diagrammed in Figure 7.4.

Let's assume that there is some 'extent of reaction' parameter, n, which has a value of zero for the encounter complex $AH \cdots B^-$ and a value of unity for the encounter complex $A^- \cdots HB$, as shown on the right side of Figure 7.4. We further assume that the energy of the system can be written as a sum of two terms, one identified with A and the other

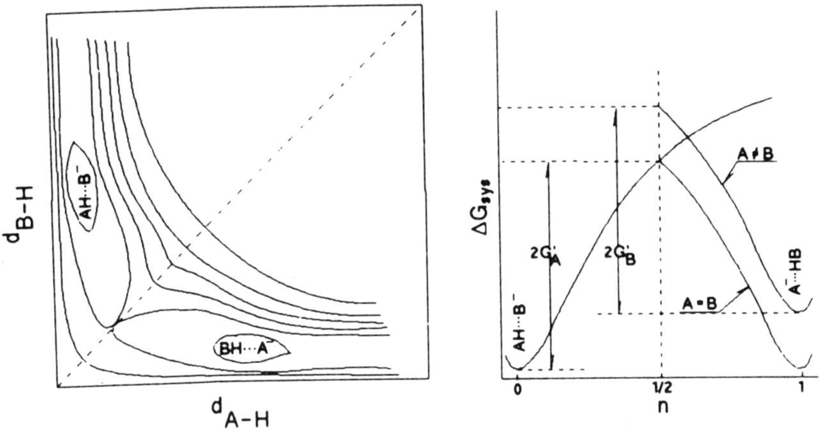

Figure 7.4: *Simplified potential energy contour diagram and the corresponding diagram of energy as a function of extent of reaction for a proton transfer between HA and* B^-.

identified with B including their interactions with the proton, such that the free energy of the system as a function of n is given by

$$\Delta G_{sys} = n\Delta G^{\circ} + G_A + G_B, \tag{7.25}$$

where ΔG° is the standard free energy change for the conversion of reactant complex to product complex[1]. Both G_A and G_B are functions of n. For $n = 0$, both G_A and G_B are equal to zero. As we distort the geometry of B from that of the anion toward that of the product, HB, G_B increases to some point before it can decrease due to the bonding interactions with the proton. Similarly, as the geometry of HA is distorted from that of the reactant, G_A will increase to some point before decreasing as A becomes the stable anion. Eq. (7.25) essentially says that the 'reaction path' curve for a cross reaction of HA and B^- is the average of the curves that would result from an HA + A^- reaction and from an HB + B^- reaction, but with a vertical displacement, $n\Delta G^{\circ}$, that is proportional to the 'extent of reaction'.

It must be remembered, as discussed in Section 4.3, that the distortions about which we are speaking involve the coordinates which are not shown in Figure 7.4. For example, if HA is nitromethane, the C-N and N-O bond lengths, and the H-C-N bond angles are considerably different in HA and in A^-, but none of these are shown in the simplified potential energy diagram. The basic idea being developed is that these types of distortions require more energy than is gained from the interaction of acid and base in the initial stages of the proton transfer. This can probably be most easily appreciated by starting to think about an 'identity reaction' in which A and B are identical. This is the one type of system about which we have some *a priori* knowledge. If A and B are identical, then the dashed diagonal lines in Figure 7.4 represents a plane of symmetry perpendicular to the page. If there is a single transition state for the reaction, it must lie on the diagonal line, and the two A moieties must be indistinguishable from one another at this point. In order to go from reactant to product, A^- must change its geometry

---

[1] In some discussions [11] of the Marcus treatment, the encounter complexes are defined to be such that the 'mean distance of separation' of the species is that found at the transition state. In this case, the ΔG° appropriate to our further discussions is that for conversion of one encounter complex to the other which differs from the overall free energy of the reaction by the difference in the energy required to form the reactant complex and that required to form the product complex; these are called the 'work terms', w_r and w_p, respectively. We shall implicitly assume that $w_r = w_p$ in our discussions.

toward that of the A in HA, while the A in HA changes its geometry toward that of A^-, until, at the transition state the two have the same geometry. Unless there is a strong enough interaction of HA and A^-, for example, by hydrogen bonding of the proton with a **localized** lone pair of electrons, to compensate the energy required for the distortions, the transition state will be considerably higher in energy than the reactants.

The understanding of Eq. (7.25) and the following development may be aided by Figure 7.5.

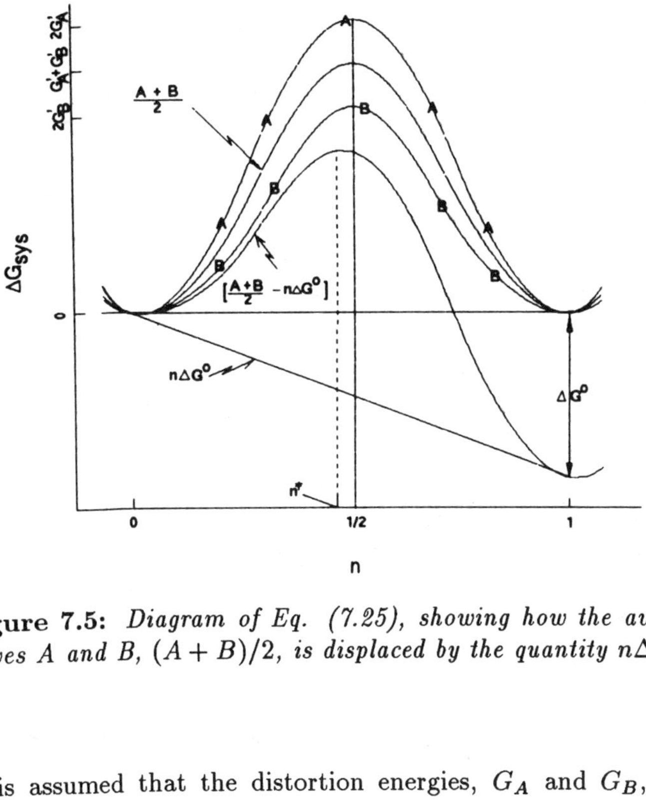

Figure 7.5: *Diagram of Eq. (7.25), showing how the average of the curves A and B, $(A + B)/2$, is displaced by the quantity $n\Delta G^\circ$.*

It is assumed that the distortion energies, G_A and G_B, can be written generally as some smooth functions of the extent of reaction parameter, n

$$G_A = G'_A g_1(n) \quad \text{and} \quad G_B = G'_B g_2(1-n),$$

where G'_A and G'_B are the energy contributions at the point where $g_1(n) = g_2(1-n) = 1$ and $g_1(n)$ and $g_2(1-n)$ are smooth functions of

n with values of zero at $n = 0$ and at $n = 1$. For an identity reaction, $\Delta G^\circ = 0$ and $n = 1/2$ at the transition state, so that Eq. (7.25) gives $G'_A = \Delta G^\ddagger/2$ for the HA, A^- reaction, and $G'_B = \Delta G^\ddagger/2$ for the HB, B^- reaction, as identified in Figure 7.5. Note that the G''s are, at least in principle, measurable by using isotopically labelled species.

We now need to know what happens when we go to some reactions other than identity reactions. As shown in Figure 7.5, the maximum point for the curves of a non-identity reaction, marked as $n^\ddagger$, is generally not at $n = 1/2$, and the energy at the maximum point will depend on ΔG° and on the shapes of the curves. First, we find the value of n in Eq. (7.25) for the transition state of the particular reaction.

Let's write Eq. (7.25) as a Taylor expansion about $n = 0.5$, truncated at second-order terms with the assumption that the reactions considered do not have transition states very far from $n = 1/2$. We have

$$\Delta G_{sys} \simeq n\Delta G^\circ + G'_A\left[1 + g'_1(n-0.5) + g''_1\frac{(n-0.5)^2}{2}\right] + G'_B\left[1 - g'_2(n-0.5) + g''_2\frac{(n-0.5)^2}{2}\right],$$

where

$$g'_1 = \left(\frac{dg_1}{dn}\right)_{n=1/2}, \quad g''_1 = \left(\frac{d^2g_1}{dn^2}\right)_{n=1/2},$$

$$g'_2 = \left(\frac{dg_2}{dn}\right)_{n=1/2}, \quad \text{and} \quad g''_2 = \left(\frac{d^2g_2}{dn^2}\right)_{n=1/2}.$$

Since the free energy of the system is at a maximum with respect to n at the transition state, we can find the value of n at the transition state, $n^\ddagger$, by taking the derivative of ΔG_{sys} with respect to n, setting it equal to zero, and solving for $n^\ddagger$. From the Taylor expansion, we get

$$\frac{d\Delta G_{sys}}{dn} = \Delta G^\circ + (G'_A g'_1 - G'_B g'_2) + (G'_A g''_1 + G'_B g''_2)(n^\ddagger - 0.5) = 0,$$

or, rearranging,

$$(n^\ddagger - 0.5)(G'_A g''_1 + G'_B g''_2) = -\Delta G^\circ - G'_A g'_1 + G'_B g'_2.$$

Substituting this last equation into the Taylor expansion, we obtain

$$\Delta G^\ddagger_{sys} = 0.5\Delta G^\circ + (G'_A + G'_B) - \frac{(n^\ddagger - 0.5)^2}{2}(G'_A g''_1 + G'_B g''_2). \quad (7.26)$$

This is the most general form of the Marcus equation [11]. It is worth noting that the equation contains no first-order terms in $(n^{\ddagger} - 0.5)$; to first-order approximation, $\Delta G^{\ddagger}_{sys} = 0.5\Delta G^{\circ} + G'_A + G'_B$.

We've used the Taylor expansion so that we could keep the development general to this point. No assumptions have been made about the shapes of the curves in arriving at Eq. (7.26). In order to go beyond the first-order approximation, we need to at least know the second derivatives of g_1 and g_2 with respect to n at $n = 0.5$. Let's assume that the g's are parabolic functions of n,

$$g_1(n) = g_2(n) = 4n(1-n), \quad \text{so that} \quad g_1'' = g_2'' = -8.$$

Note that these functions have the required values of unity at $n = 0.5$, and that both g_1' and g_2' are equal to zero. Substitution of these values into the equation for $n^{\ddagger} - 0.5$, and further substitution into Eq. (7.26) gives the simplest form of the Marcus equation:

$$\Delta G^{\ddagger} = 0.5\Delta G^{\circ} + (G'_A + G'_B) + \frac{(\Delta G^{\circ})^2}{16\,(G'_A + G'_B)}.$$

It is obvious that the Brønsted slopes for the exchange reactions can be equated to the terms multiplying ΔG° in the Marcus equation:

$$\beta = 0.5 + \frac{\Delta G^{\circ}}{16\,(G'_A + G'_B)},$$

for a series of bases, B^-. Thus, if G'_B is constant for a series of bases, the Marcus equation gives a variable Brønsted slope that is equal to 0.5 when $\Delta G^{\circ} = 0$, greater than 0.5 for positive ΔG°, and less than 0.5 for negative ΔG°. This, of course, is in the direction expected from our discussion of Figure 7.3. The curvatures of the Brønsted plots will depend on the magnitudes of $\Delta G^{\circ}/16(G'_A + G'_B)$; those reaction series having large values of $(G'_A + G'_B)$ will have small curvatures and Brønsted slopes different from zero or unity over wide ranges of ΔG°, while those with small values of $(G'_A + G'_B)$ will have large curvatures and small ranges of slopes different from zero or unity.

In essence, the Marcus equation is nothing more than an estimate of the shift in the maximum of the average curve shown in Figure 7.5 as the ΔG° for the reaction changes, thereby skewing the symmetrical average curve. The final form of the equation is dependent on the shapes of these curves and upon the assumption that one doesn't get too far away from the symmetrical situation which exists for the identity reactions. The

Marcus equation was actually developed for electron transfer reactions and later applied to other types of transfers, including methyl transfer reactions [12]. It should be kept in mind that the equation is highly approximate, applies only to near-symmetric cases where the Taylor series can be truncated to square terms, and requires the existence of identity reactions to evaluate the *intrinsic* terms, $(G'_A + G'_B)$, which may, or may not, be constant for a series of reactions [13]. For fairly complex acids or bases involving changes in polar, resonance, and solvation interactions on going from reactant to product, the assumption of a single extent of reaction parameter, n, is particularly questionable, as we discussed at the end of Section 3.4 in connection with ρ values and in Section 7.7, above, in connection with the Brønsted slope greater than unity for reactions of arylnitromethanes.

7.10 Enforced Acid-Base Catalysis [15]

We have already stated that general acid-base catalysis occurs in cases where a proton transfer is part of the rate-determining step of the reaction[2]. When the actual proton transfer is fairly slow, as is the case when the existence of resonance forms of the acid or conjugate base result in appreciable geometry changes between the two and in the absence of localized lone pairs at the basic site of the conjugate base [14], acid-base catalysis obviously results. Many reactions of carbonyl compounds involving formation of enolates or of enols, which often subsequently react with another reagent to form products, such as base catalyzed brominations of ketones, are of the 'slow' proton transfer type.

Reactions of carbonyl compounds involving additions to the carbonyl carbon, or the reverse formation of the carbonyl group, are not expected to have 'slow' proton transfers, and yet many exhibit general acid-base catalysis. Invariably, these reactions are such that very large pK_a changes occur on going from reactants to products [15]. A typical case is that of the reaction of methoxylamine with p-methoxybenzaldehyde,

---

[2] We exclude those reactions which actually involve protonation followed by nucleophilic attack by the conjugate bases in solution and show kinetics indicative of general-acid catalysis.

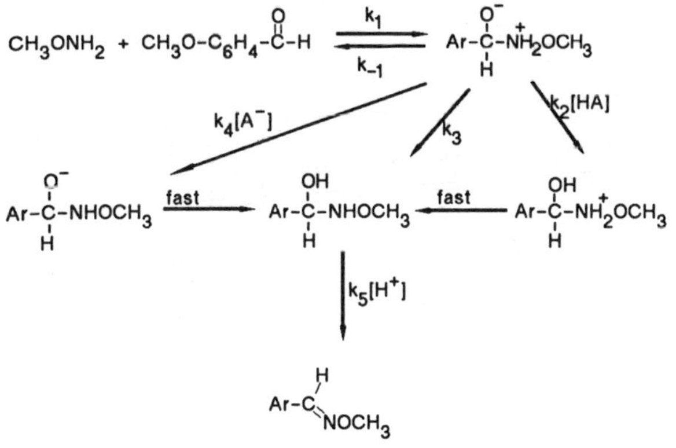

Methyloxime Formation Scheme

in which the $pK_a \simeq 25$ of the methoxylamine changes to the $pK_a \simeq 9$ of the nitrogen bound proton of the zwitterionic intermediate, $T^{\pm}$, and the $pK_a \simeq -4$ of the conjugate acid of the carbonyl oxygen changes to the $pK_a \simeq 8$ of the conjugate acid of the oxide oxygen of the zwitterion [16]. Such reactions frequently involve 'unstable' intermediates, such as those shown for methyloxime formation, and catalysis can result from the fact that extremely fast reactions of these intermediates can compete with fast proton transfers. For the reaction shown, k_{-1} has been estimated equal to $5.3 \times 10^8 sec^{-1}$ [16]. In other such reactions, proton transfers simultaneous with addition to the carbonyl group can avoid formation of the unstable intermediates, and in some cases, the intermediates are so unstable that they have no appreciable lifetime and force proton transfers to occur simultaneously with the additions.

If the reverse of the addition of a nucleophile, such as the methoxylamine shown above, to a carbonyl compound is faster than proton transfers to and from solvent and intramolecularly, then the reaction must show general acid or base catalysis. The rates of the proton transfers from the intermediate to bases, or from acids to the intermediate, are expected to be diffusion controlled if the equilibrium constants for the proton transfer are much greater than unity. If catalysis by a series of bases, for example, were studied, the observed rate constants would be equal to the equilibrium constant for formation of the intermediate times the rate constant for the proton transfer. For bases whose conjugate acids are stronger than the intermediate, the observed rate constants should show a Brønsted β near unity. For bases whose conjugate acids are weaker than the intermediate, a β near zero should be observed. We may reverse the argument to conclude that a Brønsted plot having a slope of unity for weak bases, a slope of zero for strong

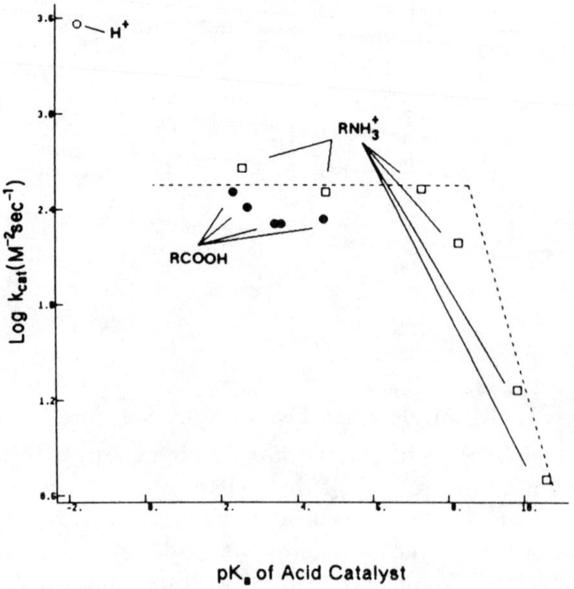

Figure 7.6: *Brønsted plot for observed rate constant for acid catalyzed formation of carbinolamine from methoxylamine and p-methoxybenzaldehyde.* k_{cat} *is equal to* K_1k_2 *in the Methyloxime Formation Scheme. Data are from [16].*

bases, and a small region of curvature in between for bases whose conjugate acids have pK$_a$'s approximately equal to that of the intermediate, is indicative of this type of trapping mechanism.

Although the kinetics of the reaction of methoxylamine with p-methoxybenzaldehyde are fairly complex, it was possible to observe the rates of acid-catalyzed formation of the neutral carbinolamine intermediate with a fairly wide range of acid catalysts. The Brønsted plot of the observed rate constants corresponding to K_1k_2 in the Methyloxime Formation Scheme above is shown in Figure 7.6. The family of ammonium ion catalysts conforms very closely to the expectations discussed in the last paragraph. The dashed lines shown in the figure are the ones with slope of zero and unity, with the break at $pK_a = 8.6$, which is in reasonable agreement with the $pK_a = 9.0$ for the conjugate acid of $T^{\pm}$ estimated by other means.

For the trapping mechanism shown, it should also be possible to increase the concentration of the catalyzing acid to a point where the proton transfer becomes faster than the loss of nucleophile since the proton transfer is a second-order reaction while the reverse of the addition is first-order. The resulting curvature toward a slope of zero at high concentrations in a plot of pseudo-first-order rate constants against concentration of catalyzing acid was observed for the case of cyanoethylammonium ion catalysis of the methyloxime formation reaction.

When the intermediates in these types of reactions become so unstable that they have no appreciable lifetimes, the transition states for the reactions must shift toward the center of the potential energy diagrams and the reactions become concerted. The transition from the step-wise trapping mechanism, through the pre-association mechanism, to the fully concerted mechanism as intermediates become less and less stable is exactly analogous to the situation that we discussed in Chapter 4 for SN1 and SN2 mechanisms. A detailed discussion of these mechanisms, and some other variations, as applied to reactions of carbonyl compounds is given in ref. [15].

7.11 REFERENCES

[1]. W. P. Jencks, **Prog. Phys. Org. Chem., 2**, 63 (**1964**).

[2]. A. A. Frost and R. G. Pearson, **Kinetics and Mechanism**, 2nd Ed., Wiley and Sons, New York, **1961**, pp. 213–223.

[3]. P. E. Sørensen and W. P. Jencks, **J. Amer. Chem. Soc., 109**, 4675 (**1987**).

[4]. R. P. Bell, **The Proton in Chemistry**, 1st Ed., Cornell Univ. Press, Ithaca, New York, **1959**, pg. 165.

[5]. E. M. Arnett, **Prog. Phys. Org. Chem., 1**, 223 (**1963**).

[6]. F. G. Bordwell and W. J. Boyle, Jr., **J. Amer. Chem. Soc., 94**, 3907 (**1972**).

[7]. W. P. Jencks, M. T. Haber, D. Herschlag, and K. T. Nazaretian, **J. Amer. Chem. Soc., 108**, 6100 (**1986**).

[8]. M. Eigen, **Angew. Chem., International Ed., 3**, 1 (**1964**).

[9]. M. M. Kreevoy and Y. Wang, **J. Phys. Chem., 81**, 1924 (**1977**); C. F. Bernasconi, J. P. Fox, and S. Fornarini, **J. Amer. Chem. Soc., 102**, 2810 (**1980**); C. D. Ritchie, **J. Amer. Chem. Soc., 105**, 3573 (**1983**).

[10]. C. D. Ritchie and H. F. King, **J. Amer. Chem. Soc., 88**, 1069 **(1966)**; **90**, 825, 833, 838 **(1968)**. S. Scheiner and Z. Latajka, **J. Phys. Chem., 91**, 724 **(1987)**.

[11]. R. A. Marcus, **J. Phys. Chem., 72**, 891 **(1968)**.

[12]. W. J. Albery and M. M. Kreevoy, **Adv. Phys. Org. Chem., 16**, 87 **(1978)**.

[13]. E. S. Lewis, T. A. Douglas, and M. L. McLaughlin in **Nucleophilicity**, Edited by J. M. Harris and S. P. McManus, ACS Advances in Chemistry Series #215, Wash., D. C., **1987**, pp. 35–49.

[14]. C. D. Ritchie, **J. Amer. Chem. Soc., 91**, 6749 **(1969)**.

[15]. W. P. Jencks, **Accts. Chem. Research, 9**, 425 **(1976)**; **13**, 161 **(1980)**.

[16]. S. Rosenberg, S. M. Silver, J. M. Sayer, and W. P. Jencks, **J. Amer. Chem. Soc., 96**, 7986 **(1974)**.

7.12 PROBLEMS

Problem 7.1

Derive the expression for the pseudo-first-order rate constant for specific base catalysis that is analogous to Eq. (7.11) for specific acid catalysis.
Hint: Just follow the same steps as those given for specific acid catalysis.

Problem 7.2

Show that Eq. (7.17) leads to an expression for the pseudo-first-order rate constant that is experimentally indistinguishable from that for general base catalysis.
Hint: Follow the same steps shown for Eq. (7.16).

Problem 7.3

The reaction of bromine with ketones to form α-bromoketones:

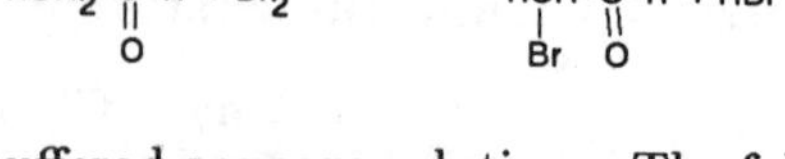

was studied in buffered aqueous solutions. The following data were obtained using $H_2PO_4^-/HPO_4^{-2}$ buffers at a constant ionic strength of 0.1 M:

$[HPO_4^{-2}]$	$[H_2PO_4^-]$	k_ψ, sec^{-1}
2.0×10^{-3}	4.0×10^{-3}	2.05×10^{-2}
2.0×10^{-3}	2.0×10^{-3}	2.46×10^{-2}
2.0×10^{-3}	1.0×10^{-3}	3.30×10^{-2}
2.0×10^{-3}	5.0×10^{-4}	4.96×10^{-2}
3.0×10^{-3}	3.0×10^{-3}	2.78×10^{-2}
4.0×10^{-3}	4.0×10^{-3}	3.10×10^{-2}
6.0×10^{-3}	6.0×10^{-3}	3.73×10^{-2}
1.0×10^{-2}	1.0×10^{-2}	4.99×10^{-2}
1.0×10^{-2}	2.0×10^{-2}	4.58×10^{-2}
1.0×10^{-2}	5.0×10^{-3}	5.82×10^{-2}

The effective pK_a of $H_2PO_4^-$ is 7.12 at an ionic strength of 0.1 M. Determine whether the reaction is general acid or general base catalyzed. Calculate rate constants for each of the catalyzing species in the solutions. Do the rate constants obey the Brønsted relationship?

Hint: Look first at cases where $[HPO_4^{-2}]$ is kept constant, then at cases where $[H_2PO_4^-]$ is kept constant, to determine whether the reaction rate is increasing with base concentration or with acid concentration. After deciding which increases rate, $[OH^-]$ or $[H^+]$, make plots of k_ψ against the appropriate one for cases where the other possible catalyst concentrations are constant; then make plots of buffer catalyst concentration against k_ψ for cases where pH is kept constant. The slopes of these plots should give you the catalytic rate constants for everything but water. This latter constant can be obtained by difference.

Problem 7.4

The data shown in Figure 7.1 for the reaction of hydroxylamine with acetone are from the following table. Each solution contained precisely $0.0167M$ total hydroxylamine and $1.0 \times 10^{-4}M$ acetone. The effective pK_a of hydroxylammonium ion under the conditions used is 6.00.

pH	$10^2 k_\psi$ sec^{-1}	pH	$10^2 k_\psi$ sec^{-1}
2.0	0.548	5.6	2.18
2.2	0.693	5.8	1.87
2.4	0.891	6.0	1.53
2.6	1.14	6.2	1.18
2.8	1.44	6.4	0.871
3.0	1.76	6.6	0.614
3.2	2.07	6.8	0.418
3.4	2.34	7.0	0.278
3.6	2.55	7.2	0.181
3.8	2.71	7.4	0.117
4.0	2.81	7.6	0.075
4.2	2.87	7.8	0.048
4.4	2.89	8.0	0.030
4.6	2.89	8.2	0.019
4.8	2.84	8.4	0.012
5.0	2.76	8.6	0.0077
5.2	2.63	8.8	0.0048

Show that the data are consistent with the mechanism:

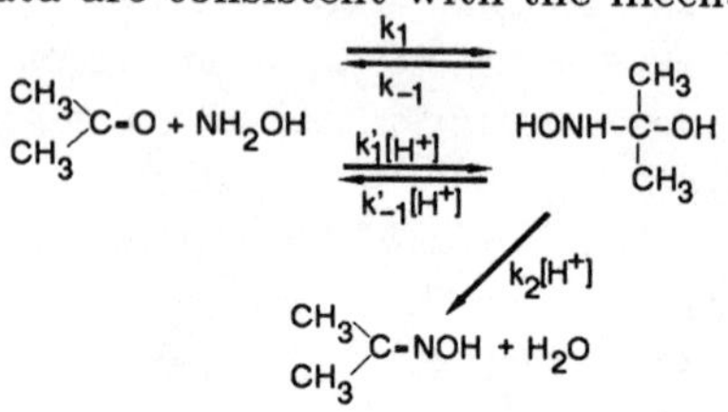

in which formation of the neutral tetrahedral intermediate takes place by an uncatalyzed pathway, k_1, and by a concerted acid catalyzed pathway simultaneously. Evaluate as many of the rate constants, or rate constant ratios, as possible. What argument can you give that the k_1 pathway is **not** a water catalyzed path? What rationalization can you provide for the fact that this reaction gives an uncatalyzed formation of the neutral tetrahedral intermediate while the methoxylamine reaction with p-methoxybenzaldehyde, discussed in Section 7.10, exhibited acid catalysis by a trapping mechanism?

Hint: Begin this problem by using the Bodenstein approximation that the tetrahedral intermediate is an unstable species in order to obtain an expression for k_ψ *in terms of the rate constants in the mechanistic scheme, and concentrations of* H^+ *and* NH_2OH. *Make a table showing* $[H^+]$, k_ψ, $[NH_2OH]$, $k_\psi/[NH_2OH]$, *and* $k_\psi/[H^+][NH_2OH]$ *for each pH. From the data for pH of ca. 4.0–8.8, evaluate* k_1k_2/k_{-1}. *For the three highest concentrations of* H^+, *note that a plot of* $k_\psi/[NH_2OH]$ *against* $[H^+]$ *is nearly a straight line. Use this to evaluate* $k_1'k_2/(k_{-1}' + k_2)$ *and* $k_1k_2/(k_{-1}' + k_2)$. *Note particularly the ratio of* k_1/k_1' *that you can get from the above ratios of rate constants. If* k_1 *represented a water catalyzed pathway, what would you expect for* k_1/k_1'*?*
In order to understand why this reaction is different from the reaction of methoxylamine with aromatic aldehydes, consider the differences in the equilibrium constants for formation of the zwitterionic tetrahedral intermediate, and what this means with respect to the relative values of the rate constants for the reverse of the zwitterion formation in the two reactions.

Problem 7.5

It was stated in Section 7.7 that the Brønsted α for reaction of hydroxide ion with a series of arylnitromethanes is 1.54. What is the Brønsted β for the reaction of water with a series of arylnitronate anions?
The Brønsted β for the reactions of a series of bases with phenylnitromethane is 0.65. The Hammett rho values for reactions of a series of arylnitromethanes with bases are ca. 1.3. If the $\beta = 0.65$ is indicative of the extent of proton transfer, and if the negative charge of the arylnitromethyl moiety at the transition state is localized on the carbon, what would be the Hammett rho value for formation of hypothetical nitronate ions with charge localized on carbon? Is this the magnitude that you expect for such a species?
Hints: The first part is too easy; $\alpha + \beta = ?$. *For the second part, go back to Chapter 4 and look up some rho values for reactions involving equilibrium formation of charges on the atoms attached to the aryl ring.*

Problem 7.6

In the reaction of methoxylamine with p-methoxybenzaldehyde, trapping of the zwitterionic intermediate by an acid produces the positively charged tetrahedral intermediate which must then lose a proton in order to get to the neutral tetrahedral intermediate. Why is that proton loss fast enough that it is not kinetically significant. *Hint: Consider the pK_a's of the -OH and the -NH_2^+- groups of the positively charged intermediate, and what this means for the rates of removal of the two different protons by any of the bases present.*

Problem 7.7

Rate constants for hydroxide ion catalyzed breakdown of hemiorthoesters are frequently near the diffusion controlled limit of $\simeq 10^{10} M^{-1} sec^{-1}$ (R. A. McClelland and L. J. Santry, **Accts. Chem. Res., 16**, 394 **(1983)**.). The pK_a's of hemiorthoesters are $\simeq 11$. What is the approximate upper limit on the lifetime of the conjugate base of such hemiorthoesters?

Hint: If the observed rate is near the diffusion controlled limit, then the breakdown of the conjugate base of the hemiorthoester cannot be the rate-limiting step; the rate of breakdown must be greater than the rate of re-formation of the hemiorthoester.

Problem 7.8

Elimination reactions of some alkyl halides and sulfonates show general base catalysis while others show specific base catalysis. Discuss the structural features, and resulting acid-base properties, of the alkyl halides which are likely to lead to the two types of catalysis. *Hint: This is just like the reverse of the carbonyl addition reactions. If you have difficulty, see the 1980 part of ref. [15].*

Problem 7.9

Some addition reactions of carbonyl compounds which produce only moderately basic zwitterionic intermediates and which show general acid catalysis with Brønsted slopes near unity are unusually well catalyzed by 'bifunctional' catalysts, such as bicarbonate ion. Suggest a reasonable explanation.

Hint: The slope of unity must mean that the protonation of the intermediate is thermodynamically unfavorable. Since conversion of the zwitterion to the neutral intermediate is thermodynamically favorable, the general acid catalysis must go uphill, then downhill. Think how can this be avoided.

CHAPTER 8

Equilibrium and Secondary Kinetic Isotope Effects

8.1 Introduction

In our discussions of structural effects on reactivities, we have usually made arguments about the origins of the effects that are based on potential energy curves or surfaces. Free energy, however, is the actual energy that is directly connected with rate or equilibrium constants. Even in those cases where we can be reasonably sure that entropies of reaction are not appreciably changed by changes in structures, changes in reactivities are directly connected with enthalpies, which are still not equivalent to potential energies.

Isotope effects on rates and equilibria of reactions are the most direct examples of the differences between thermodynamic energy quantities, such as free energies and enthalpies, and potential energies. Our development of molecular orbital theory in Chapter 5 made use of the Born-Oppenheimer approximation that nuclear and electronic wavefunctions can be separated so that we were able to calculate electronic wave functions for fixed nuclear positions. The electronic Hamiltonian for any molecule contains no terms having to do with nuclear mass. The potential energy, which is the electronic plus nuclear–nuclear repulsion energy, of a system is, therefore, completely independent of nuclear masses and does not change when isotopic substitutions are made in the system.

Nevertheless, free energies, enthalpies, and entropies, of a system do change when isotopic substitutions are made. These changes are the result of changes in the kinetic energies associated with nuclear motions such as vibrations, rotations, and translations of molecules. These motions, particularly the vibrations, do not cease even at absolute zero of temperature, so that the total energy of any system, even at absolute zero, is made up partly of kinetic energy and is nuclear mass dependent.

Just how the kinetic energies of systems are associated with thermodynamic quantities is the subject matter of statistical thermodynamics.

8.2 The Boltzmann Distribution

The Boltzmann distribution law,

$$n_i = \frac{N_0}{P} \exp\left(-\frac{\varepsilon_i}{\kappa T}\right), \tag{8.1}$$

where n_i is the number of systems having the energy ε_i at a temperature T, N_0 is the total number of systems, κ is Boltzmann's constant, and P is a normalization constant chosen such that $\sum_i n_i = N_0$, is the most important equation of statistical mechanics for our purposes. Although the Boltzmann distribution may be derived from slightly more basic postulates [1], [2], we can regard it as an assumption that, like the basic postulates of quantum mechanics, leads to results in agreement with experiment.

It might make things seem more reasonable to note that the exponential form of the Boltzmann distribution is directly related to the general behavior of probabilities. If p_i is the probability of an event i and p_j is the probability of an event j, then $p_i p_j$ is the probability of event i and event j simultaneously. If the probability of a system being in a particular state is a function only of the energy of the state, and the total energy of a system, ε, is the sum of the energies, ϵ, of any component parts, we would have to have, for example, that

$$p_{\epsilon_1} p_{\epsilon_2} = p_{(\epsilon_1+\epsilon_2)}.$$

The simplest form of p's satisfying equations of this type is an exponential function of the energies. The Boltzmann distribution law, which has the exponential form, gives

$$\begin{aligned} P\frac{n(\varepsilon)}{N_0} &= \exp\left(-\frac{\varepsilon}{\kappa T}\right) = \exp\left(-\frac{\epsilon_1+\epsilon_2+\cdots+\epsilon_r}{\kappa T}\right) \\ &= \left[\exp\left(-\frac{\epsilon_1}{\kappa T}\right)\right]\cdot\left[\exp\left(-\frac{\epsilon_2}{\kappa T}\right)\right]\cdots\left[\exp\left(-\frac{\epsilon_r}{\kappa T}\right)\right] \\ &= P\frac{n_1 n_2\cdots n_r}{N_0}, \end{aligned} \tag{8.2}$$

$$\text{when} \quad \varepsilon = \epsilon_1+\epsilon_2+\cdots+\epsilon_r.$$

The total energy of a molecule is usually the sum of the electronic, vibrational, rotational, and translational energies. For a given molecule, the energies of each type are quantized and only certain discrete values are possible. These allowed energy levels form a sort of 'space' that can be occupied according to the Boltzmann distribution; the lower energy spaces are more 'desirable' than the higher energy ones. If a molecule rearranges to an isomer, the allowed discrete energy levels will be different from those of the original isomer and will provide a different amount and desirability of 'space'. Suppose we have a molecule A and an isomer, B. Let's define a system to be a collection of atoms of the proper number and type to make one molecule of A or one molecule of B, and let's take a large number of systems, such as a mole, 10^{23}, and ask how many systems there are with an arrangement of the component atoms that we recognize as that of an A molecule. Let's label the energy levels that are available to the A molecules as $\varepsilon_1^A, \varepsilon_2^A, \ldots, \varepsilon_r^A$, so that the total number of A molecules according to the Boltzmann distribution law is

$$N_A = \sum_{i=1}^{r_A} n_i = \frac{N_0}{P} \sum_{i=1}^{r_A} \exp\left(-\frac{\varepsilon_i^A}{\kappa T}\right).$$

Similarly, the Boltzmann distribution gives the number of B molecules as

$$N_B = \sum_{i=1}^{r_B} n_i = \frac{N_0}{P} \sum_{i=1}^{r_B} \exp\left(-\frac{\varepsilon_i^B}{\kappa T}\right).$$

If we know the number of molecules of each type, then we also know the equilibrium constant for the isomerization reaction,

$$K = \frac{N_B}{N_A} = \frac{\sum_{i=1}^{r_A} \exp\left(-\frac{\varepsilon_i^A}{\kappa T}\right)}{\sum_{i=1}^{r_B} \exp\left(-\frac{\varepsilon_i^B}{\kappa T}\right)}. \tag{8.3}$$

The summations in the numerator and denominator of Eq. (8.3) are called *sums over states*, or *partition functions*. In terms of the 'space' analogy, the sum over states for a molecule is a measure of the 'volume' available to the molecule. The sums over states determine the partitioning of systems between the two different atomic arrangements in an isomerization reaction.

For the type of exchange reaction in which we have been interested in all of our discussions of structural effects on equilibria,

$$A + B \rightleftharpoons C + D, \tag{8.4}$$

the same reasoning as above gives [1]:

$$K = \frac{\sum_{i=1}^{r_C} \exp\left(-\frac{\varepsilon_i^C}{\kappa T}\right) \sum_{i=1}^{r_D} \exp\left(-\frac{\varepsilon_i^D}{\kappa T}\right)}{\sum_{i=1}^{r_A} \exp\left(-\frac{\varepsilon_i^A}{\kappa T}\right) \sum_{i=1}^{r_B} \exp\left(-\frac{\varepsilon_i^B}{\kappa T}\right)}. \tag{8.5}$$

If we wish to obtain expressions for the thermodynamic quantities pertinent to the reactions in terms of the sums over states, we can obtain them from the relationships:

$$\Delta G^\circ = -RT \ln K, \quad \Delta H^\circ = -R\frac{d\ln K}{d(1/T)},$$

$$\text{and} \quad \Delta S^\circ = R\ln K - \frac{R}{T}\frac{d\ln K}{d(1/T)}.$$

It is really remarkable that the sums over states for the molecules completely determine the thermodynamics of the equilibria.

8.3 The Resolution of Sums Over States

The first task necessary for the use of the equations given above is to resolve the sums over states into their component parts of electronic, vibrational, rotational, and translational energy levels. We will assume that these component parts are independent one of another in the sense that, for example, the same set of rotational energy levels is available to a molecule regardless of which vibrational energy level is occupied. For most organic molecules at ordinary temperatures, this assumption should be valid. The assumption would be false if, for example, a vibrational energy level had an average bond length different enough from another vibrational energy level that moments of inertia were affected, and if these levels were both appreciably occupied.

With this assumption of independence of contributions, we can write the total molecular energy, ε, of any molecule as

$$\varepsilon_{evrt} = \epsilon_e + \epsilon_v + \epsilon_r + \epsilon_t,$$

where ϵ_e is the occupied electronic energy level, ϵ_v is the occupied vibrational energy level, ϵ_r is the occupied rotational energy level, and ϵ_t is the occupied translational energy level. For virtually all organic molecules at ordinary temperatures, only one electronic energy level needs to be considered since the difference in energies between the ground and first excited electronic states is large in comparison to κT. The sum over states for a molecule can then be written as

$$\sum_{evrt} \exp\left(\frac{-\varepsilon_{evrt}}{\kappa T}\right) = \exp\left(\frac{-\epsilon_0^\circ}{\kappa T}\right)\left[\sum_{v=1}^{\infty} \exp\left(\frac{-\epsilon_v}{\kappa T}\right)\right]$$
$$\times \left[\sum_{r=1}^{\infty} \exp\left(\frac{-\epsilon_r}{\kappa T}\right)\right]\left[\sum_{t=1}^{\infty} \exp\left(\frac{-\epsilon_t}{\kappa T}\right)\right],$$

where ϵ_0° is the ground state electronic energy, and v, r, and t are the indices over the vibrational, rotational, and translational energy states, respectively. It will be convenient to define vibrational, rotational, and translational partition functions, f'_{vib}, f_{rot}, and f_{trans} as

$$f'_{vib} = \sum_{v=1}^{\infty} \exp\left(\frac{-\epsilon_v}{\kappa T}\right), \quad f_{rot} = \sum_{r=1}^{\infty} \exp\left(\frac{-\epsilon_r}{\kappa T}\right),$$

$$\text{and} \quad f_{trans} = \sum_{t=1}^{\infty} \exp\left(\frac{-\epsilon_t}{\kappa T}\right),$$

so that

$$\sum_{evrt} \exp\left(\frac{-\varepsilon_{evrt}}{\kappa T}\right) = \exp\left(\frac{-\epsilon_0^\circ}{\kappa T}\right) f'_{vib} f_{rot} f_{trans}. \tag{8.6}$$

The electronic energy levels of a molecule can, of course, be obtained from the Schrödinger equation for the electrons in the presence of the nuclei fixed at their equilibrium positions as described in Chapter 5. The other energy levels are obtained by solutions of the quantum mechanical equations for the various types of motion. Some of the energy levels can be obtained from experimental data; the vibrational levels from infrared and Raman spectra, or the rotational levels from infrared fine structure or microwave spectra. The quantum mechanical equations are particularly useful since they provide information about how molecular structure affects the various partition functions.

8.4 Translational Energy Levels

We will obtain the translational partition function, f_{trans}, by starting with the 'particle in a box' problem. We have a particle of mass m in a box having 'impenetrable' walls and dimensions $a \times b \times c$. The potential energy within the box is zero, and becomes infinite at the walls. Within the box, the Schrödinger equation is

$$-\frac{\hbar^2}{2m}\left(\frac{\partial^2}{\partial x^2} + \frac{\partial^2}{\partial y^2} + \frac{\partial^2}{\partial z^2}\right) \Psi_I(x,y,z) = \epsilon_I \Psi_I(x,y,z).$$

It is fairly obvious that the wave function is separable: $\Psi_{ijk}(x,y,z) = \Psi_i(x)\Psi_j(y)\Psi_k(z)$, so that the Schrödinger equation can be written as three independent equations

$$-\frac{\hbar^2}{2m}\frac{\partial^2 \Psi_i(x)}{\partial x^2} = \epsilon_i \Psi_i(x), \quad -\frac{\hbar^2}{2m}\frac{\partial^2 \Psi_j(y)}{\partial y^2} = \epsilon_j \Psi_j(y),$$

$$\text{and} \quad -\frac{\hbar^2}{2m}\frac{\partial^2\Psi_k(z)}{\partial z^2} = \epsilon_k \Psi_k(z),$$

where the total translational energy is the sum of the three components, $\epsilon_{ijk}(x,y,z) = \epsilon_i(x)+\epsilon_j(y)+\epsilon_k(z)$. The most general solution of this type of equation is an exponential function of the pertinent coordinate (which can be written as a linear combination of sine and cosine functions of the coordinate), but we must also have the wave function equal to zero at the two walls in order for it to be continuously zero outside the walls. If we choose the walls to be located at the origin of the Cartesian coordinate system and at the coordinate value of the dimension for each coordinate, we can choose sine functions that satisfy the Schrödinger equations and the continuity requirement by setting

$$\Psi_i(x) = C_i \sin\frac{n_i x}{a}\pi, \qquad \Psi_j(y) = C_j \sin\frac{n_j y}{b}\pi,$$

$$\text{and} \quad \Psi_k(z) = C_k \sin\frac{n_k z}{c}\pi,$$

where a, b, and c are, as stated above, the dimensions of the box, and the n's are any integers. Substitution of these functions back into the one-dimensional wave equations gives:

$$\epsilon_i(x) = \frac{n_i^2 h^2}{8ma^2}, \quad \epsilon_j(y) = \frac{n_j^2 h^2}{8mb^2}, \quad \text{and} \quad \epsilon_k(z) = \frac{n_k^2 h^2}{8mc^2},$$

so that the allowed total translational energies are

$$\epsilon_{ijk} = \frac{n_i^2 h^2}{8ma^2} + \frac{n_j^2 h^2}{8mb^2} + \frac{n_k^2 h^2}{8mc^2}.$$

Substitution of these values into the expression for f_{trans} gives:

$$f_{trans} = \sum_{n_i=0}^{\infty}\sum_{n_j=0}^{\infty}\sum_{n_k=0}^{\infty} \exp\left\{\frac{-h^2}{8m\kappa T}\left(\frac{n_i^2}{a^2}+\frac{n_j^2}{b^2}+\frac{n_k^2}{c^2}\right)\right\}$$

$$= \left[\sum_{n_i=0}^{\infty} \exp\left(\frac{-n_i^2 h^2}{8m\kappa T a^2}\right)\right] \times \left[\sum_{n_j=0}^{\infty} \exp\left(\frac{-n_j^2 h^2}{8m\kappa T b^2}\right)\right]$$

$$\times \left[\sum_{n_k=0}^{\infty} \exp\left(\frac{-n_k^2 h^2}{8m\kappa T c^2}\right)\right].$$

For molecules of any reasonable mass, the translational energy levels are so closely spaced that we may consider the n's to be continuous variables and replace the summations by integrations over the quantum numbers:

$$\sum_{n_i=0}^{\infty} \exp\left(\frac{-n_i^2 h^2}{8m\kappa T a^2}\right) = \int_0^{\infty} \exp\left(\frac{-n_i^2 h^2}{8m\kappa T a^2}\right) dn_i$$

$$= \frac{a(2\pi m\kappa T)^{1/2}}{h}.$$

The last line can be found in any handbook from the standard form of the integral. Completely analogous expressions for the sums over n_j and n_k are obvious. Substitution into the expression for the translational partition function finally gives:

$$f_{trans} = \frac{(2\pi m\kappa T)^{3/2} V}{h^3}, \quad \text{where} \quad V = abc. \tag{8.7}$$

Before proceeding with the other partition functions, it is interesting to note that the translational partition functions, alone, can lead to effects on equilibria. For the exchange reaction of Eq. (8.4), even if all other terms in Eq. (8.5) cancel, we are left with

$$K = \frac{f_{trans}(C) f_{trans}(D)}{f_{trans}(A) f_{trans}(B)} = \left(\frac{m_C m_D}{m_A m_B}\right)^{3/2}.$$

Suppose we have a reaction in which a very light molecule A, $m_A = 15$, and a very heavy molecule B, $m_B = 615$, produce two medium molecules C and D, $m_C = m_D = 315$; then the translational partition functions, alone, contribute a factor of $(315^2/15 \cdot 615)^{3/2} \simeq 35$ to the equilibrium constant. There is a clear danger in trying to interpret structural effects on equilibria in terms of the usual electronic, steric, and conformational factors unless the products of masses are nearly equal on the two sides of the reaction.

8.5 Rotational Energy Levels

The rotational and vibrational motions of molecules are most efficiently treated by the use of some operator algebra, which avoids the problem of actually obtaining the rather messy wave functions. The needed energy levels can be obtained without ever solving the Schrödinger equations. The operator algebra to be used deals with the commutation properties of quantum mechanical operators pertinent to the motions.

The basic theorem to be used is that if any two operators, say $\mathcal{R}$ and $\mathcal{S}$, commute, $\mathcal{RS} = \mathcal{SR}$, then any eigenfunction of one operator may also be an eigenfunction of the other. Conversely, if two operators do not commute, then they cannot have common eigenfunctions. Although we won't prove the theorem, the following should make it seem reasonable, and could be developed into a proof without too much trouble. Suppose we have

$$\mathcal{R}\Psi = r\Psi,$$

and then left-multiply by the commuting operator, $\mathcal{S}$, to obtain

$$\mathcal{SR}\Psi = \mathcal{S}r\Psi = r(\mathcal{S}\Psi) = \mathcal{R}(\mathcal{S}\Psi),$$

using the fact that any operator commutes with a constant. This equation shows that if Ψ is an eigenfunction of $\mathcal{R}$, then $\mathcal{S}\Psi$ is also an eigenfunction of $\mathcal{R}$ with the same eigenvalue as Ψ. This must mean that Ψ is either an eigenfunction of $\mathcal{S}$, or is transformed by $\mathcal{S}$ into a linear combination of degenerate eigenfunctions of $\mathcal{R}$.

In Chapter 5, particularly in connection with Table 5.1, we discussed how one can construct quantum mechanical operators by substituting basic operators, such as that for momentum, into the classical expressions for the desired observables. For rotation problems, the observables will have to do with angular momentum.

If $\vec{\wp}$ is the linear momentum of a particle of mass m located at a point (x, y, z),

$$\vec{\wp} = m\left(\frac{\partial x}{\partial t}\vec{\imath} + \frac{\partial y}{\partial t}\vec{\jmath} + \frac{\partial z}{\partial t}\vec{k}\right),$$

where $\vec{\imath}, \vec{\jmath}, \vec{k}$ are unit vectors along the three cartesian coordinates, then the angular momentum, $\vec{\omega}$, of the particle about the origin is defined as the vector cross product of the position vector, $\vec{r} = x\vec{\imath} + y\vec{\jmath} + z\vec{k}$, and $\vec{\wp}$:

$$\vec{\omega} = \vec{r} \times \vec{\wp} = \begin{vmatrix} \vec{\imath} & \vec{\jmath} & \vec{k} \\ x & y & z \\ p_x & p_y & p_z \end{vmatrix},$$

$$= [yp_z - zp_y]\vec{\imath} + [zp_x - xp_z]\vec{\jmath} + [xp_y - yp_x]\vec{k}$$

where $p_x = m\partial x/\partial t$, $p_y = m\partial y/\partial t$, and $p_z = m\partial z/\partial t$.

We can now write the quantum mechanical operators corresponding to angular momentum, $\mathcal{M}$, the three components of angular momentum, $\mathcal{M}_x$, $\mathcal{M}_y$, and $\mathcal{M}_z$, and the operator corresponding to the square of the length of the angular momentum vector, $\mathcal{M}^2$, by substituting the quantum mechanical operators for linear momentum, $-i\hbar\partial/\partial x$, $-i\hbar\partial/\partial y$, and $-i\hbar\partial/\partial z$, into the classical expressions. We obtain:

$$\mathcal{M} = -i\hbar \begin{vmatrix} \vec{i} & \vec{j} & \vec{k} \\ x & y & z \\ \frac{\partial}{\partial x} & \frac{\partial}{\partial y} & \frac{\partial}{\partial z} \end{vmatrix},$$

$$\mathcal{M}_x = -i\hbar \left[y\frac{\partial}{\partial z} - z\frac{\partial}{\partial y} \right],$$

$$\mathcal{M}_y = -i\hbar \left[z\frac{\partial}{\partial x} - x\frac{\partial}{\partial z} \right],$$

$$\mathcal{M}_z = -i\hbar \left[x\frac{\partial}{\partial y} - y\frac{\partial}{\partial x} \right],$$

and

$$\mathcal{M}^2 = \mathcal{M}_x^2 + \mathcal{M}_y^2 + \mathcal{M}_z^2.$$

Now, let's look at the commutation properties of these operators with one another. In order to develop the operator algebra, we will apply the operators to a general function, $f(x, y, z)$; if relationships exist for the operators applied to this general function, then we will write them as relationships of the operators. The process should become obvious as we proceed.

We'll start with $\mathcal{M}_x$ and $\mathcal{M}_y$:

$$\mathcal{M}_x f(x,y,z) = -i\hbar \left[y\frac{\partial f(x,y,z)}{\partial z} - z\frac{\partial f(x,y,z)}{\partial y} \right],$$

$$\begin{aligned} \mathcal{M}_y\mathcal{M}_x f(x,y,z) &= -\hbar^2 \left[z\frac{\partial}{\partial x}y\frac{\partial f}{\partial z} - z\frac{\partial}{\partial x}z\frac{\partial f}{\partial y} - x\frac{\partial}{\partial z}y\frac{\partial f}{\partial z} + x\frac{\partial}{\partial z}z\frac{\partial f}{\partial y} \right] \\ &= -\hbar^2 \left[zy\frac{\partial^2 f}{\partial x \partial z} - z^2\frac{\partial^2 f}{\partial x \partial y} - xy\frac{\partial^2 f}{\partial z^2} + x\frac{\partial f}{\partial y} + xz\frac{\partial^2 f}{\partial y \partial z} \right]. \end{aligned}$$

Note the last term in the next-to-last line, $x\frac{\partial}{\partial z}z\frac{\partial f}{\partial y}$, where the operator $\partial/\partial z$ operates on everything to the right of it, giving the two terms in the last line of the equation. Applying the operators in the opposite order, we get

$$\begin{aligned} \mathcal{M}_x\mathcal{M}_y f(x,y,z) &= -\hbar^2 \left[y\frac{\partial}{\partial z}z\frac{\partial f}{\partial x} - y\frac{\partial}{\partial z}x\frac{\partial f}{\partial z} - z\frac{\partial}{\partial y}z\frac{\partial f}{\partial x} + z\frac{\partial}{\partial y}x\frac{\partial f}{\partial z} \right] \\ &= -\hbar^2 \left[y\frac{\partial f}{\partial x} + yz\frac{\partial^2 f}{\partial x \partial z} - xy\frac{\partial^2 f}{\partial z^2} - z^2\frac{\partial^2 f}{\partial x \partial y} + xz\frac{\partial^2 f}{\partial y \partial z} \right]. \end{aligned}$$

Subtracting one from another, we have

$$\begin{aligned}\mathcal{M}_x\mathcal{M}_y f(x,y,z) - \mathcal{M}_y\mathcal{M}_x f(x,y,z) =& \hbar^2\left[x\frac{\partial f}{\partial y} - y\frac{\partial f}{\partial x}\right],\\ =& i\hbar\mathcal{M}_z f(x,y,z).\end{aligned}$$

Since the equation is true for any function, we write it as an operator equation:

$$\mathcal{M}_x\mathcal{M}_y - \mathcal{M}_y\mathcal{M}_x = i\hbar\mathcal{M}_z. \tag{8.8}$$

The developments of the following commutation relationships are similar to that just given and are left as an exercise at the end of the Chapter:

$$\mathcal{M}_y\mathcal{M}_z - \mathcal{M}_z\mathcal{M}_y = i\hbar\mathcal{M}_x, \tag{8.9}$$

$$\mathcal{M}_z\mathcal{M}_x - \mathcal{M}_x\mathcal{M}_z = i\hbar\mathcal{M}_y, \tag{8.10}$$

$$\mathcal{M}^2\mathcal{M}_x - \mathcal{M}_x\mathcal{M}^2 = 0, \tag{8.11}$$

$$\mathcal{M}^2\mathcal{M}_y - \mathcal{M}_y\mathcal{M}^2 = 0, \tag{8.12}$$

$$\mathcal{M}^2\mathcal{M}_z - \mathcal{M}_z\mathcal{M}^2 = 0, \tag{8.13}$$

Equations (8.8)–(8.10) show that the three component angular momentum operators cannot have shared eigenfunctions since they do not commute. If a system is in a state such that a definite value for one component of angular momentum is observed, the other two components have indeterminate values. The last three equations show that $\mathcal{M}^2$ can have eigenfunctions in common with the component operators. Since the three components cannot have common eigenfunctions, any eigenfunction of $\mathcal{M}^2$ can also be an eigenfunction of one and only one of the three components; this is arbitrarily chosen to be the z component, so that we write:

$$\mathcal{M}^2\Psi_{lm} = \omega_l^2\Psi_{lm}, \quad \text{and} \quad \mathcal{M}_z\Psi_{lm} = \xi_m\Psi_{lm}.$$

Now, we can begin to consider the relationship between the eigenvalues of the two operators. Obviously, the eigenvalue ω_l^2 is positive since it corresponds to the observable square of the angular momentum. The eigenvalue ξ_m does not need to be positive, but its square is certainly positive, and is the eigenvalue of $\mathcal{M}_z^2$:

$$\mathcal{M}_z\mathcal{M}_z\Psi_{lm} = \mathcal{M}_z^2\Psi_{lm} = \mathcal{M}_z\xi_m\Psi_{lm} = \xi_m^2\Psi_{lm}.$$

Now, subtracting $\mathcal{M}_z^2$ from $\mathcal{M}^2$, we have

$$[\mathcal{M}^2 - \mathcal{M}_z^2]\,\Psi_{lm} = [\mathcal{M}_x^2 + \mathcal{M}_y^2]\,\Psi_{lm} = (\omega_l^2 - \xi_m^2)\Psi_{lm},$$

showing that Ψ_{lm} is an eigenfunction of $[\mathcal{M}_x^2 + \mathcal{M}_y^2]$ with the eigenvalue of $(\omega_l^2 - \xi_m^2)$. Since the sum of the squares of the two components of angular momentum must be positive, we have

$$(\omega_l^2 - \xi_m^2) \geq 0, \quad \text{therefore} \quad -\sqrt{\omega_l^2} \leq \xi_m \leq +\sqrt{\omega_l^2}. \tag{8.14}$$

Consider now the two operators defined by

$$\mathcal{R} \equiv [\mathcal{M}_x + i\mathcal{M}_y]\,, \quad \text{and} \quad \mathcal{L} \equiv [\mathcal{M}_x - i\mathcal{M}_y]\,. \tag{8.15}$$

You should be able to show that the commutation relationships

$$\mathcal{M}_z\mathcal{R} - \mathcal{R}\mathcal{M}_z = \hbar\mathcal{R}, \quad \text{and} \quad \mathcal{M}_z\mathcal{L} - \mathcal{L}\mathcal{M}_z = -\hbar\mathcal{L}, \tag{8.16}$$

hold by using Eqs. (8.8)–(8.10). Starting with $\mathcal{M}_z\Psi_{lm} = \xi_m\Psi_{lm}$, we left-multiply by $\mathcal{R}$ to obtain

$$\begin{aligned} \mathcal{R}\mathcal{M}_z\Psi_{lm} &= \mathcal{R}\xi_m\Psi_{lm} = \xi_m\mathcal{R}\Psi_{lm} \\ &= [\mathcal{M}_z\mathcal{R} - \hbar\mathcal{R}]\,\Psi_{lm} = \mathcal{M}_z\mathcal{R}\Psi_{lm} - \hbar\mathcal{R}\Psi_{lm} \\ \mathcal{M}_z\,(\mathcal{R}\Psi_{lm}) &= [\xi_m + \hbar]\,(\mathcal{R}\Psi_{lm})\,, \end{aligned} \tag{8.17}$$

where the first line uses the fact that any operator commutes with a constant, the second line results from substitution for $\mathcal{R}\mathcal{M}_z$ from Eq. (8.16), and the third line simply collects terms. Similarly, left multiplication of $\mathcal{M}_z\Psi_{lm} = \xi_m\Psi_{lm}$ by $\mathcal{L}$ gives

$$\begin{aligned} \mathcal{L}\mathcal{M}_z\Psi_{lm} &= \mathcal{L}\xi_m\Psi_{lm} = \xi_m\mathcal{L}\Psi_{lm} \\ &= [\mathcal{M}_z\mathcal{L} + \hbar\mathcal{L}]\,\Psi_{lm} = \mathcal{M}_z\mathcal{L}\Psi_{lm} + \hbar\mathcal{L}\Psi_{lm} \\ \mathcal{M}_z\,(\mathcal{L}\Psi_{lm}) &= [\xi_m - \hbar]\,(\mathcal{L}\Psi_{lm})\,. \end{aligned} \tag{8.18}$$

In words, these equations show that if Ψ_{lm} is an eigenfunction of $\mathcal{M}_z$ with eigenvalue ξ_m, then $\mathcal{R}\Psi_{lm}$ is another eigenfunction of $\mathcal{M}_z$ with eigenvalue of $\xi_m + \hbar$. Similarly, operation on an eigenfunction with $\mathcal{L}$ produces another eigenfunction whose eigenvalue is $\hbar$ less than the original eigenvalue. The operators $\mathcal{R}$ and $\mathcal{L}$ are called *raising* and *lowering operators*, respectively, or *ladder operators*, collectively. Notice that both $\mathcal{L}$ and $\mathcal{R}$ commute with $\mathcal{M}^2$; this means that $\mathcal{L}\Psi_{lm}$ and $\mathcal{R}\Psi_{lm}$ have the same eigenvalues, ω_l^2, of $\mathcal{M}^2$ as did the original Ψ_{lm}. The ladder operators only change the eigenvalues of $\mathcal{M}_z$.

Since we know, from Eq. (8.14), that the ξ's are bounded, the ladder operators cannot continue to generate higher or lower eigenfunctions without limit. There must then be some lowest eigenfunction, Ψ_{lo}, which is annihilated by $\mathcal{L}$, and some highest eigenfunction, Ψ_{ln}, which is annihilated by $\mathcal{R}$:

$$\mathcal{L}\Psi_{lo} = 0, \quad \text{and} \quad \mathcal{R}\Psi_{ln} = 0.$$

If the eigenvalue of Ψ_{lo} is ξ_o, then, since each application of $\mathcal{R}$ adds $\hbar$ to the eigenvalue, the eigenvalue of the m'th eigenfunction, Ψ_{lm}, is $\xi_o + m\hbar$. We can go further by left-multiplying $\mathcal{L}\Psi_{lo}$ by $\mathcal{R}$, and then substituting

$$\begin{aligned}\mathcal{R}\mathcal{L} &= [\mathcal{M}_x + i\mathcal{M}_y]\,[\mathcal{M}_x - i\mathcal{M}_y] \\ &= \left[\mathcal{M}_x^2 + \mathcal{M}_y^2 - i\hbar\mathcal{M}_x\mathcal{M}_y + i\hbar\mathcal{M}_y\mathcal{M}_x\right] \\ &= \left[\mathcal{M}^2 - \mathcal{M}_z^2 + \hbar\mathcal{M}_z\right],\end{aligned}$$

as follows:

$$\begin{aligned}\mathcal{R}\mathcal{L}\Psi_{lo} = 0 &= \left[\mathcal{M}^2 - \mathcal{M}_z^2 + \hbar\mathcal{M}_z\right]\Psi_{lo} \\ &= \left[\omega_l^2 - \xi_o^2 + \hbar\xi_o\right]\Psi_{lo}.\end{aligned}$$

Since Ψ_{lo} exists, we must have

$$\omega_l^2 - \xi_o^2 + \hbar\xi_o = 0, \quad \text{or} \quad \omega_l^2 = \xi_o(\xi_o - \hbar).$$

Working analogously with $\mathcal{L}\mathcal{R}\Psi_{ln} = 0$, we can find that

$$\omega_l^2 - \xi_n^2 - \hbar\xi_n = 0, \quad \text{or} \quad \omega_l^2 = \xi_n(\xi_n + \hbar).$$

We also know that $\xi_n = \xi_o + n\hbar$, so we have the result that

$$\omega_l^2 = (\xi_o + n\hbar)\,(\xi_o + (n+1)\hbar) = \xi_o(\xi_o - \hbar),$$

which can be solved to give

$$\xi_o = -\frac{n}{2}\hbar, \quad \text{and} \quad \omega_l^2 = -\frac{n}{2}\left(-\frac{n}{2} - 1\right)\hbar^2 = \ell(\ell+1)\hbar^2,$$

where n is any integer and $\ell = n/2$ is defined by the last equality to be any integer or half-integer. There will be $2\ell + 1$ states for each ℓ having the same eigenvalue, ω_l^2 for the $\mathcal{M}^2$ operator, and eigenvalues of $\xi_o = -\ell\hbar$ to $\xi_n = +\ell\hbar$ for the $\mathcal{M}_z$ operator.

The classical expression for the energy of a linear system is given by

$$\epsilon_{rot} = \frac{\vec{\omega} \cdot \vec{\omega}}{2I},$$

where I is the moment of inertia ($I = \sum_i m_i r_i^2$ where m_i is the mass of the i'th nucleus located at a distance of r_i from the rotational axis.), about the rotational axis, so that the quantum mechanical operator for rotational energy can be formulated from $\mathcal{M}^2$ to give the Schrödinger equation as

$$\frac{\mathcal{M}^2}{2I}\Psi_{lm} = \epsilon_l \Psi_{lm} = \frac{\ell(\ell+1)}{2I}\hbar^2 \Psi_{lm}.$$

Remembering that there are $2\ell + 1$ states for each energy level, we can write the sum over states for the linear rotating system as

$$f_{rot} = \sum_{\ell=0}^{\infty} (2\ell + 1) \exp\left(\frac{-\ell(\ell+1)\hbar^2}{2I\kappa T}\right).$$

There is a considerable complication at this point. We have assumed that all existing states can be occupied by any particle, the only restriction being that the Boltzmann distribution law holds. If a molecule posesses any elements of symmetry, the rotational wavefunctions will be either symmetric or anti-symmetric with respect to each such element. The complication arises because of the quantum restrictions on the ability of a molecule to go from one rotational state to another. The situation is similar to that discussed in Chapter 5 for the *non-crossing rule*; for the transitions between rotational states, symmetric states can undergo transitions only to other symmetric states and anti-symmetric states can undergo transitions only to anti-symmetric states. This restricts the rotational energy states that can be occupied by any particular molecule to some fraction of the total number of states, with the fraction depending on the number of symmetry operations. In somewhat loose language[1] that will suffice for our discussions, if a molecule has σ indistinguishable orientations in space, only $1/\sigma$ of the total rotational states may be occupied by any particular molecule. Taking this complication into account, and recognizing that the rotational states are closely enough spaced that the summation may be replaced by integration over

---

[1] The looseness of language arises from the fact that a precise definition of σ requires a consideration of conformations of most complex molecules and a more rigorous handling of symmetry groups [3].

the quantum number ℓ, we obtain

$$f_{rot} = \frac{1}{\sigma}\int_0^{\infty}(2\ell+1)\exp\left(\frac{-\ell(\ell+1)\hbar^2}{2I\kappa T}\right)d\ell$$
$$= \frac{2I\kappa T}{\sigma\hbar^2},$$

where the integral was put into a standard form by the change of variable $x = \ell(\ell+1)$, $dx = (2\ell+1)d\ell$.

For a general, non-linear, molecule, there are three principal axes of rotation with moments of inertia, I_x, I_y, and I_z. The total rotational partition function in the classical limit is then given by [4].

$$f_{rot} = \left(\frac{2\kappa T}{\hbar^2}\right)^{3/2}\frac{(\pi I_x I_y I_z)^{1/2}}{\sigma}, \tag{8.19}$$

which differs by the factor of $\pi^{1/2}$ from what one might have expected by extension of the two dimensional linear system to three dimensions. The derivation of this expression for the case of a *spherical top* molecule, one having three equal principal moments of inertia, is left as a problem at the end of this chapter.

Most of the cases in which the rotational partition functions make large contributions to the equilibrium constants for the exchange reactions of Eq. (8.4) are ones involving changes in symmetry numbers. For example, proton exchanges of dicarboxylic acids according to the reaction

$$^{-}OOCRCOO^{-} \quad + \quad HOOCRCOOH \xrightleftharpoons{K_{ex}} 2HOOCRCOO^{-},$$

where the two carboxyl groups are equivalently attached to the R group, have both reactants with symmetry numbers of 2, and products with symmetry numbers of unity. In the absence of any other effects, this symmetry change will result in $K = 4$.

8.6 Energy Levels of a 1-Dimensional Oscillator

We will see in the next section that complex molecules can be treated as if they were composed of a number of one-dimensional harmonic oscillators. We discussed the Hamiltonian for a single one-dimensional harmonic oscillator in Section 5.6, pg. 177 and found

$$\mathcal{H} = -\frac{\hbar^2}{2m_r}\frac{\partial^2}{\partial x^2} + \frac{k}{2}x^2, \tag{8.20}$$

where m_r is the reduced mass of the oscillator and k is the Hooke's Law force constant. The harmonic oscillator may be diagrammed as two masses joined by a Hooke's law spring

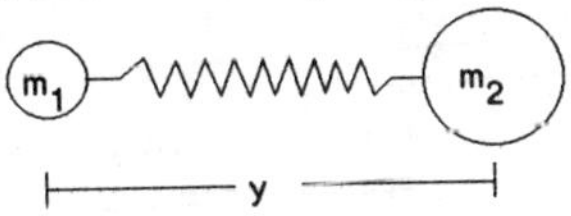

$$Force = k(y_0 - y) = kx$$

and the reduced mass is given by

$$m_r = \frac{m_1 m_2}{m_1 + m_2}.$$

Classical mechanics finds that such an oscillator vibrates with a frequency, ν_o, given by

$$\nu_o = \frac{1}{2\pi}\sqrt{\frac{k}{m_r}},$$

and we will use the symbol ν_o as shorthand for the quantities on the right.

We can construct two operators, $\mathcal{U}$, and $\mathcal{D}$,

$$\mathcal{U} = \frac{\partial}{\partial x} - 2\alpha x, \quad \text{and} \quad \mathcal{D} = \frac{\partial}{\partial x} + 2\alpha x,$$

$$\text{where} \quad \alpha = \frac{\sqrt{km_r}}{2\hbar} = \frac{\pi m_r}{\hbar}\nu_o,$$

and derive the commutation relationships of these with each other and with $\mathcal{H}$ just as we did with the operators in the previous section. You should be able to obtain

$$\mathcal{HU} - \mathcal{UH} = \frac{2\alpha\hbar^2}{m_r}\mathcal{U} = h\nu_o\mathcal{U}, \tag{8.21}$$

$$\mathcal{HD} - \mathcal{DH} = -h\nu_o\mathcal{D}, \tag{8.22}$$

$$\text{and} \quad \mathcal{UD} - \mathcal{DU} = 4\alpha. \tag{8.23}$$

Suppose we have some eigenfunction of $\mathcal{H}$

$$\mathcal{H}\Psi_i = \epsilon_i\Psi_i.$$

Left-multiplying both sides by $\mathcal{D}$, and using the commutation relationships, we obtain

$$\mathcal{DH}\Psi_i = (\mathcal{HD} + h\nu_o\mathcal{D})\,\Psi_i = \epsilon_i\mathcal{D}\Psi_i,$$

$$\mathcal{H}(\mathcal{D}\Psi_i) = (\epsilon_i - h\nu_o)(\mathcal{D}\Psi_i),$$

showing that the operation of $\mathcal{D}$ on an eigenfunction of $\mathcal{H}$ produces a new eigenfunction whose eigenvalue is $h\nu_o$ lower than that of the original eigenfunction. Since the observable energy must always be greater than zero, the operation of $\mathcal{D}$ cannot go on indefinitely; there must be some lowest eigenfunction of $\mathcal{H}$, Ψ_0, which is annihilated by $\mathcal{D}$, just as we found for the lowering operator for the rotational problem. We can, therefore, write

$$\mathcal{D}\Psi_0 = \frac{\partial \Psi_0}{\partial x} + 2\alpha x \Psi_0 = 0,$$

$$\text{or, rearranging} \quad \frac{\partial \Psi_0}{\partial x} = -2\alpha x \Psi_0.$$

Operating on this last equation with $\partial/\partial x$, we obtain

$$\frac{\partial^2 \Psi_0}{\partial x^2} = -2\alpha\Psi_0 - 2\alpha x \frac{\partial \Psi_0}{\partial x} = -2\alpha\Psi_0 + 4\alpha^2 x^2 \Psi_0.$$

Rearrangement of this last equation, and multiplication of both sides by $-\hbar^2/2m_r$, gives

$$-\frac{\hbar^2}{2m_r}\frac{\partial^2 \Psi_0}{\partial x^2} + \frac{4\hbar^2\alpha^2}{2m_r} x^2 \Psi_0 = \frac{2\hbar^2 \alpha}{2m_r}\Psi_0,$$

and, finally, substitution of $\alpha = \sqrt{km_r}/2\hbar$, shows that

$$-\frac{\hbar^2}{2m_r}\frac{\partial^2 \Psi_0}{\partial x^2} + \frac{k}{2}x^2\Psi_0 = \mathcal{H}\Psi_0 = \frac{h}{2}\nu_o. \tag{8.24}$$

Thus, the lowest eigenfunction has an eigenvalue of $h\nu_o/2$. Successive application of the *raising operator*, $\mathcal{U}$, to the lowest eigenfunction generates the higher eigenfunctions, with each application producing an eigenvalue incremented by $h\nu_o$. We have, then,

$$\mathcal{H}\Psi_n = \left(n + \frac{1}{2}\right) h\nu_o \Psi_n \tag{8.25}$$

as the general solution to the one dimensional harmonic oscillator problem.

8.7 Vibrations of Complex Molecules

For an N-atom molecule, the potential energy is a function of the $3N$ cartesian coordinates of the atoms $x_1, y_1, z_1, x_2, \cdots, z_N$. Instead of labeling the coordinates in this way, they are more commonly labelled as $q'_1, q'_2, \cdots, q'_{3N}$, where $q'_1 = x_1$, $q'_2 = y_1$, etc. At the equilibrium geometry of the molecule, energy is at a minimum with respect to changes in any of the coordinates of any of the atoms, so that we have

$$V = V_\circ + \frac{1}{2}\sum_{i=1}^{3N}\sum_{j=1}^{3N} k_{ij} q_i q_j \quad + \text{higher order terms},$$

$$\text{with } q_i = q'_i - q'_{io},$$

where q'_{io} is the equilibrium value of q'_i. Neglect of the higher order terms gives the potential energy in the quadratic approximation used for the derivation of harmonic frequencies.

Matrix notation is virtually indispensable for the discussion of vibrations of complex molecules. We can define the vector

$$\mathcal{Q}^\dagger = \begin{pmatrix} q_1 \\ q_2 \\ \vdots \\ q_{3N} \end{pmatrix} \quad \text{and} \quad \mathcal{Q} = (\, q_1 \quad q_2 \quad \cdots \quad q_{3N}\,),$$

and the matrix

$$\mathcal{K} = \begin{pmatrix} k_{11} & k_{12} & \cdots & k_{13N} \\ k_{21} & k_{22} & \cdots & k_{23N} \\ \vdots & \vdots & \ddots & \vdots \\ k_{3N1} & k_{3N2} & \cdots & k_{3N3N} \end{pmatrix},$$

and write

$$V - V_\circ = \frac{1}{2}\mathcal{Q}\mathcal{K}\mathcal{Q}^\dagger .$$

We can also write the kinetic energy T as a matrix equation

$$T = \frac{1}{2}\dot{\mathcal{Q}}\Omega\dot{\mathcal{Q}}^\dagger$$

$$= \frac{1}{2}(\,\dot q_1 \quad \dot q_2 \quad \cdots \quad \dot q_{3N}\,)\begin{pmatrix} m_1 & 0 & 0 & 0 & \cdots & 0 \\ 0 & m_1 & 0 & 0 & \cdots & 0 \\ 0 & 0 & m_1 & 0 & \cdots & 0 \\ 0 & 0 & 0 & m_2 & \cdots & 0 \\ \vdots & \vdots & \vdots & \vdots & \ddots & \vdots \\ 0 & 0 & 0 & 0 & \cdots & m_N \end{pmatrix}\begin{pmatrix} \dot q_1 \\ \dot q_2 \\ \vdots \\ \dot q_{3N} \end{pmatrix}$$

where the second line defines the matrices in the first, and $\dot{q}_i = \partial q_i/\partial t$.

The mass matrix, Ω, is a positive definite diagonal matrix so that the matrices $\Omega^{1/2}$, and $\Omega^{-1/2}$ exist and are obviously defined to have diagonal elements of $m_i^{1/2}$ and $1/m_i^{1/2}$, respectively. For reasons that will quickly become obvious, we now write the equations for potential energy and for kinetic energy as

$$V - V_\circ = \mathcal{Q}\Omega^{1/2}\Omega^{-1/2}\mathcal{K}\Omega^{-1/2}\Omega^{1/2}\mathcal{Q}^\dagger,$$

and

$$T = \dot{\mathcal{Q}}\Omega^{1/2}\Omega^{1/2}\dot{\mathcal{Q}}^\dagger.$$

We now define the *mass weighted coordinates*, $r_i \equiv q_i\sqrt{m_i}$, the vector $\mathcal{R}$ whose elements are the r_i, and a matrix $\mathcal{F}$ whose elements are $f_{ij} = k_{ij}/\sqrt{m_i m_j}$:

$$\mathcal{R} \equiv \mathcal{Q}\Omega^{1/2}, \quad \text{and} \quad \mathcal{F} \equiv \Omega^{-1/2}\mathcal{K}\Omega^{-1/2},$$

and write

$$2T = \dot{\mathcal{R}}\dot{\mathcal{R}}^\dagger, \tag{8.26}$$

and

$$2(V - V_\circ) = \mathcal{R}\mathcal{F}\mathcal{R}^\dagger. \tag{8.27}$$

The matrix $\mathcal{F}$ is obviously Hermitian since $\mathcal{K}$ is Hermitian, and can be diagonalized by a Unitary transformation

$$\mathcal{U}^\dagger\mathcal{F}\mathcal{U} = \Lambda.$$

We now insert $\mathcal{I} = \mathcal{U}\mathcal{U}^\dagger$, at two places in Eq.(8.27) to obtain

$$\begin{aligned} 2(V - V_\circ) &= \mathcal{R}\mathcal{U}\mathcal{U}^\dagger\mathcal{F}\mathcal{U}\mathcal{U}^\dagger\mathcal{R}^\dagger \\ &= (\mathcal{R}\mathcal{U})\left(\mathcal{U}^\dagger\mathcal{F}\mathcal{U}\right)(\mathcal{R}\mathcal{U})^\dagger \\ &= \mathcal{N}\Lambda\mathcal{N}^\dagger, \end{aligned} \tag{8.28}$$

where the last line defines the vector $\mathcal{N} = \mathcal{R}\mathcal{U}$ whose components are the *normal coordinates* or *normal vibrational modes* for the molecule. Note also that

$$2T = \dot{\mathcal{R}}\mathcal{U}\mathcal{U}^\dagger\dot{\mathcal{R}}^\dagger = \dot{\mathcal{N}}\dot{\mathcal{N}}^\dagger. \tag{8.29}$$

In terms of the normal coordinates, we can write the $3N$ *independent* Schrödinger equations

$$-\frac{\hbar^2}{2}\frac{\partial^2}{\partial n_i^2}\Psi_k + \frac{\lambda_i}{2}n_i^2\Psi_k = (k + \frac{1}{2})h\nu_\circ^i$$

where $2\pi\nu_0^i = \sqrt{\lambda_i}$, since each of the independent equations is of exactly the form of the one-dimensional harmonic oscillator that we treated above.

The matrix Λ in Eq. (8.28) will have six of its diagonal elements equal to zero for a non-linear molecule. The normal coordinates corresponding to these six zero eigenvalues are those for the three degrees of translational and three degrees of rotational freedom.

Each of the $3N-6$ vibrations of a non-linear molecule will have its particular fundamental frequency, and the vibrational partition function for the molecule will be a product of the sums over states for each of the vibrations:

$$f'_{vib} = \prod_{j=1}^{3N-6} \sum_{n_j=0}^{\infty} \exp\left(\frac{-\left(n_j + \frac{1}{2}\right) h\nu_0^j}{\kappa T}\right). \tag{8.30}$$

Since $h\nu_0^j/2$ of energy is present for each vibration even at absolute zero of temperature, it is customary to collect these energies into a sum,

$$\epsilon_{zp} = \sum_{j=1}^{3N-6} \frac{h\nu_0^j}{2},$$

and write

$$f'_{vib} = f_{vib} \exp\left(\frac{-\epsilon_{zp}}{\kappa T}\right),$$

where f_{vib} is now given by

$$f_{vib} = \prod_{j=1}^{3N-6} \sum_{n_j=1}^{\infty} \exp\left(\frac{-n_j h\nu_0^j}{\kappa T}\right). \tag{8.31}$$

The *zero point energy*, ϵ_{zp}, is added to the electronic energy, ϵ_0°, of the molecule, and the sum over states for the molecule, Eq. (8.6), is written as

$$\sum_{evrt} \exp\left(\frac{-\varepsilon_{evrt}}{\kappa T}\right) = \exp\left(\frac{-\varepsilon_0}{\kappa T}\right) f_{vib} f_{rot} f_{trans} \tag{8.32}$$

$$\text{where} \quad \varepsilon_0 = \varepsilon_0^\circ + \epsilon_{zp}.$$

Each of the sums in Eq. (8.31) can be written in a more convenient form,

$$\sum_{n_j=1}^{\infty} \exp\left(\frac{-n_j h\nu_0^j}{\kappa T}\right) = \left[1 - \exp\left(\frac{-h\nu_0^j}{\kappa T}\right)\right]^{-1},$$

which can be verified by multiplication of both sides by the quantity enclosed in the square brackets. The vibrational partition function is then:

$$f_{vib} = \prod_{j=1}^{3N-6} \left[1 - \exp\left(\frac{-h\nu_o^j}{\kappa T} \right) \right]^{-1} . \tag{8.33}$$

Although the force constants and the related fundamental frequencies for vibrations of a molecule can, in principle, be obtained from basic theory by solving the electronic Schrödinger equation for all sets of nuclear positions corresponding to the various distortions from equilibrium geometry, they are most often obtained from experimental infrared and Raman spectroscopic studies. The observed absorbances are due to transitions from one vibrational level to the next higher so that the energy of an absorbed photon is equal to $h\nu_o$ for the vibrational mode. The matter of which of the normal modes are infrared active, which are Raman active, and the far-from-trivial problem of going from the observed spectra to a complete set of normal mode frequencies are discussed in reference [5].

8.8 Isotope Effects on Equilibria

The partition functions, and the formulation of equilibrium constants in terms of them, that we have derived above may be used for any exchange reaction of the type given in Eq. (8.4). Their uses for cases of isotope exchange are particularly frequent.

The effect of isotopic substitution in a molecule on the equilibrium constant for any reaction of the molecule can always be formulated as an exchange reaction. Suppose we have some molecule, R, involved in a reaction

$$R + A + B + \cdots \xrightleftharpoons{K_1} P + C + D + \cdots ,$$

and an isotopically substituted molecule, R', undergoes the same reaction, with the isotope being in the P molecule, P', of the products

$$R' + A + B + \cdots \xrightleftharpoons{K_1'} P' + C + D + \cdots .$$

The isotope effect on the equilibrium constant is usually taken to mean the value of K_1/K_1', which, of course, is equal to the equilibrium constant for the exchange reaction

$$R + P' \xrightleftharpoons{K_{ex}} R' + P, \tag{8.34}$$

which may be written

$$K_{ex} = \frac{f_{trans}(R')f_{rot}(R')f_{vib}(R')f_{trans}(P)f_{rot}(P)f_{vib}(P)}{f_{trans}(R)f_{rot}(R)f_{vib}(R)f_{trans}(P')f_{rot}(P')f_{vib}(P')} \times \exp\left(-\frac{\epsilon_{zp}(R') - \epsilon_{zp}(R) + \epsilon_{zp}(P) - \epsilon_{zp}(P')}{\kappa T}\right), \tag{8.35}$$

where the exponential part has only zero-point energy terms because the electronic energies of the normal and isotopically substituted molecules are the same. In the ratios of partition functions which occur for any exchange reaction, all constant terms will cancel. We can write the ratios for the R and R' molecules as

$$\frac{f_{trans}(R')f_{rot}(R')f_{vib}(R')}{f_{trans}(R)f_{rot}(R)f_{vib}(R)} = \left(\frac{M_{R'}}{M_R}\right)^{3/2} \left(\frac{I_x(R')I_y(R')I_z(R')}{I_x(R)I_y(R)I_z(R)}\right)^{1/2} \left(\frac{\sigma(R)}{\sigma(R')}\right) \frac{f_{vib}(R')}{f_{vib}(R)}.$$

There is another simplification that can be made in the ratio of the partition functions for the normal and isotopically substituted molecules that comes from what is called the Redlich-Teller product rule [6]. In order to derive this rule, let's look again at the matrices that were used to arrive at Eq. (8.28). In particular, note that the matrix $\mathcal{K}$ has no mass dependencies, and is, therefore, the same for a normal and any isotopically substituted molecule. The matrices $\mathcal{F}$ and Λ, however, are mass dependent. We had

$$\Lambda = \mathcal{U}^\dagger \mathcal{F} \mathcal{U} = \mathcal{U}^\dagger \Omega^{-1/2} \mathcal{K} \Omega^{-1/2} \mathcal{U}. \tag{8.36}$$

In order to avoid the zero's in Λ corresponding to rotational and translational modes, let's imagine that the K matrix has some force constants causing these values to be non-zero. In the limit of vanishing force constants for the rotational modes, the λ's will be inversely proportional to the moments of inertia for the modes, and for the translational modes, the λ's will be inversely proportional to the molecular mass since these parameters are the effective masses for motions along the pertinent normal coordinates. Taking determinants of the matrices in Eq. (8.36), we have

$$\lambda_1 \lambda_2 \cdots \lambda_{3N} = \left(\frac{1}{m_1}\frac{1}{m_2}\cdots\frac{1}{m_N}\right)^3 \det(\mathcal{K}),$$

for the normal molecule, and

$$\lambda_1' \lambda_2' \cdots \lambda_{3N}' = \left(\frac{1}{m_1'}\frac{1}{m_2'}\cdots\frac{1}{m_N'}\right)^3 \det(\mathcal{K}),$$

for the isotopically substituted molecule. Dividing one equation by the other, and noting the behaviors of the λ's for the rotational and translational modes stated above, we obtain

$$\left(\frac{I'_x I'_y I'_z}{I_x I_y I_z}\right)\left(\frac{M'}{M}\right)^3\left(\prod_{i=1}^{3N-6}\frac{\lambda_i}{\lambda'_i}\right)=\prod_{j=1}^{N}\left(\frac{m'_j}{m_j}\right)^3,$$

where M and M' are the molecular masses of the normal and isotopically substituted molecules, respectively. Noting that $\lambda_i = 4\pi^2(\nu_o^i)^2$, we can take square roots of both sides of the last equation, and rearrange to obtain the usual form of the product rule

$$\left(\frac{M}{M'}\right)^{3/2}\left(\frac{I_x I_y I_z}{I'_x I'_y I'_z}\right)^{1/2}=\left(\frac{m_i}{m'_i}\right)^{3/2}\prod_{j=1}^{3N-6}\left(\frac{\nu_o^j}{\nu_o'^j}\right), \tag{8.37}$$

for a single isotopic substitution of the i'th atom. Obviously, for a large molecule many of the frequencies will be the same in the substituted and unsubstituted molecules, leading to many cancellations in the last term on the right of Eq. (8.37). It is not possible, however, to make any general rules as to which of the normal modes will be affected by the isotopic substitution, so we carry the entire product over all vibrational modes.

Substitution of Eq. (8.37) for both R and P into Eq. (8.35) gives

$$\begin{aligned} K_{ex} &= \exp\left(\frac{-\Delta\epsilon_{zp}}{\kappa T}\right) \\ &\times \prod_{i=1}^{3N_R-6}\left(\frac{\nu_o'^i\left[1-\exp\left(\frac{-h\nu_o^i}{\kappa T}\right)\right]}{\nu_o^i\left[1-\exp\left(\frac{-h\nu_o'^i}{\kappa T}\right)\right]}\right) \\ &\times \prod_{j=1}^{3N_P-6}\left(\frac{\nu_o^j\left[1-\exp\left(\frac{-h\nu_o'^j}{\kappa T}\right)\right]}{\nu_o'^j\left[1-\exp\left(\frac{-h\nu_o^j}{\kappa T}\right)\right]}\right) \\ &\times \frac{\sigma_R\sigma_{P'}}{\sigma_{R'}\sigma_P}, \end{aligned} \tag{8.38}$$

where

$$\Delta\epsilon_{zp} = \sum_{i=1}^{3N_R-6}\left(\frac{h\left[\nu_o'^i-\nu_o^i\right]}{2}\right)-\sum_{j=1}^{3N_P-6}\left(\frac{h\left[\nu_o'^j-\nu_o^j\right]}{2}\right).$$

Note that the isotopic mass terms, $(m_i/m_i')^{3/2}$, in Eq. (8.37) have cancelled between R and P.

Eq. (8.38) might be termed the 'master equation' for the calculation of isotope effects on equilibria. As a simple example of its use, let's consider the exchange reaction

$$^{13}CO \quad + \quad ^{12}CO_2 \xrightleftharpoons{K_{ex}} {}^{12}CO \quad + \quad ^{13}CO_2.$$

Carbon-12 monoxide has a stretching frequency of 2155 cm^{-1}; the reduced mass is $12 \cdot 16/(12+16) = 6.86$ for the C-12, and 7.17 for the C-13, species, so that the stretching frequency for carbon-13 monoxide is $2155\sqrt{6.86/7.17} = 2108cm^{-1}$. Carbon-12 dioxide has a symmetric stretch at 1337 cm^{-1}, an anti-symmetric stretch at 2349 cm^{-1}, and a doubly degenerate bending mode at 676 cm^{-1}. The symmetric stretching mode involves motion only of the two oxygens so that the frequency is unchanged by the isotopic substitution of carbon. We will approximate the effective mass for both the anti-symmetric stretching mode and the two bending modes to be equal to $m_O m_C/(2m_O + m_C)$ [7], so that the frequencies for carbon-13 dioxide are $2349\sqrt{(4.0/4.24)} = 2281cm^{-1}$, and $656cm^{-1}$. The zero-point energy difference is

$$\begin{aligned} \Delta\epsilon_{zp} &= \frac{1}{2}h\,(2281 - 2349 + 2155 - 2108 - 2(676 - 656))\,cm^{-1} \\ &= -\,h \cdot 30.5cm^{-1}, \end{aligned}$$

and the exponential term in the zero-point energy difference is[2]

$$\exp\left(\frac{h \cdot 30.5cm^{-1}}{\kappa T}\right) = \exp\left(4.77 \cdot 10^{-3} \cdot 30.5\right) = 1.16,$$

at $298°K$. The term from the ratio of frequencies, which comes from the translational and rotational partition functions, is

$$\frac{2281 \cdot 656 \cdot 656 \cdot 2155}{2349 \cdot 676 \cdot 676 \cdot 2108} = 0.935.$$

Finally, the terms from the vibrational partition functions are

$$\begin{aligned} &\frac{(1 - \exp(-2349h/\kappa T))\,(1 - \exp(-676h/\kappa T))^2}{(1 - \exp(-2281h/\kappa T))\,(1 - \exp(-656h/\kappa T))^2} \\ &\quad \times \frac{(1 - \exp(-2108h/\kappa T))}{(1 - \exp(-2155h/\kappa T))} = 1.008, \end{aligned}$$

2 The value of the quantity $h/\kappa T$ at $25°C$ and in units of cm is 4.77×10^{-3}.

so that we calculate

$$K_{ex} = 1.16 \cdot 0.935 \cdot 1.008 = 1.09$$

for the exchange reaction. The experimental value is 1.086 [8].

There are several points that should be noted in the calculation, since they are quite general. First, the largest factor contributing to K_{ex} is the exponential in the zero-point energy term. This term will always work in such a direction that the heavier isotope is preferentially in the molecule having the highest **sum** of frequencies involving the isotopically substituted atom. Second, the rotational and translational partition functions will generally work in the opposite direction to the zero-point energy term, but will be considerably smaller in magnitude. Last, the vibrational partition functions will generally reinforce the zero-point energy term, but will be quite small.

Some other examples of isotope exchange equilibria, all at near room temperature, are [8], [9]

$$HT \quad + \quad H_2O \quad \overset{K = 6.3}{\rightleftharpoons} \quad H_2 \quad + \quad HTO$$

$$H_2O \quad + \quad D_2O \quad \overset{K = 3.94}{\rightleftharpoons} \quad 2HDO$$

$$D_2 \quad + \quad 2HCl \quad \overset{K = 1.89}{\rightleftharpoons} \quad H_2 \quad + \quad 2DCl$$

$$H \cdot {}^{12}CN \quad + \quad {}^{13}CN^- \quad \overset{K = 1.03}{\rightleftharpoons} \quad H \cdot {}^{13}CN \quad + \quad {}^{12}CN^-$$

$${}^{15}NH_3 \quad + \quad {}^{14}NH_4^+ \quad \overset{K = 1.02}{\rightleftharpoons} \quad {}^{14}NH_3 \quad + \quad {}^{15}NH_4^+$$

$$S({}^{16}O)_2 \quad + \quad 2H_2{}^{18}O \quad \overset{K = 1.03}{\rightleftharpoons} \quad S({}^{18}O)_2 \quad + \quad 2H_2{}^{16}O.$$

A few other similar exchanges are left for exercises at the end of this chapter.

8.9 Solvent Isotope Effects

A slightly different type of equilibrium isotope effect from those discussed above is involved in comparisons of reactions in H_2O and in D_2O solutions. In most of the reactions of interest, the solute reactants and products contain rapidly exchanged hydrogens so that the actual reactions studied involve isotopic substitutions in the solute species as well as the change of solvent. Most attention has been devoted to Brønsted acids and bases and to Lewis acid-base reactions in which the solvent or its conjugate base is the base of interest.

Ionization constants for a large number of Brønsted acids have been determined in D_2O solution. Some of the reported data [10] are shown in Table 8.1.

TABLE 8.1
Solvent Isotope Effects on Acid Ionizations, 25°C

Acid	pK_{H_2O}	pK_{D_2O}	ΔpK
Maleic Acid, K_1	1.98	2.53	0.55
Cyanoacetic Acid	2.50	2.94	0.44
Fluoroacetic Acid	2.72	3.13	0.41
2-Chlorobenzoic Acid	2.91	3.38	0.47
Nitrous Acid	2.99	3.53	0.54
m-Nitrobenzoic Acid	3.62	4.12	0.50
2,4-Dinitrophenol	4.02	4.54	0.52
Benzoic Acid	4.16	4.65	0.49
Anilinium Ion	4.55	5.13	0.58
Hydrazoic Acid	4.69	5.15	0.46
Acetic Acid	4.74	5.26	0.52
Malonic Acid, K_2	5.69	6.19	0.43
3,5-Dinitrophenol	6.70	7.31	0.61
p-Nitrophenol	7.26	7.74	0.48
Ammonium Ion	9.26	9.87	0.61
Trimethylammonium Ion	9.78	10.37	0.59
2,2,2-Trifluoroethanol	12.37	13.02	0.65
Water	15.74	16.60	0.86

Measurements of the equilibrium hydrations of aldehydes in deuterium oxide have also been reported [10]. Values of K_{H_2O}/K_{D_2O} are generally about 0.85 near room temperature.

These solvent isotope effects on equilibria are usually discussed as involving several solvent molecules hydrogen bonded to reactants and products as well as the hydrogens formally bonded to reactants and products. *Fractionation factors* for the exchange equilibria of various types of solute-H species with solvent have been defined, tabulated, and used for estimating solvent isotope effects on equilibria [11], particularly in mixtures of H_2O and D_2O. Usually, the zero-point energy effects and symmetries are assumed to be the only significant terms in Eq. (8.38).

8.10 Secondary Kinetic Isotope Effects

Isotope effects on rates of reactions are generally divided into two classes. *Primary isotope effects* are those which occur for reactions in which bonds directly to the isotopically substituted atom are made or broken. *Secondary isotope effects* are those which occur for reactions only indirectly involving bonds to the isotopically substituted atom. These latter, secondary, isotope effects on rates are quite similar in origin and in magnitude to isotope effects on equilibria which we have been discussing.

The most commonly studied secondary kinetic isotope effects are those involving substitution of deuterium for hydrogen, and most common of these are cases in which the substitution is at a carbon undergoing a change in hybridization. Additions of nucleophiles to carbonyl compounds, Diels-Alder type reactions of alkenes, and SN1 type nucleophilic substitutions are examples of interconversions of sp^2 and sp^3 hybridized carbons. Comparisons of the isotope effects on rates with those on corresponding, or similar, equilibria should provide some measure of the change in hybridization that has occurred in going from reactant to transition state for a reaction.

In order to estimate the equilibrium effect, we need to consider the vibrations of the two different C–H bonds. An sp^3-C–H bond has a stretching mode at about 2900 cm^{-1} and a doubly degenerate bending bending mode at about 1250 cm^{-1}. A typical sp^2-C–H bond has a stretching mode at about 2900 cm^{-1}, an in-plane bending mode at about 1200 cm^{-1}, and an out-of-plane bending mode at about 900 cm^{-1}. Obviously, the greatest difference is between the out-of-plane bending mode and the bending of the sp^3-C–H; a difference of *ca.* 400 cm^{-1}.

Since the masses of hydrogen and deuterium are much less than that of carbon, the reduced mass for vibration of a C–H bond is essentially that of the isotope alone. This means that the frequencies of vibration of the C–H and C–D bonds are related by the expression $\nu_o(CD) = \nu_o(CH)\sqrt{\frac{1}{2}}$, and that the difference in zero-point energy for a vibration is given by

$$\Delta\epsilon_{zp} = \frac{1}{2}h\nu_o(CH)\left(1 - \sqrt{1/2}\right) = 0.146h\nu_o(CH).$$

For the equilibrium change from an sp^2 to sp^3 carbon, then, we expect that

$$K_D/K_H \simeq \exp\left(0.146 \cdot h \cdot 400cm^{-1}/\kappa T\right) = 1.32$$

near 25°C, neglecting the small contributions from partition functions. This valuc is in reasonable agreement with estimates of 1.18–1.30 for equilibrium additions of nucleophiles to CH_3CDO based on experiment [12].

The secondary deuterium kinetic isotope effect on the rate of reaction of hydroxide ion with p-nitrophenyl formate is $k_D/k_H = 1.24$, for substitution of D for H at the carbonyl carbon [13]. This indicates that the hybridization of the carbonyl carbon at the transition state for the reaction has progressed almost all the way to sp^3. For the addition of mercaptides to the same ester, $k_D/k_H = 1.03$, indicating very little change in hybridization at the transition state from that in the reactant [13].

Kinetic solvent isotope effects on reactions not involving proton transfers can be interpreted in much the same way by comparisons of the kinetic and corresponding equilibrium effects on reactions.

Effects of deuterium substitutions at carbons β to a reaction center can also frequently be detected. For example, The substitution of CD_3 for CH_3 in acetate esters leads to estimated equilibrium effects of $K_{CH_3}/K_{CD_3} \simeq 0.87$ for nucleophilic additions to the carbonyl carbon [14]. Even larger effects are seen in β-substitutions of compounds undergoing SN1 type reactions, and are believed to be due to a decrease in the frequency of the β-C–H bonds caused by hyperconjugation with the developing carbocation sp^2-carbon. The solvolysis of 2-methyl, 3,3-dideuterio, 2-chlorobutane, for example, shows $k_{H_2}/k_{D_2} = 1.4$.

In the discussions of general properties of transition states in Chapters 2 and 4, it was pointed out that transition states are **almost** like ordinary species. The characteristic of a transition state is that the force constant matrix that we discussed above for molecular vibrations has one negative eigenvalue. This leads to a negative λ in Eq. (8.28), and to an *imaginary* frequency in the normal mode analysis. The particular normal mode having negative λ must be handled in a completely different way from the other modes. If this normal mode has any contributions from the mass of the isotopically substituted atom, the isotopic substitution cannot be treated by Eq. (8.38). We will see in the next chapter how the special mode is handled.

8.11 REFERENCES

[1]. H. Eyring, D. Henderson, B. J. Stover, and E. M. Eyring, **Statistical Mechanics and Dynamics**, Wiley and Sons, New York, **1964**, pgs. 86–89.

[2]. F. C. Andrews, **Equilibrium Statistical Mechanics**, Wiley and Sons, New York, **1963**.

[3]. J. E. Kilpatrick, K. S. Pitzer, and R. Spitzer, **J. Amer. Chem. Soc., 69**, 2483 (**1947**) gives a treatment of cyclopentane that shows the complication and how to handle it.

[4]. H. Eyring, *et al*, **Statistical Mechanics and Dynamics**, *loc. cit.*, pgs. 49–59.

[5]. E. B. Wilson, J. C. Decius, and P. C. Cross, **Molecular Vibrations**, McGraw-Hill, New York, **1955**.

[6]. E. B. Wilson, *et al*, **Molecular Vibrations**, *loc. cit.*, pgs. 183–184.

[7]. K. Wiberg, **Physical Organic Chemistry**, Wiley and Sons, New York, **1964**, pgs. 145–149.

[8]. T. Moeller, **Inorganic Chemistry**, Wiley and Sons, New York, **1952**, pg. 50.

[9]. J. Bigeleisen, M. W. Lee, and F. Mandel, **Ann. Rev. Phys. Chem., 24**, 407 (**1973**).

[10]. P. M. Laughton and R. E. Robertson in **Solute Solvent Interactions**, Vol. 1, Ed. J. F. Coetzee and C. D. Ritchie, M. Dekker Inc., New York, **1969**, Chap. 7.

[11]. R. L. Schowen, **Prog. Phys. Org. Chem., 9**, 275 (**1972**).

[12]. S. L. Shames And L. D. Byers, **J. Amer. Chem. Soc., 103**, 6170 (**1981**).

[13]. E. R. Pohl and D. J. Hupe, **J. Amer. Chem. Soc., 102**, 2763 (**1980**).

[14]. I. M. Kovach, J. P. Elrod, and R. L. Schowen **J. Amer. Chem. Soc., 102**, 7530 (**1980**).

8.12 PROBLEMS

Problem 8.1

Derive Equations (8.9)–(8.13) following the same procedure as that for the derivation of Eq. (8.8).

Problem 8.2

Derive the commutation relationships, Eq. (8.16), between the ladder operators and $\mathcal{M}_z$.

Problem 8.3

Show that $\mathcal{L}\Psi_{lm}$ and $\mathcal{R}\Psi_{lm}$ are eigenfunctions of $\mathcal{M}^2$, both with eigenvalues of ω_ℓ^2, if $\mathcal{L}$ and $\mathcal{R}$ are defined in Eq. (8.15) and Ψ_{lm} is an eigenfunction of $\mathcal{M}^2$ with eigenvalue of ω_ℓ^2.
Hint: If you show that the ladder operators commute with $\mathcal{M}^2$, the rest is easy.

Problem 8.4

The wave functions for a spherical top molecule (one having three equal principal moments of inertia), $\Psi_{\ell km}$ have eigenvalues of the rotational Hamiltonian operator of $\epsilon_\ell = \ell(\ell+1)\hbar^2$, just as was the case for a linear system as developed in Section 8.5. For the spherical top, however, there are two quantum numbers other than ℓ, k and m, each of which can take on values from $-\ell$ to $+\ell$. This means that each energy level is $(2\ell+1)^2$-fold degenerate. The rotational partition function, then, is

$$f_{rot} = \frac{1}{\sigma}\sum_{\ell=0}^{\infty}(2\ell+1)^2 \exp\left(\frac{\ell(\ell+1)\hbar^2}{2I\kappa T}\right).$$

Show that the replacement of the summation by integration over ℓ gives the same result as Eq. (8.19).
Hint: Write the integral just as was done for the linear rotor. The whole problem is to put the integral into standard form. First, note that

$$\exp\left(-\alpha\ell(\ell+1)\right) = \exp\left(-\alpha(\ell+\frac{1}{2})^2\right)\exp(+\alpha/4).$$

Substitution of $z = \ell + 1/2$ *now puts the integrand into a standard form of definite integral,*

$$\int_0^\infty z^2 \exp\left(-\alpha z^2\right) dz = \frac{\sqrt{\pi}}{4\alpha^{3/2}},$$

but the limits are from $z = 1/2$ *to* $z = \infty$ *rather than from* $z = 0$ *to* $z = \infty$. *Look back at the summation, and notice that the addition of a term with* $\ell = -1/2$ *doesn't change anything; so that the integration can be taken from* $z = 0$. *After evaluating the definite integral, write* $\exp(\alpha/4)$ *as a series:* $1 + (\alpha/4) + \ldots + (\alpha/4)^n/n!$. *A typical moment of inertia for a fairly small organic molecule is about* $1. \times 10^{-38} g \cdot cm^2$, *and* $\hbar = 1.04 \times 10^{-27} erg \cdot secs$, $\kappa = 1.37 \times 10^{-16} erg \cdot deg^{-1}$. *Show that the series can safely be terminated at the first term, 1, for reasonable temperatures.*

Problem 8.5

Derive the commutation relationships among the operators for the harmonic oscillator shown in Equations (8.21)–(8.23).

Hint: Follow the same general method that you used in Problem 1, above; the algebra is just a little more involved in the present case. Pay particular attention to the fact that

$$\frac{\partial^2}{\partial x^2} x \cdot f(x) = \frac{\partial}{\partial x}\left(x \frac{\partial}{\partial x} f(x) + f(x)\right).$$

Problem 8.6

Calculate the equilibrium constant at $25°C$ for the reaction

$$H_2 + D_2 \overset{K_{ex}}{\rightleftharpoons} 2HD$$

from the fact that the stretching frequency of H_2 is 4405 cm^{-1}. Use both Eq. (8.35) and Eq. (8.38) and verify that they give the same result.

Hint: For a diatomic molecule, the moment of inertia is equal to $r^2 \cdot m_1 m_2/(m_1 + m_2)$ *where r is the bond length. Also, the vibrational frequency is proportional to the square root of* $(m_1 + m_2)/m_1 m_2$. *In using Eq. (8.35), remember that the diatomic molecules have two equal moments of inertia. Don't forget the symmetry numbers in the rotational partition functions. The quantity* $h/\kappa T$ *is about* $4.77 \times 10^{-3} cm$ *at room temperature.*

Problem 8.7

Calculate the equilibrium constant for the reaction

$$CH_4 + DBr \xrightleftharpoons{K_{ex}} CH_3D + HBr$$

Methane has a triply degenerate mode at 3019 cm^{-1}, a triply degenerate mode at 1306 cm^{-1}, a doubly degenerate mode at 1534 cm^{-1}, and a non-degenerate mode at 2917 cm^{-1}. Monodeuteriomethane has three doubly degenerate modes at 3021, 1471, and 1155 cm^{-1}, and three non-degenerate modes at 2945, 2200, and 1310 cm^{-1}. The stretching mode for HBr is at 2650 cm^{-1}, and that for DBr is at 1880 cm^{-1}. Methane has a symmetry number of 12, and CH_3D has a symmetry number of 3.

Hint: Use Eq. (8.38). If you wish to check your results, the problem is worked out in K. Wiberg, **Physical Organic Chemistry,** *loc.cit., pgs. 273–276..*

Problem 8.8

For a simple isomerization reaction $A \xrightleftharpoons{K_i} B$, derive expressions for ΔG°, ΔH°, and ΔS° in terms of electronic + zero-point energy differences, $\Delta\varepsilon_\circ$ (see Eq. (8.32)), moments of inertia, and vibrational frequencies.

Hint: Start by re-writing Eq. (8.3), pg. 259 in terms of $\Delta\varepsilon_\circ$, and the rotational and vibrational partition functions; the translational functions cancel. Then use the equations at the top of pg. 260 giving the thermodynamic quantities in terms of the equilibrium constant. You'll probably save a lot of time if you work by pieces; for example, you should write the equations for $d\ln K/d(1/T)$ in terms of $d\ln f_{rot}/d(1/T) = (1/f_{rot})df_{rot}/d(1/T) =$, etc. and expand the derivatives one at a time. You will probably also find it helpful to define $\beta \equiv (1/T)$ right at the start.

Problem 8.9

Consider the Diels-Alder reaction of 2-butene with butadiene to form 4,5-dimethylcyclohexene, and the same reactions of 2-deuterio, 2-butene, and of 2,3-dideuterio, 2-butene. Derive expressions for the ratio of the rate constants of the di-deuterio/di-hydrogen reactions in terms of the measured mono-deuterio/di-hydrogen reactions for the following possible mechanisms: (1) both bonds are formed synchronously in a single step; (2) one bond is formed in a rate determining step to produce a di-radical (with the radical centers hybridized sp^2), which then forms the second bond in a fast step.

Hint: If you have trouble, see S. Seltzer, **Tetrahedron Letters, 1962**, *457. A discussion and critique of this particular type of study is given by B. K. Carpenter,* **Determination of Organic Reaction Mechanisms**, *Wiley-Interscience, New York,* **1984**, *pgs. 107–110.*

Problem 8.10

Because real potential wells for bond distortions are not precisely quadratic, it is thought that hydrogen and deuterium might actually differ in equilibrium bond-length, and, as a result, that a steric isotope effect might exist. Would hydrogen or deuterium appear to be the 'larger' isotope? One might also expect a steric isotope effect to arise in 'crowded' environments where the force constants for the vibrations might be increased from usual values. What direction would this factor have?

Hint: Draw a reasonable potential curve and draw in the zero-point energy levels for the two isotopes. What would happen if the curve were made narrower?

CHAPTER 9

Transition State Theory and Primary Isotope Effects

9.1 Introduction

In our discussions up to the last section of the last chapter, we have treated transition states just as if they were ordinary chemical species. It has been remarked several times that this isn't quite true, but the fiction caused no real problems until we began to look at vibrational behaviors of molecules. For ordinary chemical species, the force constant matrix, $\mathcal{K}$, with elements equal to $k_{ij} = \partial^2 V/\partial q_i \partial q_j$, has all eigenvalues greater than or equal to zero. This leads to the vibrational problem discussed in the last chapter where all of the motions other than translational and rotational had restoring forces acting for any displacement away from equilibrium geometry. Transition states, as we discussed in Chapter 4, have a $\mathcal{K}$ matrix with one and only one negative eigenvalue. This leads to a Λ matrix in Eq. (8.28) which has one and only one negative element. The normal mode corresponding to this negative λ_{ii} is the *reaction coordinate*. Motion along this normal mode corresponds to the transformation of reactants into products, or products into reactants, and requires special consideration in setting up the sums over states for transition states.

9.2 Sums Over States for Transition States [1]

Let's consider a simple triatomic system represented by

$$A - B \quad + \quad C \quad \rightleftharpoons \quad A \quad + \quad B - C, \tag{9.1}$$

having a potential energy surface as shown in Fig. 9.1.

Remember from our discussions in Chapter 4 that Fig. 9.1 is a simplification of the many-dimensional surface; for three particles, we have nine coordinates, and three or four of these are for other than rotational and translational motions, depending on whether the system is non-linear or linear, respectively. We assume that the angle formed by the $A - B - C$ system is such that the potential energy is minimized

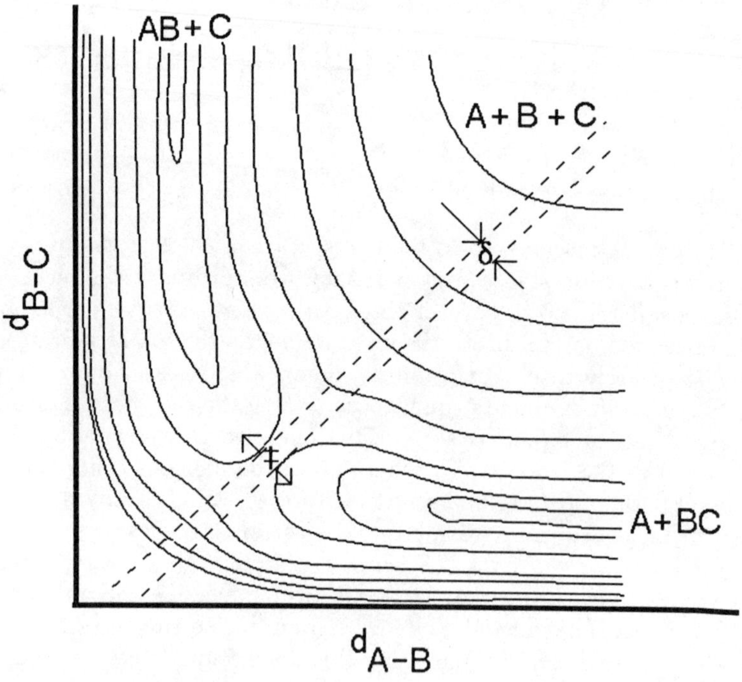

Figure 9.1: *A Potential Energy Surface for the System Shown in Eq. (9.1).*

for each specified point corresponding to given distances between A and B and between B and C.

In order for a system to undergo the transformation from reactants, at the upper left of Fig. 9.1, to products, at the lower right, it must traverse some path that passes over a potential energy maximum. The lowest maximum that can be attained is, of course, the saddle point marked by the double dagger in the lower left region of Fig. 9.1. We define a **reaction coordinate**, q_{rc}, to be the *normal coordinate* for the vibrational system, as discussed in the last Chapter, which has a negative eigenvalue at the saddle point. Note that the **reaction coordinate** is defined with respect to curvatures at the saddle point; there is no 'reaction path' that can be uniquely defined for points other than the saddle point.

We **define a transition state** as a species which lies within some distance δ of the saddle point along q_{rc}. The transition state can have any value for the coordinates orthogonal to the reaction coordinate, and these will be determined by the Boltzmann distribution just as for any other species. Any system that passes from reactants to products along any path on the surface must, somewhere, take on the value of q_{rc} corresponding to what we have defined as a transition state; this would be anywhere within the region of width δ marked by the two dashed lines in Fig. 9.1, and with **any** value of the $A - B - C$ angle.

We can now see that the rate of a reaction is the rate at which systems move from one of the dashed-line boundaries to the opposite one. The motion from reactants to products is motion along the reaction coordinate, and we need to solve the problem of the average absolute velocity along this coordinate. If we know the concentration of transition states at any time, and the average velocity at which they pass from one boundary to the other, then we know the rate of reaction:

$$\text{rate} = [TS]\left(\frac{\bar{v}^{\ddagger}}{2\delta}\right),$$

where $\bar{v}^{\ddagger}$ is the root-mean-square velocity of systems along the reaction coordinate, and the factor of 2 gives the rate in one direction which is equal to that in the opposite direction.

The average velocity can be obtained from the Boltzmann distribution:

$$n_v = N_o \exp\left(-\frac{m^{\ddagger}v^2}{2\kappa T}\right),$$

which gives the number of systems n_v having a velocity v as a function of kinetic energy, and the usual expression for averages:

$$\bar{v}^{\ddagger} = \frac{\int_0^{\infty} v \cdot n_v \cdot dv}{\int_0^{\infty} n_v \cdot dv} = \left(\frac{2\kappa T}{\pi m^{\ddagger}}\right)^{1/2}. \tag{9.2}$$

Just as we did for any ordinary species, we assume that the concentration of transition states is determined by the Boltzmann distribution, and, therefore, by the partition function expression for an equilibrium constant between reactants and transition states:

$$K_{TS} = \frac{[TS]}{[AB][C]},$$

and we obtain

$$\text{rate} = K_{TS}[AB][C]\left(\frac{\bar{v}^{\ddagger}}{2\delta}\right),$$

or

$$k_f = K_{TS}\left(\frac{\bar{v}^{\ddagger}}{2\delta}\right) = K_{TS}\frac{(2\kappa T)^{1/2}}{(\pi m^{\ddagger})^{1/2}2\delta}, \tag{9.3}$$

where k_f is the forward rate constant for the reaction under discussion. The partition functions for the transition state, which determine K_{TS}, however, differ in one respect from those for ordinary molecules; there is one less vibrational mode for transition states since one of the normal vibrational modes corresponds to translation along the reaction coordinate. The vibrational partition function of a non-linear transition state will have $3N-7$ product terms, rather than the usual $3N-6$ terms, and the motion along the reaction coordinate generates another partition function, $f_{rc}^{\ddagger}$, which must be treated differently; we define:

$$f_{vib}^{\ddagger} \equiv \prod_{j=1}^{3N-7}\left[1-\exp\left(-\frac{h\nu_o^j}{\kappa T}\right)\right]^{-1},$$

$$\epsilon_{zp}^{\ddagger} = \sum_{j=1}^{3N-7}\frac{h\nu_o^j}{2},$$

and write

$$K_{TS} = \frac{f_{trans}(ts)f_{rot}(ts)f_{vib}^{\ddagger}f_{rc}^{\ddagger}}{\prod_r f_{trans}(r)f_{rot}(r)f_{vib}(r)} \times \exp\left(-\frac{\epsilon_o^o(ts)-\sum_r\epsilon_o^o(r)+\epsilon_{zp}^{\ddagger}-\sum_r\epsilon_{zp}(r)}{\kappa T}\right), \tag{9.4}$$

where r is an index over the reactant molecules.

If we choose the length δ to be very small, potential energy is essentially constant along the reaction coordinate since $\partial V/\partial q_{rc} = 0$ at the saddle point. This means that motion of a system along q_{rc}, within a small distance of the maximum, is just the same as translational motion of a particle in a one-dimensional box. The sum over states for this motion, $f^{\ddagger}_{rc}$, is exactly that which we calculated for translation of a molecule in one-dimension in the last Chapter:

$$f^{\ddagger}_{rc} = \left(\frac{\delta}{h}\right)(2\pi m^{\ddagger}\kappa T)^{1/2},$$

where $m^{\ddagger}$ is the effective mass of the transition state as determined by the reduced mass for the normal coordinate.

It is now convenient to define a pseudo-equilibrium constant, $K^{\ddagger}$, as

$$K^{\ddagger} = \frac{K_{TS}}{f^{\ddagger}_{rc}} = \frac{hK_{TS}}{\delta(2\pi m^{\ddagger}\kappa T)^{1/2}},$$

so that substitution into Eq. (9.3) for K_{TS} gives

$$k_f = K^{\ddagger}\frac{\delta(2\pi m^{\ddagger}\kappa T)^{1/2}}{h}\frac{(2\kappa T)^{1/2}}{(\pi m^{\ddagger})^{1/2}2\delta} = \left(\frac{h}{\kappa T}\right)K^{\ddagger}. \tag{9.5}$$

Note particularly that the width, δ, in the sum over states for motion along the reaction coordinate has cancelled out of the equation along with some constants, and that the term $h/\kappa T$ is the residue from the sum over states for motion along the reaction coordinate. The pseudo-equilibrium constant, $K^{\ddagger}$, is missing one of the contributions from the $3N$ degrees of freedom of the transition state. In an ordinary molecule, the missing degree of freedom would be a vibration, and the sum over states for that vibration would have a zero-point energy term, as well as the the term in f_{vib}.

9.3 Isotope Effects

If we have a reaction

$$A \;+\; B \;+\ldots+C \;\overset{K_{TS}}{\rightleftharpoons}\; TS \;\longrightarrow\; \text{Products},$$

and a corresponding reaction of an isotopically substituted reactant, A',

$$A' \;+\; B \;+\ldots+C \;\overset{K'_{TS}}{\rightleftharpoons}\; TS' \;\longrightarrow\; \text{Products}',$$

with forward rate constants k_f and k'_f, respectively, we can formulate the ratio of rate constants,

$$\frac{k_f}{k'_f} = \frac{K^{\ddagger}}{K'^{\ddagger}},$$

as a *pseudo-equilibrium constant* for the hypothetical exchange reaction

$$A \quad + \quad TS' \quad \overset{K_{ex}^{\ddagger}}{\rightleftharpoons} \quad A' \quad + \quad TS,$$

which is similar to the type of exchange that we discussed for equilibrium isotope effects in the last chapter. It is worth pointing out, however, that this exchange reaction between ordinary molecules and transition states is not completely analogous to ordinary equilibria[1] because one of the degrees of freedom of the transition states, the one for motion along the reaction coordinate, is not included in the pseudo-equilibrium constant, $K_{ex}^{\ddagger}$.

The pseudo-equilibrium constant may be written in terms of partition functions to obtain an equation analogous to Eq. (8.35), and the Redlich-Teller product rule, Eq. (8.37), may be used to write the ratios of translational and rotational partition functions as ratios of frequencies and masses of the substituted atom and its isotope. Note that the Redlich-Teller product rule involves all $3N-6$ vibrational coordinates for the transition states, so that we obtain:

$$\frac{f_{rot}(ts)f_{trans}(ts)}{f_{rot}(ts')f_{trans}(ts')} = \left(\frac{\sigma(ts')}{\sigma(ts)}\right)\left(\frac{m_i}{m_i'}\right)^{3/2}\left(\frac{\nu^{\ddagger}}{\nu'^{\ddagger}}\right)\prod_{j=1}^{3N^{\ddagger}-7}\left(\frac{\nu_o^j}{\nu_o'^j}\right),$$

where $\nu^{\ddagger}$ is the imaginary frequency (*ie: the square root of the negative* λ) for the transition state, and

$$\frac{\nu^{\ddagger}}{\nu'^{\ddagger}} = \left(\frac{m'^{\ddagger}}{m^{\ddagger}}\right)^{1/2}.$$

---

[1] The actual thermodynamic exchange equilibrium constant, K_{ex}, is given by

$$K_{ex} = K_{ex}^{\ddagger}\left(\frac{m^{\ddagger}}{m'^{\ddagger}}\right)^{1/2},$$

where $m^{\ddagger}$ is the reduced mass of the transition state.

The full equation for primary kinetic isotope effects is

$$\frac{k_f}{k'_f} = K^{\ddagger}_{ex} = \exp\left(-\frac{\Delta\epsilon_{zp}}{\kappa T}\right)\left(\frac{m'^{\ddagger}}{m^{\ddagger}}\right)^{1/2} \times \prod_{i=1}^{3N^{\ddagger}-7} \frac{\nu^i_o\left[1-\exp(-h\nu'^i_o/\kappa T)\right]}{\nu'^i_o\left[1-\exp(-h\nu^i_o/\kappa T)\right]} \times \prod_{j=1}^{3N-6} \frac{\nu'^j_o\left[1-\exp(-h\nu^j_o/\kappa T)\right]}{\nu^j_o\left[1-\exp(-h\nu'^j_o/\kappa T)\right]} \times \frac{\sigma(ts')\sigma(A)}{\sigma(ts)\sigma(A')}, \tag{9.6}$$

where: $$\Delta\epsilon_{zp} = \frac{h}{2}\left(\sum_{i=1}^{3N^{\ddagger}-7}(\nu^i_o - \nu'^i_o) - \sum_{j=1}^{3N-6}(\nu^j_o - \nu'^j_o)\right),$$

and the i index is over transition state vibration modes while the j index is for reactant A. The major difference between Eq. (9.6) and Eq. (8.35) is the 'missing' vibrational frequency term in both the zero-point energy difference and in the partition functions in Eq. (9.6).

For heavy isotopes, kinetic effects are larger than those usually found for equilibria, but are still fairly small. Typical kinetic effects for moderately heavy isotopes, such as carbon and nitrogen, are those in SN2 type reactions [2]. The $^{12}C/^{14}C$ kinetic isotope effect for the reaction of isotopically substituted methyl brosylate with dimethylaniline in acetonitrile at 55°C is $k_{12}/k_{14} = 1.16$, and that for the reaction of isotopically substituted methyl iodide with pyridine in 2-nitropropane at 25°C is $k_{12}/k_{14} = 1.07$. The nitrogen isotope effect on the rate of reaction of isotopically substituted dimethyltoluidine with benzyl benzenesulfonate in acetone at 35°C is $k_{14}/k_{15} = 1.003$. All of these effects have appreciable contributions in Eq. (9.6) from both the exponential zero-point energy term and the term in the ratio of effective masses of the transition state [2].

9.4 Hydrogen, Deuterium and Tritium Kinetic Effects

For light isotopes, particularly those of hydrogen, the dominant term in Eq. (9.6) is usually the exponential term in zero-point energy. If a high frequency vibration of the isotopically substituted atom in the reactant becomes, essentially, the motion along the reaction coordinate at the transition state, then the entire zero-point energy difference between isotopes for this vibration in the reactant disappears at the transition state. Some approximate frequencies and associated zero-point energies for various vibrations involving hydrogen are given in Table 9.1.

The last column of Table 9.1 shows that, at normal temperatures, the terms in $[1 - \exp(-h\nu_o/\kappa T)]$ make negligible contributions to the isotope effects except for the very lowest frequency vibrations.

Because of the small masses of the isotopes of hydrogen, the reduced masses for most vibrations involving these isotopes are essentially equal to the mass of the isotope. The vibrational frequencies for the three isotopes, then, are usually related by the reciprocal square root of the isotopic masses:

$$\nu_o(H) : \nu_o(D); \nu_o(T) = 1 : \sqrt{\frac{1}{2}} : \sqrt{\frac{1}{3}},$$

so that we can calculate the frequencies of vibrations involving D or T from the frequencies of the corresponding vibrations involving H. As a rough indication of the magnitudes of kinetic isotope effects to be expected for substitution of H by D, let's just look at the zero-point energy term in Eq. (9.6) for a reaction in which an R_3C–H stretching frequency of a reactant becomes the motion along the reaction coordinate at the transition state. The $\Delta\epsilon_{zp}$ will have the single term from this reactant vibration, so that

$$\Delta\epsilon_{zp} = \frac{h}{2}\left(2900 - \frac{2900}{\sqrt{2}}\right) = 425h = 1.18 \text{ kcal/mole},$$

and

$$\exp\left(\frac{1.18}{\kappa T}\right) = 7.6 \quad \text{at } 25°C.$$

For tritium substitution we can calculate the exponential term to equal ≈ 18.6 at $25°C$.

If the hydrogen of an R_3C–H reactant became completely free at the transition state, so that the stretching and both bending modes had no zero-point energies, the exponential term would have a value of ≈ 24 at room temperature for deuterium substitution. This represents the maximum deuterium kinetic isotope effect for reaction of an R_3C–H hydrogen that can be calculated from Eq. (9.6).

TABLE 9.1
Approximate Frequencies of Some X-H Vibrations

Vibration	ν_o, cm^{-1}	ϵ_{zp}^{a}	$\exp(-h\nu_o/\kappa T)^b$
R_3C–H stretch	2900	4.1	9.7×10^{-7}
R_3C–H bend	1400	2.0	1.3×10^{-3}
RO–H stretch	3600	5.1	3.5×10^{-8}
RO–H in-plane bend	1400	2.0	1.3×10^{-3}
RO–H out-of-plane bend	1000	1.4	8.5×10^{-3}
R_2N–H stretch	3500	4.9	5.6×10^{-8}
RC(=O)–H stretch	2800	3.9	1.6×10^{-6}
RC(=O)–H in-plane bend	1400	2.0	1.3×10^{-3}
RC(=O)–H out-of-plane bend	900	1.3	1.4×10^{-2}
C=C–H stretch	3050	4.3	4.7×10^{-7}
C=C–H in-plane bend	1100	1.6	5.3×10^{-3}
C=C–H out-of-plane bend	800	1.1	2.2×10^{-2}
C≡C–H stretch	3300	4.7	1.4×10^{-7}
C≡C–H bend	625	0.9	5.1×10^{-2}

[a] In units of kcal/mole. Values of the constants in various units are:

$$h = 6.547 \times 10^{-27}\text{erg-sec} = 2.826 \times 10^{-3}\text{cm-kcal/mole.}$$

$$\kappa = 1.371 \times 10^{-16}\text{erg/deg} = 1.974 \times 10^{-3}\text{kcal/mole-deg.}$$

$$h/\kappa T = 4.77 \times 10^{-3}\text{cm, at } 300^\circ K.$$

[b] At 25°C.

9.5 Vibrational Analysis of Simple Transition States

As a more complex, and more realistic, example of the factors contributing to light isotope effects, let's consider a proton or hydrogen atom transfer reaction for which we can write the pseudo-exchange reaction

$$AH \quad + \quad [A \cdots D \cdots B] \quad \overset{K_{ex}^{\ddagger}}{\rightleftharpoons} \quad AD \quad + \quad [A \cdots H \cdots B],$$

where charges are omitted to maintain generality. If A represents a fairly complex group, the normal mode vibrations of the reactant which involve hydrogen will be a stretching and two bending modes. We will assume that the transition state has a linear geometry, and, therefore, has four vibrational modes that need to be considered. In order to see exactly what is involved, let's work through the problem of three particles in two dimensions.

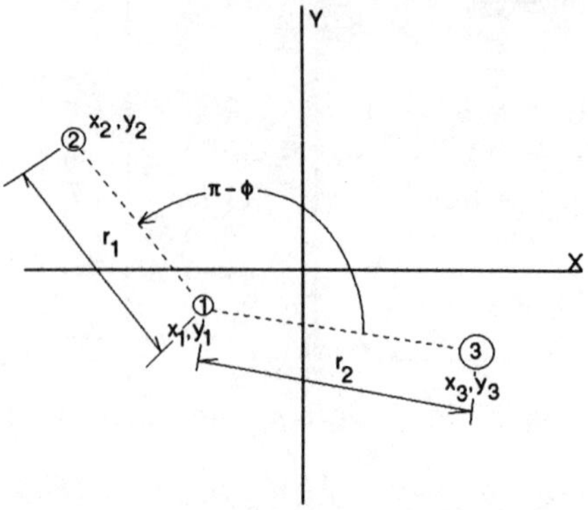

Figure 9.2: *Three Particles in Two Dimensions. The equilibrium bond distances are ℓ_1 and ℓ_2, and the equilibrium value of ϕ is zero.*

The basic layout of the problem is shown in Fig. 9.2. Particle 1 is attached to particle 2 by a Hooke's law force with force constant of k_1, and to particle 3 by a Hooke's law force with force constant of k_2. In addition, there is an interaction between the two forces with an interaction constant of k_{12}. Bending of the system is associated with a Hooke's law force constant of k_ϕ. With the three particles located at the positions $x_1, y_1; x_2, y_2; x_3, y_3$, we define

$$r_1 = \left[(x_1 - x_2)^2 + (y_1 - y_2)^2\right]^{1/2},$$

$$r_2 = \left[(x_3 - x_1)^2 + (y_3 - y_1)^2\right]^{1/2}, \quad \text{and,}$$

$$\sin(\pi - \phi) = \sin\phi = \frac{(x_3 - x_1)(y_1 - y_2) - (x_1 - x_2)(y_3 - y_1)}{r_1 r_2},$$

so that the potential energy, V, of the system is given as

$$V = \frac{k_1}{2}(r_1 - \ell_1)^2 + \frac{k_2}{2}(r_2 - \ell_2)^2 + k_{12}(r_1 - \ell_1)(r_2 - \ell_2) + k_\phi \phi^2,$$

where ℓ_1 and ℓ_2 are the equilibrium values of r_1 and r_2, respectively, and, at equilibrium, $\phi = 0$.

As discussed in Section 8.7, we wish to write Newton's law,

$$Force = mass \times acceleration,$$

as the matrix equation:

$$\begin{pmatrix}
\frac{\partial^2 V}{\partial x_1^2} & \frac{\partial^2 V}{\partial x_1 \partial x_2} & \frac{\partial^2 V}{\partial x_1 \partial x_3} & \frac{\partial^2 V}{\partial x_1 \partial y_1} & \frac{\partial^2 V}{\partial x_1 \partial y_2} & \frac{\partial^2 V}{\partial x_1 \partial y_3} \\
\frac{\partial^2 V}{\partial x_2 \partial x_1} & \frac{\partial^2 V}{\partial x_2^2} & \frac{\partial^2 V}{\partial x_2 \partial x_3} & \frac{\partial^2 V}{\partial x_2 \partial y_1} & \frac{\partial^2 V}{\partial x \partial y_2} & \frac{\partial^2 V}{\partial x_2 \partial y_3} \\
\frac{\partial^2 V}{\partial x_3 \partial x_1} & \frac{\partial^2 V}{\partial x_3 \partial x_2} & \frac{\partial^2 V}{\partial x_3^2} & \frac{\partial^2 V}{\partial x_3 \partial y_1} & \frac{\partial^2 V}{\partial x_3 \partial y_2} & \frac{\partial^2 V}{\partial x_3 \partial y_3} \\
\frac{\partial^2 V}{\partial y_1 \partial x_1} & \frac{\partial^2 V}{\partial y_1 \partial x_2} & \frac{\partial^2 V}{\partial y_1 \partial x_3} & \frac{\partial^2 V}{\partial y_1^2} & \frac{\partial^2 V}{\partial y_1 \partial y_2} & \frac{\partial^2 V}{\partial_1 \partial y_3} \\
\frac{\partial^2 V}{\partial y_2 \partial x_1} & \frac{\partial^2 V}{\partial y_2 \partial x_2} & \frac{\partial^2 V}{\partial y_2 \partial x_3} & \frac{\partial^2 V}{\partial y_2 \partial y_1} & \frac{\partial^2 V}{\partial y_2^2} & \frac{\partial^2 V}{\partial y_1 \partial y_3} \\
\frac{\partial^2 V}{\partial y_3 \partial x_1} & \frac{\partial^2 V}{\partial y_3 \partial x_2} & \frac{\partial^2 V}{\partial y_3 \partial x_3} & \frac{\partial^2 V}{\partial y_3 \partial y_1} & \frac{\partial^2 V}{\partial y_3 \partial y_2} & \frac{\partial^2 V}{\partial y_3^2}
\end{pmatrix}
\begin{pmatrix} x_1 - x_1^o \\ x_2 - x_2^o \\ x_3 - x_3^o \\ y_1 - y_1^o \\ y_2 - y_2^o \\ y_3 - y_3^o \end{pmatrix}$$

$$= \begin{pmatrix}
m_1 & 0 & 0 & 0 & 0 & 0 \\
0 & m_2 & 0 & 0 & 0 & 0 \\
0 & 0 & m_3 & 0 & 0 & 0 \\
0 & 0 & 0 & m_1 & 0 & 0 \\
0 & 0 & 0 & 0 & m_2 & 0 \\
0 & 0 & 0 & 0 & 0 & m_3
\end{pmatrix}
\begin{pmatrix} \ddot{x}_1 \\ \ddot{x}_2 \\ \ddot{x}_3 \\ \ddot{y}_1 \\ \ddot{y}_2 \\ \ddot{y}_3 \end{pmatrix},$$

in which we now need to evaluate the second partials of V with respect to the cartesian coordinates. We could substitute the expressions for r_1, r_2, and ϕ in terms of the cartesian coordinates into the equation for V, but it is much easier in the present case to use the general equations that can be derived from the chain rule for differentiating a function of a function [3]. Let r_1, r_2, and ϕ, be represented as s_1, s_2, and s_3, respectively, and let the cartesian coordinates be represented as $q_1, q_2, \ldots, q_6$. We can then write

$$\frac{\partial V}{\partial q_i} = \sum_{k=1}^{3} \frac{\partial V}{\partial s_k} \frac{\partial s_k}{\partial q_i},$$

and,

$$\frac{\partial^2 V}{\partial q_i \partial q_j} = \sum_{k=1}^{3}\sum_{l=1}^{3}\left[\left(\frac{\partial^2 V}{\partial s_k \partial s_l}\right)\left(\frac{\partial s_k}{\partial q_i}\right)\left(\frac{\partial s_l}{\partial q_j}\right) + \left(\frac{\partial V}{\partial s_k}\right)\left(\frac{\partial^2 s_k}{\partial q_i \partial q_j}\right)\right],$$

but the second term within the summation on the right-hand-side vanishes because $\partial V/\partial s_k$ is zero for all k at the equilibrium geometry. Writing $\phi \approx \sin\phi$, it is easy to obtain:

$$\frac{\partial^2 V}{\partial r_1^2} = k_1, \quad \frac{\partial^2 V}{\partial r_2^2} = k_2, \quad \frac{\partial^2 V}{\partial r_1 \partial r_2} = \frac{\partial^2 V}{\partial r_2 \partial r_1} = k_{12}, \quad \frac{\partial^2 V}{\partial \phi^2} = k_\phi.$$

We wish to evaluate the $\mathcal{K}$ matrix elements for equilibrium geometry, and we will lose no generality by taking the equilibrium position to have the three masses on the X axis, so that we get

$$\frac{\partial r_1}{\partial x_1} = \frac{x_1 - x_2}{r_1} \stackrel{\text{eq}}{=} \frac{\ell_1}{r_1^\circ - r_2^\circ} = 1,$$

$$\frac{\partial r_1}{\partial x_2} \stackrel{\text{eq}}{=} -1, \quad \frac{\partial r_1}{\partial x_3} \stackrel{\text{eq}}{=} 0 \stackrel{\text{eq}}{=} \frac{\partial r_1}{\partial y_1} \stackrel{\text{eq}}{=} \frac{\partial r_1}{\partial y_2} \stackrel{\text{eq}}{=} \frac{\partial r_1}{\partial y_3},$$

$$\frac{\partial r_2}{\partial x_1} \stackrel{\text{eq}}{=} -1, \quad \frac{\partial r_2}{\partial x_2} \stackrel{\text{eq}}{=} 0, \quad \frac{\partial r_2}{\partial x_3} \stackrel{\text{eq}}{=} 1, \quad \frac{\partial r_1}{\partial y_1} \stackrel{\text{eq}}{=} \frac{\partial r_1}{\partial y_1} \stackrel{\text{eq}}{=} \frac{\partial r_1}{\partial y_1} \stackrel{\text{eq}}{=} 0,$$

$$\frac{\partial \phi}{\partial x_1} \stackrel{\text{eq}}{=} \frac{\partial \phi}{\partial x_2} \stackrel{\text{eq}}{=} \frac{\partial \phi}{\partial x_3} \stackrel{\text{eq}}{=} 0, \quad \frac{\partial \phi}{\partial y_1} \stackrel{\text{eq}}{=} -\frac{\ell_1 + \ell_2}{\ell_1 \ell_2},$$

$$\frac{\partial \phi}{\partial y_2} \stackrel{\text{eq}}{=} \frac{1}{\ell_1}, \quad \frac{\partial \phi}{\partial y_3} \stackrel{\text{eq}}{=} \frac{1}{\ell_2}.$$

Note that the matrix elements $\partial^2 V/\partial x_i \partial y_j$ are all zero, so that we can write the 6×6 matrix equation as two 3×3 equations:

$$\begin{pmatrix} k_1 + k_2 - 2k_{12} & -k_1 + k_{12} & -k_2 + k_{12} \\ -k_1 + k_{12} & k_2 & -k_{12} \\ -k_2 & -k_{12} & k_2 \end{pmatrix} \begin{pmatrix} x_1 - x_1^\circ \\ x_2 - x_2^\circ \\ x_3 - x_3^\circ \end{pmatrix}$$

$$= \begin{pmatrix} m_1 & 0 & 0 \\ 0 & m_2 & 0 \\ 0 & 0 & m_3 \end{pmatrix} \begin{pmatrix} \ddot{x}_1 \\ \ddot{x}_2 \\ \ddot{x}_3 \end{pmatrix}, \quad \text{and}$$

$$\frac{k_\phi}{\ell_1^2 \ell_2^2} \begin{pmatrix} (\ell_1 + \ell_2)^2 & -\ell_2(\ell_1 + \ell_2) & -\ell_1(\ell_1 + \ell_2) \\ -\ell_2(\ell_1 + \ell_2) & \ell_2^2 & \ell_1 \ell_2 \\ -\ell_1(\ell_1 + \ell_2) & \ell_1 \ell_2 & \ell_1^2 \end{pmatrix} \begin{pmatrix} y_1 \\ y_2 \\ y_3 \end{pmatrix}$$

$$= \begin{pmatrix} m_1 & 0 & 0 \\ 0 & m_2 & 0 \\ 0 & 0 & m_3 \end{pmatrix} \begin{pmatrix} \ddot{y}_1 \\ \ddot{y}_2 \\ \ddot{y}_3 \end{pmatrix}.$$

We now transform to the *mass weighted coordinates* by inserting the identity matrix, written as $\Omega^{-1/2}\Omega^{1/2}$, between the $\mathcal{K}$ and $\mathcal{Q}^\dagger$ matrices and left-multiplying by $\Omega^{-1/2}$ to obtain

$$\begin{pmatrix} \frac{k_1+k_2-2k_{12}}{m_1} & \frac{-k_1+k_{12}}{\sqrt{m_1m_2}} & \frac{-k_2+k_{12}}{\sqrt{m_1m_3}} \\ \frac{-k_1+k_{12}}{\sqrt{m_1m_2}} & \frac{k_1}{m_2} & \frac{-k_{12}}{\sqrt{m_2m_3}} \\ \frac{-k_2+k_{12}}{\sqrt{m_1m_3}} & \frac{-k_{12}}{\sqrt{m_2m_3}} & \frac{k_2}{m_3} \end{pmatrix} \begin{pmatrix} (x_1 - x_1^\circ)\sqrt{m_1} \\ (x_2 - x_2^\circ)\sqrt{m_2} \\ (x_3 - x_3^\circ)\sqrt{m_3} \end{pmatrix}$$

$$= \begin{pmatrix} \ddot{x}_1\sqrt{m_1} \\ \ddot{x}_2\sqrt{m_2} \\ \ddot{x}_3\sqrt{m_3} \end{pmatrix}, \quad \text{and}$$

$$\frac{k_\phi}{\ell_1^2\ell_2^2} \begin{pmatrix} \frac{(\ell_1+\ell_2)^2}{m_1} & -\frac{\ell_2(\ell_1+\ell_2)}{\sqrt{m_1m_2}} & -\frac{\ell_1(\ell_1+\ell_2)}{\sqrt{m_1m_3}} \\ -\frac{\ell_2(\ell_1+\ell_2)}{\sqrt{m_1m_2}} & \frac{\ell_2^2}{m_2} & \frac{\ell_1\ell_2}{\sqrt{m_2m_3}} \\ -\frac{\ell_1(\ell_1+\ell_2)}{\sqrt{m_1m_3}} & \frac{\ell_1\ell_2}{\sqrt{m_2m_3}} & \frac{\ell_1^2}{m_3} \end{pmatrix} \begin{pmatrix} y_1\sqrt{m_1} \\ y_2\sqrt{m_2} \\ y_3\sqrt{m_3} \end{pmatrix}$$

$$= \begin{pmatrix} \ddot{y}_1\sqrt{m_1} \\ \ddot{y}_2\sqrt{m_2} \\ \ddot{y}_3\sqrt{m_3} \end{pmatrix}.$$

Eigenvalues of the $\mathcal{F}$ matrices can be obtained most simply by expanding the determinants $|\mathcal{F} - \lambda\mathcal{I}|$, to obtain

$$\lambda^3 - \lambda^2\left(\frac{k_1 + k_2 - 2k_{12}}{m_1} + \frac{k_1}{m_2} + \frac{k_2}{m_3}\right) + \lambda\left(k_1k_2 - k_{12}^2\right)\left(\frac{m_1 + m_2 + m_3}{m_1m_2m_3}\right) = 0,$$

from the X equation, and

$$\lambda^3 - \lambda^2\frac{k_\phi}{\ell_1^2\ell_2^2}\left(\frac{\ell_1^2}{m_3} + \frac{\ell_2^2}{m_2} + \frac{(\ell_1 + \ell_2)^2}{m_1}\right) = 0,$$

from the Y equation. The X equations have one zero eigenvalue, corresponding to translation of the entire system along the X-axis, and the Y equations have two zero eigenvalues, corresponding to translation along the Y-axis, and rotation about the center of gravity. Before proceeding, we can note that nothing new would be added by extending the above treatment to three dimensions; we would obtain two more zero eigenvalues corresponding to translation along the Z-axis and rotation in the XZ-plane, and one more real eigenvalue for bending in the third dimension which would be equal to the one already included for bending.

Dividing the Y equation by λ^2, we obtain the eigenvalue for bending

$$\lambda = \frac{k_\phi}{\ell_1^2 \ell_2^2}\left(\frac{\ell_1^2}{m_3} + \frac{\ell_2^2}{m_2} + \frac{(\ell_1+\ell_2)^2}{m_1}\right).$$

In the case of $m_1 << m_2$ and $m_1 << m_3$, the eigenvalue will be inversely proportional to m_1, or, stated differently, the reduced mass for the bending vibration will be that for the light atom.

Dividing the X equations by λ, we obtain the quadratic

$$\lambda^2 - \lambda\left(\frac{k_1 + k_2 - 2k_{12}}{m_1} + \frac{k_1}{m_2} + \frac{k_2}{m_3}\right) + (k_1 k_2 - k_{12}^2)\left(\frac{m_1 + m_2 + m_3}{m_1 m_2 m_3}\right) = 0, \tag{9.7}$$

for the non-zero eigenvalues. If the system is a transition state, one of the eigenvalues must be negative, and this requires that $k_{12}^2 > k_1 k_2$. Table 9.2 shows how the eigenvalues vary as functions of the masses of particles 2 and 3, for particle 1 being either hydrogen, $m_1 = 1$, or deuterium, $m_1 = 2$, and for various ratios of the stretching and interaction force constants.

The primary points of interest in the data of Table 9.2 are the ratios of the eigenvalues for corresponding cases of $m_1 = 1$ and $m_1 = 2$. A nearly symmetrical three particle system has vibrational modes that can be diagrammed as:

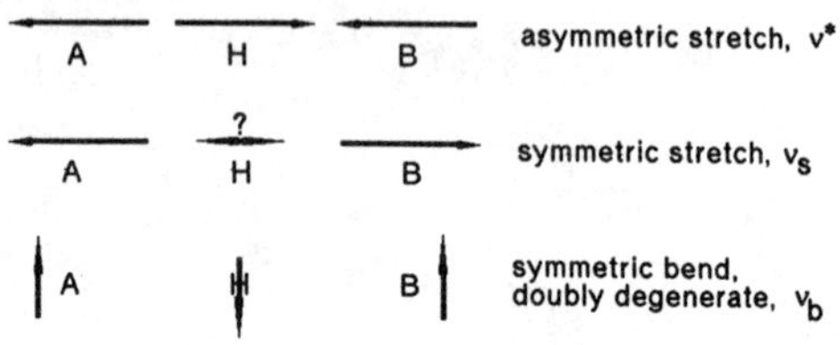

For a symmetrical system in which A=B and H is equally bonded to A and B, the symmetric stretch involves no motion of H, and, for both A and B much heavier than H, the asymmetric stretch involves the motion of H with very little motions of either A or B. This situation corresponds to all those entries in Table 9.2 for which $m_2 = m_3 = 15$ and $k_2/k_1 = 1$, and has the λ's the same, but $\lambda^\ddagger$'s differing by a factor of approximately two, for $m_1 = 1$ and $m_1 = 2$. It is worth noting that changes in the mass of A relative to B have little effect on this general behavior, as shown particularly by the last two entries in the table.

TABLE 9.2
Solutions of Eq. (9.7) For Several Masses and Ratios of Stretching Force Constants

m_1	m_2	m_3	k_2/k_1	k_{12}^2/k_1k_2	λ/k_1	$\lambda^{\ddagger}/k_1$
1	15	15	1	1.01	0.1337	-0.0103
2	15	15	1	1.01	0.1337	-0.0053
1	15	15	2	1.01	0.3650	-0.0075
2	15	15	2	1.01	0.2837	-0.0050
1	15	15	5	1.01	1.9092	-0.0036
2	15	15	5	1.01	1.1559	-0.0031
1	15	15	10	1.01	5.3798	-0.0026
2	15	15	10	1.01	3.0576	-0.0023
1	15	15	1	1.1	0.1366	-0.1009
2	15	15	1	1.1	0.1366	-0.0521
1	15	15	5	1.1	1.7490	-0.0394
2	15	15	5	1.1	1.0875	-0.0327
1	15	15	1	1.2	0.1397	-0.1973
2	15	15	1	1.2	0.1397	-0.1018
1	15	15	5	1.2	1.5878	-0.0868
2	15	15	5	1.2	1.0202	-0.0697
1	15	15	5	2.0	0.8686	-0.7931
2	15	15	5	2.0	0.7269	-0.4892
1	15	30	5	2.0	0.6708	-0.7620
2	15	30	5	2.0	0.5478	-0.4767
1	15	30	0.2	2.0	0.1473	-0.1388
2	15	30	0.2	2.0	0.1247	-0.0838
1	15	150	1	1.1	0.0746	-0.0989
2	15	150	1	1.1	0.0744	-0.0499

When the two force constants, k_1 and k_2, become unequal [4], however, the situation is drastically altered to the extent that the diagrammed vibrations have little meaning since the normal modes correspond to linear combinations of the simple modes shown[2]. The sensitivity of the eigenvalue corresponding to the real vibration to m_1 depends

[2] It is interesting that transition states differ significantly from ordinary species in this respect. If the interaction constant, k_{12}, is zero, the symmetric stretching

on both k_2/k_1 and k_{12}^2/k_1k_2 in a rather complex manner, generally increasing with increasing k_2/k_1 and decreasing with increasing k_{12}^2/k_1k_2.

Qualitatively, as the bonding of H to A and to B becomes unequal, the sensitivity of the real vibration for the transition state increases and the frequency of the real vibration increases, producing zero-point energy difference between the H and D substituted transition states. Any zero-point energy difference between the isotopically substituted transition states counteracts the zero-point energy differences of the reactants, and will cause a decrease in the observed kinetic isotope effects.

Let's consider some examples in more detail. Suppose A–H has a stretching vibration with a frequency of 3600 cm^{-1}, and doubly degenerate bending vibrations with frequency of 1200 cm^{-1}, all of which have reduced masses essentially equal to the isotope mass so that the frequencies for A–D are $1/\sqrt{2}$ times those for A–H. The total zero-point energy difference for A–H and A–D is then equal to 2.5 $kcal/mole$. For all except the extreme last case, we will assume that the bending force constant for the transition state is 80% of that for the reactants, so that the transition states all have doubly degenerate bending frequencies which are 90% ($\approx \sqrt{80\%}$) of the reactant bending frequencies. These bending frequencies, then, will contribute 0.9 $kcal/mole$ of zero-point energy difference between the H- and D-substituted transition states. Now, let's consider four possible transition states, all having $m_2 = m_3 = 15$ and $k_{12}^2/k_1k_2 = 1.01$:

Case 1: $k_1 = k_2 = 1.0 \times 10^8\ cm^{-2}$, (This is $\approx$ 78% of the stretching force constant of the reactant.). From Table 9.2, we have $\lambda_H = \lambda_D = 0.1337 \times 10^8\ cm^{-2}$, and $\nu_H^\circ = \nu_D^\circ = 582\ cm^{-1}$ The zero-point energy difference between the isotopically substituted transition states is just that from the bending frequencies, 0.9 $kcal/mole$. The terms in Eq. (9.6) from ratios of bending frequencies in the reactants will cancel those in the transition state. Note that the transition states have one real stretching frequency just as do the reactants. The ratios of these frequencies, and the ratio of the effective masses, $m^\ddagger$, in Eq. (9.6) will be given by $(\lambda_H^\ddagger/\lambda_D^\ddagger)^{1/2}(\lambda_H/\lambda_D)^{1/2}(1/2)^{1/2}$, where the λ's are those in Table 9.2 and the $(1/2)^{1/2}$ is the ratio of reactant stretching frequencies. In the present case,

---

vibration remains insensitive to change of m_1 from 1 to 2 even for extreme ratios of k_2/k_1. Even when k_{12} becomes quite large, but still less than $\sqrt{k_1k_2}$ so that we have an ordinary species, the sensitivity is small. For example, when $k_{12}^2 = 0.9k_1k_2$ and $k_2/k_1 = 5$, the symmetric stretch eigenvalue is $0.0324k_1$ for $m_1 = 1$ and is $0.0284k_1$ for $m_1 = 2$ and both other masses 15.

this product of ratios is equal to 0.9857, and we will neglect it. The terms in $1 - \exp(-h\nu/\kappa T)$ are also quite close to unity and cancel to a large extent. The isotope effect on the reaction rate is therefore given by the exponential term in zero-point energies alone, and we obtain:

$$\frac{k_H}{k_D} = \exp\left((2.5 - 0.9)/0.592\right) = 14.9, \quad \text{at } 25°C.$$

Case 2: $k_1 = 5.0 \times 10^7$, $k_2 = 1.0 \times 10^8$. Taking the eigenvalues from Table 9.2 as above, we find $\nu_H^\circ = 680\ cm^{-1}$, and $\nu_D^\circ = 600\ cm^{-1}$. This produces a zero-point energy difference of 0.11 $kcal/mole$ in addition to that from the bending modes. The exponential in zero-point energy differences then is $\exp[(2.5-1.0)/\kappa T]$. As for case 1, the product of ratios of frequencies and effective masses along the reaction coordinate is very close to unity, 0.9823, as are the partition function terms, so that we have:

$$\frac{k_H}{k_D} = \exp(1.5/0.592) = 12.6.$$

Case 3: $k_1 = 1.0 \times 10^7$, $k_2 = 1.0 \times 10^8$, so that we find $\nu_H^\circ = 1167\ cm^{-1}$, and $\nu_D^\circ = 880\ cm^{-1}$. The zero-point energy difference from this vibration is 0.4 $kcal/mole$. Adding the zero-point energy from the bending modes, and subtracting from the zero-point energy difference of the reactants, we obtain the exponential term $\exp[(2.5 - 1.3)/\kappa T]$. Again, as in the other cases, the other terms in Eq. (9.6) are very close to unity, so:

$$\frac{k_H}{k_D} = \exp(1.2/0.592) = 7.6$$

Case 4: In the extreme where the isotopically substituted transition states have no zero-point energy differences, the full zero-point energy difference of the reactants appears in the exponential term, giving:

$$\frac{k_H}{k_D} = 68.$$

These examples show that the kinetic isotope effect can, at least in theory, vary considerably as the symmetry of the transition state varies. There have been a number of studies [5] showing that kinetic isotope effects on proton transfer reactions vary as the equilibrium constant

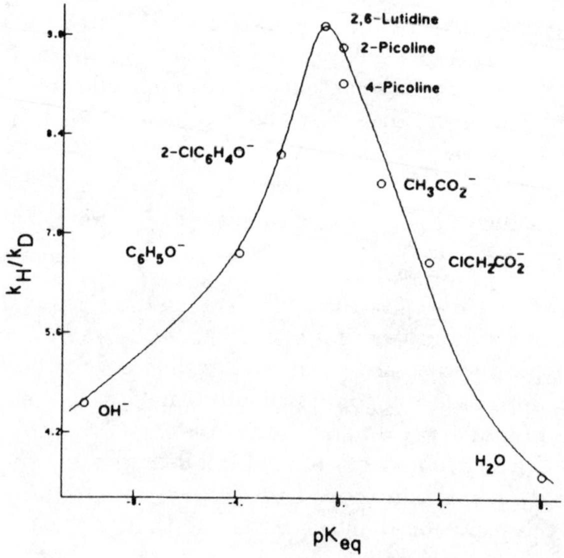

Figure 9.3: *Kinetic Isotope Effects on the Reactions of Ethylnitroacetate with Bases.* pK_{eq} *is the negative logarithm of the equilibrium constant for the proton transfer reaction. The curve drawn has no significance; it is meant only to guide the eye through the points. Data are from [5].*

for the reaction varies, with a maximum isotope effect occurring when $K_{eq} = 1$. The results of one such study are shown in Fig. 9.3.

The results are in accord with the expectation that the transition states will be most nearly symmetrical when $K = 1$ for the exchange reactions.

There is, however, some serious question as to whether the variations in isotope effects found are due to the variations in the symmetric stretching modes of the transition states [5]. The doubt is based on two primary points: first, if the negative curvature along the reaction coordinate is taken to be nearly the same magnitude as the positive curvature perpendicular to it, the change in sensitivity of the real λ with change in symmetry is fairly small, as can be seen from Table 9.2; and, second, there is another factor involved in kinetic hydrogen isotope effects that we have not considered and which could easily be responsible for the variations seen. Before becoming too serious about explaining the variation of kinetic isotope effect with change in equilibrium constant, one should take careful note of the fact that bases having entirely different structures define the shape of Fig. 9.3, so that major factors other than equilibrium constant vary along the X-axis.

9.6 Tunneling in Chemical Reactions

The factor which we have not yet considered that could contribute to kinetic isotope effects is quantum mechanical tunneling. When a quantum system is 'confined' by finite potential energy barriers, there is often a possibility that the system can appear on the other side of a barrier even though it never had enough energy to surmount the barrier. This leakage of quantum systems through barriers is a phenomenon involving time-dependent quantum mechanics, and has several associated interesting questions. Up to now, our discussions of wavefunctions have been almost exclusively of eigenfunctions. Eigenfunctions, as pointed out in Chapter 5, are time-independent functions; they are frequently called *stationary states* because a system in a state described by an eigenfunction of its Hamiltonian will remain in that state forever. Systems which can exist in more than one form, and, obviously, most chemical systems are of this type, frequently have eigenfunctions which describe the system in all forms simultaneously. If, for example we had a one-dimensional system which was confined in a symmetrical 'double box'; $V = \infty$ for $x < -x_o$, $-x_b < x > +x_b$, and $x > +x_o$, and $V = 0$ for $-x_o > x < -x_b$, $+x_b > x < +x_o$, The eigenfunctions would give equal probability of the system existing in either end of the box. The first question that we will need to answer in discussing such systems is how to describe a system located in one particular region of space. Once that question has been answered, we then need to know what the time-dependence is for the system described by a function which is not an eigenfunction of its Hamiltonian.

We can keep things reasonably general, and yet fairly simple, by considering a one-dimensional system that is bounded by infinite potential walls at $x = 0$ and $x = c$, and has two regions of low potential separated by a region of higher potential, as diagrammed in Figure 9.4. In the region from $x = 0$ to $x = a$ there is a potential, $V(1)$; in the region from $x = a$ to $x = b$, a potential of $V(2)$; and, in the region from $x = b$ to $x = c$, a potential of $V(3)$. Although this general problem cannot be solved analytically, it is really quite simple to obtain eigenvalues and eigenfunctions to any desired accuracy.

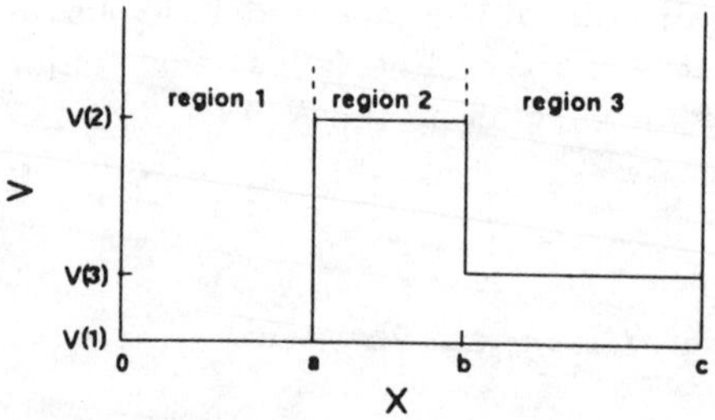

Figure 9.4: *Diagram of A General One-dimensional Double Well Problem.*

Within any region of constant potential, the most general wavefunction for a system is given by

$$\Psi = Ae^{i\alpha x} + Be^{-i\alpha x}, \quad \text{with} \quad \alpha = \frac{\sqrt{2m}}{\hbar}(E - V)^{1/2}.$$

For a system with the regions under discussion, a total energy E independent of x will require that α take on a different value in each region, as given in the last equation. Note particularly that α will become imaginary when $E < V$. The wavefunction has three different forms for the three different regions;

$$\begin{aligned} \Psi_k(x \in n) &= A_k(n)e^{i\alpha_k(n)x} + B_k(n)e^{-i\alpha_k(n)x}, \\ \alpha_k(n) &= \frac{\sqrt{2m}}{\hbar}\left(E_k - V(n)\right)^{1/2}, \quad n = 1, 2, 3. \end{aligned} \tag{9.8}$$

In order to be a suitable wavefunction, Ψ_k must go to zero at both $x = 0$ and $x = c$; in addition, both the wavefunction and its slope must be continuous at $x = a$ and at $x = b$. The slope of the wavefunction is given by

$$\frac{\partial \Psi_k(x \in n)}{\partial x} = i\alpha_k(n)\left[A_k(n)e^{i\alpha_k(n)x} - B_k(n)e^{-i\alpha_k(n)x}\right],$$

so that we can write

$$\begin{aligned} A_k(n) &= \frac{1}{2}\left(\Psi_k(x \in n) - \frac{i}{\alpha_k(n)}\frac{\partial \Psi_k(x \in n)}{\partial x}\right)e^{-i\alpha_k(n)x}, \quad \text{and} \\ B_k(n) &= \frac{1}{2}\left(\Psi_k(x \in n) + \frac{i}{\alpha_k(n)}\frac{\partial \Psi_k(x \in n)}{\partial x}\right)e^{+i\alpha_k(n)x}. \end{aligned}$$

Now, suppose that for some given E_k, Ψ_k has a value $\Psi_k(s)$, and a slope $[\partial\Psi_k(x \in n)/\partial x]_{x=s}$ at some point, $x = s \in n$; then, at any other point within the same region n, say $x = (s + r) \in n$, we find that

$$\begin{aligned}
\Psi_k(s+r) =& \frac{1}{2}\left(\Psi_k(s) - \frac{i}{\alpha_k(n)}\left[\frac{\partial\Psi_k}{\partial x}\right]_{x=s}\right)e^{+i\alpha_k(n)r} \\
&+ \frac{1}{2}\left(\Psi_k(s) + \frac{i}{\alpha_k(n)}\left[\frac{\partial\Psi_k}{\partial x}\right]_{x=s}\right)e^{-i\alpha_k(n)r} \\
=& \Psi_k(s)\left(\frac{e^{i\alpha_k(n)r} + e^{-i\alpha_k(n)r}}{2}\right) \\
&- \frac{i}{\alpha_k(n)}\left[\frac{\partial\psi_k}{\partial x}\right]_{x=s}\left(\frac{e^{i\alpha_k(n)r} - e^{-i\alpha_k(n)r}}{2}\right)
\end{aligned},$$

and

$$\begin{aligned}
\left[\frac{\partial\Psi_k}{\partial x}\right]_{x=s+r} =& i\alpha_k(n)\Psi_k(s)\left(\frac{e^{i\alpha_k(n)r} - e^{-i\alpha_k(n)r}}{2}\right) \\
&+ \left[\frac{\partial\psi_k}{\partial x}\right]_{x=s}\left(\frac{e^{i\alpha_k(n)r} + e^{-i\alpha_k(n)r}}{2}\right).
\end{aligned}$$

Now, lets define a 2×2 matrix, $\mathcal{P}_{kn}(x)$, with elements

$$p_{11} = p_{22} = \frac{1}{2}\left(e^{-i\alpha_k(n)x} + e^{+i\alpha_k(n)x}\right),$$

$$p_{12} = \frac{i}{2\alpha_k(n)}\left(e^{-i\alpha_k(n)x} - e^{+i\alpha_k(n)x}\right),$$

$$p_{21} = -\frac{i\alpha_k(n)}{2}\left(e^{-i\alpha_k(n)x} + e^{+i\alpha_k(n)x}\right),$$

and a column vector, $\vec{F}_k(x)$, with elements $f_1 = \Psi_k(x)$, and $f_2 = \partial\Psi_k(x)/\partial x$, so that we can write the matrix equation

$$\vec{F}_k(r+s) = \mathcal{P}_{kn}(r)\vec{F}_k(s), \quad r, (r+s) \in n. \tag{9.9}$$

In order to find the eigenfunctions for a system, we start at $x = 0$ where the function is required to have the value of zero. The slope of any un-normalized function at zero has a completely arbitrary non-zero value, call it g in every case. We have, then, the vector $\vec{F}(0)$ for any Ψ_k. Now guess a value for E_k, which sets the value of $\alpha_k(n)$ in each region as given in Eq. (9.8). Now, the $\mathcal{P}_{k1}(a-\delta)$ matrix can be set up and $\vec{F}_k(a-\delta)$ calculated for vanishingly small δ. Since the function and its slope must be continuous at $x = a$, $\vec{F}_k(a-\delta) = \vec{F}_k(a+\delta)$ as $\delta \to 0$. Thus, $\vec{F}_k(a)$ can now serve as the starting point in region 2, and be multiplied by $\mathcal{P}_{k2}(b)$ to obtain $\vec{F}_k(b)$, which, by the same reasoning as that at $x = a$, is the same in regions 2 and 3. Therefore, $\vec{F}_k(b)$ can now be multiplied by $\mathcal{P}_{k3}(c)$ to obtain $\vec{F}_k(c)$. Putting all of this together, we have for the present case of three regions [3],

$$\vec{F}_k(c) = \mathcal{P}_{k3}(c)\mathcal{P}_{k2}(b)\mathcal{P}_{k1}(a)\vec{F}_k(0).$$

The eigenfunctions for the system are those for which the calculated value of the function is zero at the right-hand infinite barrier, in the present case, $x = c$. In practice, we would simply program all of the above equations, then start by guessing a very small eigenvalue, then incrementally increasing it until the calculated $\Psi_k(c)$ changes sign. The eigenvalue could then be interpolated to any desired degree of accuracy.

The lowest four eigenfunctions and eigenvalues for a symmetrical double-well potential are shown, overlaid on the potential energy diagram, in Fig. 9.5. The eigenfunctions with eigenvalues below the energy of the barrier essentially occur in pairs, one symmetric and one antisymmetric about the center of the diagram, with small differences in eigenvalues for the members of a pair. For the potentials shown in Fig. 9.5, the lowest pair of eigenfunctions have a difference in eigenvalues of 0.0073 units, and the next higher pair have a difference of 0.033 units. In general, the difference in energy for a pair of functions decreases with the width of the potential barrier and with the height of the potential barrier relative to the eigenvalues. For the situation with an infinitely high barrier, the one mentioned at the start of this section, the symmetric and corresponding antisymmetric functions would have the same eigenvalue. As the barrier height is lowered and widened, the symmetric and antisymmetric states are 'split', until, for very low or very narrow barriers, the difference in eigenvalues for neighboring symmetric and antisymmetric states is comparable to that between pairs.

We are now in position to answer the first question asked in this section; how do we describe a system localized in one region of space? The general, trivial, answer to the question is that we take a linear combination of the eigenstates such that a localized state is obtained.

---

[3] It is obviously trivial to extend the machinery developed here to any number of regions with any number of different potentials. The general method, then, allows the solution of the Schrödinger equation for any bound one-dimensional system.

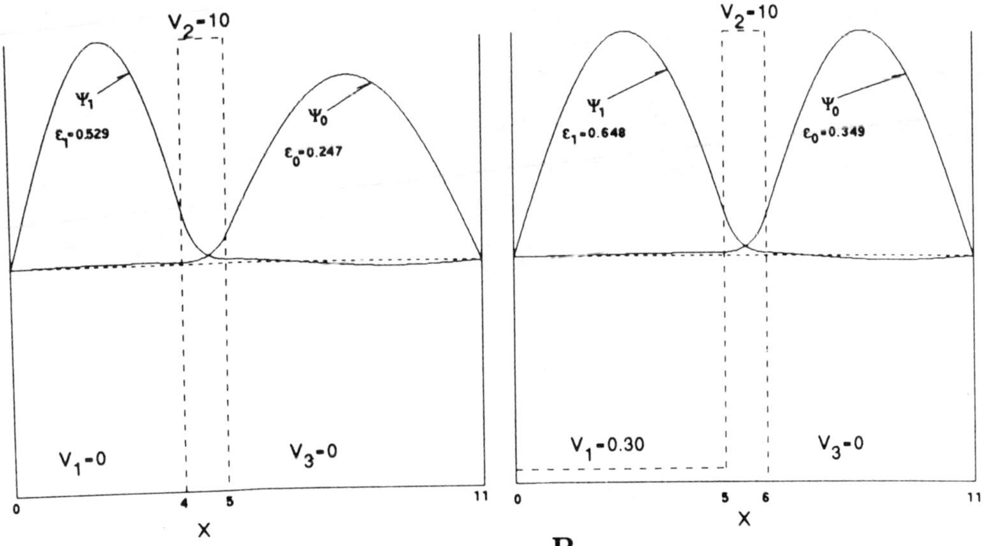

A. B.

Figure 9.5: *The Lowest Four Eigenfunctions and Eigenvalues for the Symmetrical Double-Well Potential with the Dimensions Shown.*

For systems with reasonably high barriers, such as that in Fig. 9.5, a specific answer is that we could take either a sum or a difference of the pair of symmetric and antisymmetric eigenfunctions. The simple localized functions $\Phi_1 = (\Psi_0 + \Psi_1)/\sqrt{2}$ and $\Phi_2 = (\Psi_0 - \Psi_1)/\sqrt{2}$ from the eigenfunctions given in Fig. 9.5 are shown in Fig. 9.6a.

We now need to examine the time dependence of states which are not described by an eigenfunction of the Hamiltonian. The time dependent Schrödinger equation is

$$\mathcal{H}\Psi(q,t) = i\hbar\frac{\partial\Psi(q,t)}{\partial t},$$

where $\mathcal{H}$ is the Hamiltonian operator for the system, t is time, and q represents the coordinates of the system. For a time independent Hamiltonian, we can rearrange this equation and take successive partials with respect to time to obtain

$$\frac{\partial\Psi(q,t)^n}{\partial t^n} = \left(-\frac{i}{\hbar}\mathcal{H}(q)\right)^n \Psi(q,t).$$

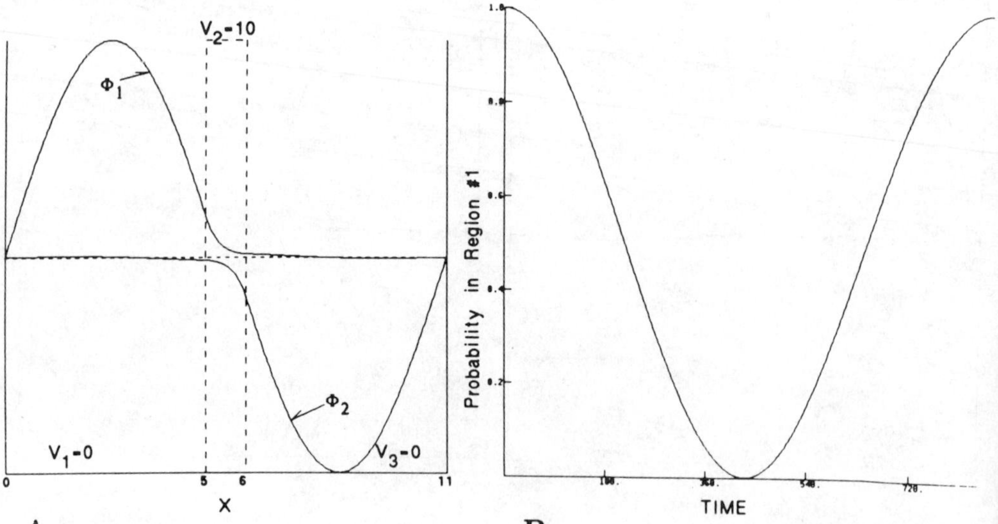

A. B.

Figure 9.6: *Localized Functions Constructed From the Two Lowest Eigenfunctions of Figure 9.5a, and The Probability of Localized Function Φ_1 Being in Region 1 as a Function of Time.*

Now, let's write $\Psi(q,t)$ as a Taylor series about an initial time, $t_\circ$:

$$\begin{aligned}\Psi(q,t) =&\Psi(q,t_\circ) + \left(\frac{\partial\Psi}{\partial t}\right)_{t=t_\circ}(t-t_\circ) + \cdots + \left(\frac{\partial\Psi^n}{\partial t^n}\right)_{t=t_\circ}\frac{(t-t_\circ)^n}{n!}\\ =&\Psi(q,t_\circ) + \left(-\frac{i}{\hbar}\mathcal{H}\right)\Psi(q,t_\circ)(t-t_\circ) + \cdots\\ &\cdots + \left(-\frac{i}{\hbar}\mathcal{H}\right)^n\Psi(q,t_\circ)\frac{(t-t_\circ)^n}{n!}.\end{aligned}$$

Using the symbolism for the series equivalent of an exponential function,

$$e^{+z} \equiv 1 + z + \cdots + \frac{z^n}{n!},$$

we can write the **formal**[4] equation

$$\Psi(q,t) = \left\{ e^{-\frac{i}{\hbar}\mathcal{H}(t-t_\circ)} \right\} \Psi(q,t_\circ).$$

A general time dependent function of a system can be written as a linear combination of the eigenfunctions, ψ_i° and $\mathcal{H}\psi_i^\circ = \epsilon_i\psi_i^\circ$, with time dependent coefficients:

$$\Psi_j(q,t) = \sum_{i=1}^{N} \psi_i^\circ(q) a_{ij}(t).$$

From here on, we will save some writing and lose no generality by always assuming that $t_\circ = 0$, and that ψ_i° and $a_{ij}(0)$ are real. The time dependence of the coefficients can be found by writing

$$\begin{aligned}\Psi_j(q,t) &= \sum_{i=1}^{N} \psi_i^\circ(q) a_{ij}(t) \\ &= \sum_{i=1}^{N} a_{ij}(0) \left\{ e^{-\frac{i}{\hbar}\mathcal{H}t} \right\} \psi_i^\circ(q) \\ &= \sum_{i=1}^{N} a_{ij}(0) e^{-i\epsilon_i t/\hbar} \psi_i^\circ(q),\end{aligned}$$

from which it should be fairly obvious that

$$a_{ij}(t) = a_{ij}(0) e^{-i\epsilon_i t/\hbar}. \tag{9.10}$$

The probability of finding a system described by some wavefunction, Ψ_j in any particular volume element of space is, of course, just the integral

$$\int_a^b \cdots \int_c^d \Psi_j^*(q,t)\Psi_j(q,t) dq_1 \ldots dq_n,$$

over the volume element. Obviously, this probability is time dependent when the wavefunction is time dependent. For the one dimensional

---

4 The expression doesn't have meaning until the exponential is expanded, because the expansion is the only definition of what we mean by the operation of the exponential of an operator. The exponential operator is sometimes called the **time evolution operator**.

bound systems that we are discussing, the time dependent wavefunctions are written as linear combinations of the eigenfunctions, ψ_i^o, of the Hamiltonian, and this probability can be expanded as

$$\begin{aligned} P(t) &\equiv \int_r^s \Psi_j^*(x,t)\Psi_j(x,t)dx \\ &= \sum_{k=1}^{N}\sum_{\ell=1}^{N} a_{kj}^*(t)a_{\ell j}(t)\int_r^s \psi_k(x)\psi_\ell(x)dx, \end{aligned}$$

which defines the probability $P(t)$ of finding the system in the region $r \leq x \leq s$. We can save some writing by defining the matrix elements

$$p_{k\ell} \equiv \int_r^s \psi_k(x)\psi_\ell(x)dx,$$

which, since the integral is not over all space, are **not** equal to zero for $k \neq \ell$. Writing $P(t)$ in terms of these matrix elements, and using Eq. (9.10), we obtain

$$\begin{aligned} P(t) &= \sum_{k=1}^{N}\sum_{\ell=1}^{N} a_{kj}^*(t)a_{\ell j}(t)p_{k\ell} \\ &= \sum_{k=1}^{N}\sum_{\ell=1}^{N} a_{kj}(0)\left[e^{+i\epsilon_k t/\hbar}\right] a_{\ell j}(0)\left[e^{-i\epsilon_\ell t/\hbar}\right] p_{k\ell} \\ &= \sum_{k=1}^{N} a_{kj}^2(0)p_{kk} \\ &\quad + 2\sum_{k=1}^{N-1}\sum_{\ell=k+1}^{N} a_{kj}(0)a_{\ell j}(0)p_{k\ell}\cos\left(\frac{(\epsilon_k-\epsilon_\ell)t}{\hbar}\right), \end{aligned} \tag{9.11}$$

where the last equivalence separates the diagonal terms into a single summation, and makes use of the facts that

$$e^{+i\epsilon_k t/\hbar}e^{-i\epsilon_k t/\hbar} = 1,$$

and, perhaps less obviously,

$$\left[e^{+i\epsilon_k t/\hbar}\right]\left[e^{-i\epsilon_\ell t/\hbar}\right] + \left[e^{-i\epsilon_k t/\hbar}\right]\left[e^{+i\epsilon_\ell t/\hbar}\right] = 2\cos\left(\frac{(\epsilon_k-\epsilon_\ell)t}{\hbar}\right).$$

Eq. (9.11) shows that $P(t)$, the probability of finding the system in the defined region, consists of the **time independent** first single summation, and the second, **time dependent**, sum of **periodic** terms. Each of the periodic terms has a maximum amplitude of $a_{kj}(0)a_{\ell j}(0)p_{k\ell}$ and a period of $h/(\epsilon_k - \epsilon_\ell)$. In order for appreciable changes in $P(t)$ to occur, there must be at least one pair of eigenfunctions, ψ_k° and ψ_ℓ°, having different eigenvalues, appreciable 'overlap' (ie. $p_{k\ell}$) in the region of interest, and appreciable coefficients in the initial wavefunction, $\Psi_j(x, 0)$.

As a simple example, lets consider the case that was developed in Figures 9.5 and 9.6a. For the initial localized function Φ_1 shown in Figure 9.6a, we have for the region $0 \leq x \leq 5$

$$a_{11}(0) = a_{22}(0) = a_{11} = a_{22} = \frac{1}{\sqrt{2}}, \qquad p_{11} \approx p_{22} \approx p_{12} \approx \frac{1}{2},$$

$$(\epsilon_2 - \epsilon_1) = 0.0073\ units,$$

so that

$$P(t) = \frac{1}{2}\left(1 + \cos\left(\frac{0.0073t}{\hbar}\right)\right),$$

which is shown in Figure 9.6b with t in arbitrary units.

It is important to note that the localized system described by Φ_1 has an expectation value for its energy (≈ 0.35 units) which is far below the barrier height (10 units), yet the system sloshes periodically back and forth between the two wells. This is the phenomenon of tunneling.

It is easy to devise a general method for constructing localized functions from eigenfunctions. We have a set of N eigenfunctions, ψ_i°, and we wish to construct a linear combination of these

$$\Psi_j = \sum_{i=1}^{N} \psi_i^\circ a_i,$$

such that

$$P_j \equiv \int_a^b \cdots \int_c^d \Psi_j^* \Psi_j dq_1 \ldots dq_n = \sum_{k=1}^{N}\sum_{\ell=1}^{N} a_{kj}^* a_{\ell j} p_{k\ell},$$

is at a maximum, and where the $p_{k\ell}$ are defined as

$$p_{k\ell} \equiv \int_a^b \cdots \int_c^d \psi_k^\circ \psi_\ell^\circ dq_1 \ldots dq_n,$$

generalizing the case discussed above. Note that P_j can be expressed as a matrix equation

$$P_j = \begin{pmatrix} a_{1j}^* & \cdots & a_{Nj}^* \end{pmatrix} \begin{pmatrix} p_{11} & \cdots & p_{1N} \\ \vdots & \ddots & \vdots \\ p_{N1} & \cdots & p_{NN} \end{pmatrix} \begin{pmatrix} a_{1j} \\ \vdots \\ a_{Nj} \end{pmatrix}.$$

Taking partial derivatives of P_j with respect to each of the a_{ij}'s and setting each equal to zero yields the matrix diagonalization problem[5]

$$\mathcal{A}^{\dagger}\mathcal{P}\mathcal{A} = \Lambda,$$

where $\mathcal{P}$ is the matrix whose elements are $p_{k\ell}$. Each column of $\mathcal{A}$ gives one possible Ψ_j for which P_j is equal to the corresponding λ_{jj}.

Since the expectation value of the energy of the system will depend on which eigenfunctions are used in the construction, we will generally want to limit the set used. Both the expectation value of the energy and the root-mean-square deviation from the expectation value can be calculated with no difficulty

$$\langle \Psi_j | \mathcal{H} | \Psi_j \rangle = \langle E \rangle = \sum_{k=1}^{N} a_{kj}^*(0) a_{kj}(0) \epsilon_k,$$

$$\langle \Delta E \rangle \equiv \left[\sum_{k=1}^{N} \left(\langle E \rangle - \epsilon_k \right)^2 a_{kj}^*(0) a_{kj}(0) \right]^{1/2},$$

since all terms with $k \neq \ell$ are zero in the double summations; for the same reason, both of these quantities are time independent.

The more eigenfunctions we use, the more localized we can make the functions, but the less precisely we know the energy. As an example, for the specific case that we have been discussing in Figures 9.5 and 9.6, there are ten eigenfunctions with eigenvalues less than the barrier height. Using all ten of these functions, two functions, shown as Φ_3 and Φ_4 in Figure 9.7, can be constructed that are initially localized to five significant figures in region 1. The expectation value for the energy of Φ_3 is 0.6565 with a root-mean-square deviation of 1.237 units. The time evolution of the probability for Φ_3 in region 1 is shown in Figure 9.7b with the same time units as in Figure 9.6b.

The two lowest eigenfunctions are dominant, with coefficients of 0.68 each, in Φ_3, so that the time behavior of the probability is roughly the same as shown by the simple function Φ_1 of Figure 9.6, but the higher frequency smaller variations in probability are due to the inclusion of the other eigenfunctions.

It has only recently been fully realized that the behavior discussed above is sensitively dependent on the relative widths and depths of the

---

[5] This minimizes P_j while retaining normalization and is just a proof of the theorem that the diagonal elements of a matrix are bounded by the largest and smallest eigenvalues of the matrix.

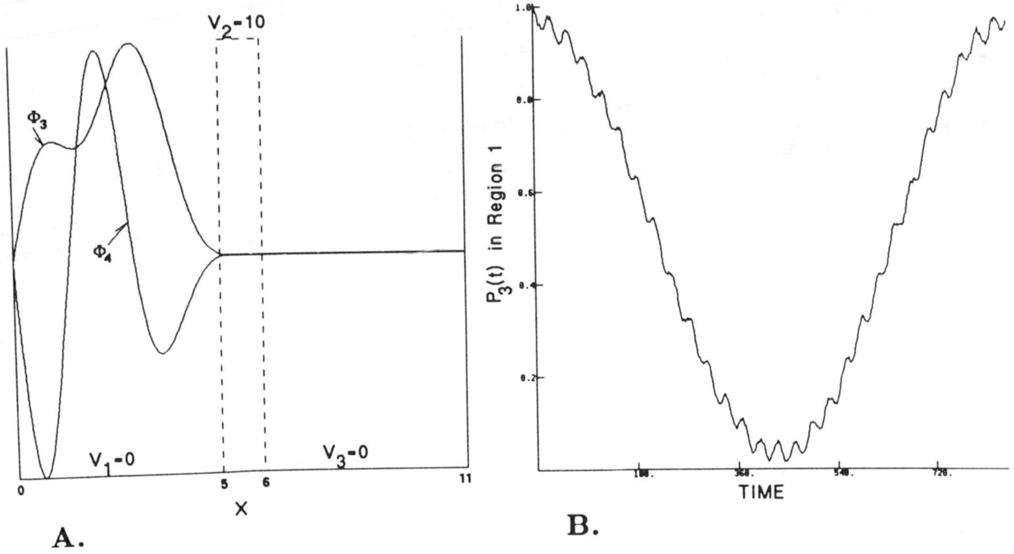

Figure 9.7: *Two Localized Functions, Φ_3 and Φ_4, Constructed From the Ten Lowest Eigenfunctions for the Symmetrical Double Well Potential Shown, and the Time Evolution of Probability for Φ_3.*

two wells [6]. Figure 9.8 shows two slightly asymmetric double wells with the two lowest eigenfunctions of each overlaid on the potential diagrams.

Since the eigenfunctions of these systems are highly localized, a localized function does not need to have any time dependence. If we construct rigorously localized wave functions, these will contain only small contributions from eigenfunctions in the other well, and the maximum change in probability will be small. For the system in Figure 9.8a, a wavefunction constructed from the ten lowest energy eigenfunctions and initially localized in region 1 to six significant figures has a coefficient of 0.916 for the lowest eigenfunction in region 1, ψ_1, and shows less than 3% maximum change in probability in region 1 with time.

Behavior similar to that for the symmetric double well is shown for situations when the width of one well becomes approximately a multiple of the other and both have the same depth, and for specific combinations of relative width and depth [6].

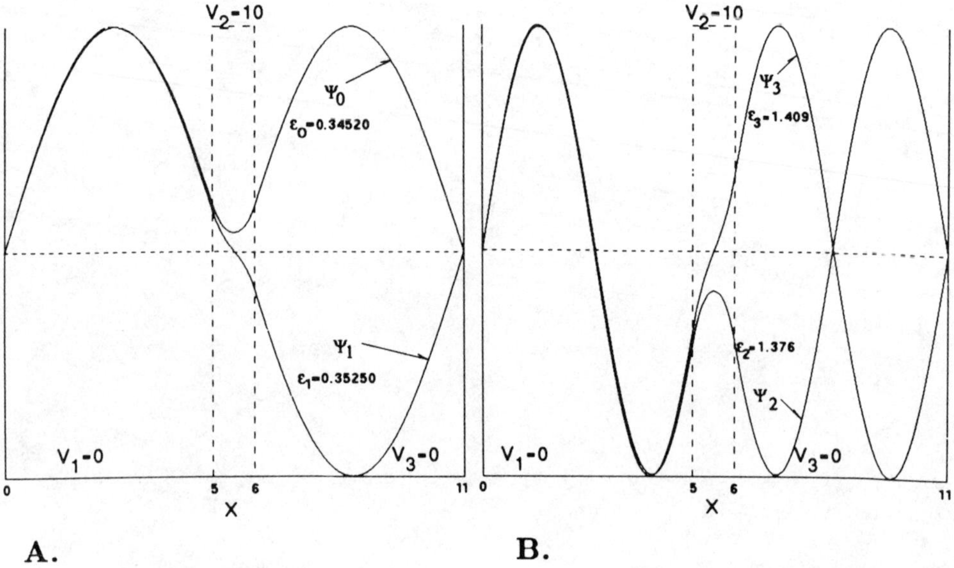

Figure 9.8: *Two Slightly Asymmetric Double Well Potentials and the Two Lowest Eigenfunctions of Each Overlaid on the Potential Diagram.*

Thus far in the discussion of tunneling, the role of mass has been hidden by the fact that we have implicitly used arbitrary units for length, energy, and time in all of the equations. The Schrödinger equation for the systems discussed can be written as

$$-\frac{\partial^2}{\partial x^2}\psi_j^\circ + \frac{2m}{\hbar^2}V(x)\psi_j^\circ = \frac{2m}{\hbar^2}\epsilon_j\psi_j^\circ,$$

and the question of units avoided by expressing energy in units of $\hbar^2/2m$ for any system under study. If we take the potential diagrams described in Figures 9.5–9.8 as being pertinent for a mass unit of $1.649 \times 10^{-24}g$, the mass of a proton, and a length unit of 1 $pm = 0.1$Å, then the energy unit is $4.23 \times 10^{-14}ergs \approx 0.61\ kcal/mole$, and the time unit in Figures 9.6b and 9.7b is $4.66 \times 10^{-14}secs$. The 10 *unit* barrier height, then, corresponds to 6.1 $kcal/mole$, the barrier width of 1 *unit* corresponds to 0.1Å, and the well width of 5 *units* corresponds to .5Å.

Since the unit of energy used is inversely proportional to mass, if we wish to consider deuterium in the same true potential diagram as that shown in Figure 9.5, we would have a barrier height of 20 arbitrary units. This 'higher' barrier reduces the energy differences between the two members of pairs of symmetric and antisymmetric eigenfunctions (For example, the two lowest eigenfunctions differ by 0.0014 *units* whereas the difference was 0.0073 *units* for the 10 *unit* barrier.) which are already two times smaller units. For the same true potentials shown in Figure 9.5 and units as given in the last paragraph, the energy difference between the two lowest eigenfunctions is 4.45 *cals* for hydrogen and 0.44 *cals* for deuterium. Since the time period for tunneling is inversely proportional to the energy differences, the tunneling frequency for hydrogen is approximately ten times that for deuterium in the present example. The differences in tunneling frequencies for different masses depend on the exact details of the potential diagram, and the differences for hydrogen and deuterium can be larger or smaller than that calculated for this one specific case. What is clear for all cases is that the lighter particle will always tunnel more than the heavier one.

Historically, the problem of tunneling was first approached for the case of unbound particles encountering barriers [7]. For one specific case, that of an *Eckart potential* [8], the quantum mechanics can be treated analytically; but numerical and approximate solutions are more frequently used. All of the calculations are aimed at evaluation of a *transmission coefficient*, $G(E)$, which is the probability that a system with an energy of V striking the barrier appears on the other side of the barrier[6]. The function $G(E)$ for a parabolic barrier can be approximated as

$$G(E) \approx \left[1 + \exp\left((V-E)(m/2V)^{1/2}a/h\right)\right]^{-1},$$

where V is the barrier height and a is the barrier width at half-height. In order to apply this to real systems having a Boltzmann distribution of energies, E, a *tunneling correction factor*, Q_t, is calculated by integrating the product of the Boltzmann factor and the transmission coefficient from zero to ∞, and dividing by the transition state theory expression for the rate constant of a reaction [7]:

$$Q_t = \exp\left(\Delta G^{\ddagger}/\kappa T\right) \int_0^{\infty} \frac{G(E)}{\kappa T} \exp(-E/\kappa T)dE,$$

6 As a matter of curiosity, it can be noted that the time dependence shown for the system in Fig.9.7b corresponds to an average transmission probability of approximately $1/28$ during the time required for P to go from 1 to zero. See Problem 9.7.

such that we can write

$$k_{corr} = k_{ts} Q_t,$$

where k_{corr} is the expected rate constant corrected for tunneling, and k_{ts} is the rate constant calculated from transition state theory not considering tunneling. Approximate equations and Tables of values for Q_t are available [7].

There is a considerable doubt about the applicability of any of the above treatments to real situations. It is quite clear that the double-well, bound, problem gives different results from the unbound problem even when averages are taken for the bound cases [6]. The more bothersome problem is the fact that the real situations involve more than a single dimension and that tunneling could involve appreciable amounts of cutting across the diagonals of the potential energy surfaces along paths other than the reaction coordinate. It is not at all clear how this problem can be handled [5].

In spite of the uncertainties of the theory, observations of kinetic deuterium isotope effects $k_H/k_D \geq 10$ are generally taken as evidence for the existence of tunneling in the reactions studied. The reaction of 2-nitropropane with 2,4,6-trimethylpyridine in aqueous ethanol at $25°C$ has $k_H/k_D = 24$ [9] which is difficult to rationalize without calling upon tunneling. Even more striking, $k_H/k_D \geq 10^4$ has been reported [10] for an intramolecular hydrogen atom transfer reaction.

Tunneling has also been held responsible for many cases where k_H/k_D is within the 'normal' range of 3–10, but where the temperature dependence of the isotope effect cannot be accomodated by Eq. (9.6). In particular, it has been argued that Arhennius pre-exponential factors, A_H/A_D, much less than unity are attributable to tunneling [5], [7].

9.7 REFERENCES

[1]. H. Eyring, **J. Chem. Phys., 3**, 107 (**1935**).

[2]. T. Ando, T. Kimura, and H. Yamatake in **Nucleophilicity**, Ed. J. M. Harris and S. P. McManus, ACS Advances in Chemistry Series, #215, Wash., D. C., **1987**.

[3]. M. L. Boas, **Mathematical Methods in the Physical Sciences**, Wiley and Sons, New York, **1966**, pgs. 131–140.

[4]. F. Westheimer, **Chem. Rev., 61**, 265 (**1961**).

[5]. R. P. Bell, **The Proton in Chemistry**, 2nd Ed., Cornell Univ. Press, Ithaca, N. Y., **1973**, pgs. 263–268.

[6]. J. H. Weiner and S. T. Tse, **J. Chem. Phys., 74**, 2419 **(1981)**; J. R. de la Vega, **Accts. Chem. Res., 15**, 185 **(1982)**.

[7]. R. P. Bell, **The Tunnel Effect in Chemistry**, Chapman and Hall, New York, **1980** reviews the 'classical' approach and its applications up to approximately 1978.

[8]. C. Eckart, **Phys. Rev., 35**, 1303 **1930**.

[9]. E. S. Lewis and L. H. Funderburk, **J. Amer. Chem. Soc., 89**, 2322 **(1967)**; E. S. Lewis and J. K. Robinson, **J. Amer. Chem. Soc., 90**, 4337 **(1968)**.

[10]. G. Bunton, D. Griller, L. R. C. Barclay, and K. U. Ingold, **J. Amer. Chem. Soc., 98**, 6803 **(1976)**.

9.8 PROBLEMS

Problem 9.1

Calculate k_H/k_D for a reaction involving a proton transfer from R_3C–H using the frequencies listed in Table 9.1 and assuming that the proton is symmetrically located at the transition state and that the bending force constant is 60% of that for R_3C–H.

Hint: Use Eq. (9.6); be sure to include the partition function terms for the low-frequency bending modes.

Problem 9.2

Calculate k_H/k_D for the same system as that in Problem 9.1, but with the following assumptions:

a. the transition state is linear and both R_3C and B can be treated as single atoms of mass 15;
b. the pertinent force constants at the transition state are $k_1 = 1.0$ mdynes/Å for the C–H stretch, $k_2 = 5.0$ mdynes/Å for the B–H stretch, and $k_{12}^2/k_1k_2 = 1.1$, and;
c. the bending force constant is the same as that of the R_3C–H reactant.

Hint: Use Table 9.2 to obtain the transition state frequencies, then plug these into Eq. (9.6).

Problem 9.3

Assuming that only zero-point energy differences contribute to the isotope effects, and that the transition states are linear and symmetric, derive an expression relating k_H/k_D to k_H/k_T and to k_D/k_T. *Hint: Just write the equations for* $\Delta\epsilon_{zp}$ *in terms of the reduced masses which may be assumed to equal the masses of the isotopes in these cases. See [5], pg 257, for a critique of the relationships.*

Problem 9.4

Show how the isotope effects on the forward and reverse rate constants for a reaction are related.
Hint: Write $k_f/k_r = K_{eq}$ *for both the normal and isotopically substituted reactants and products.*

Problem 9.5

Suppose that we have the reaction mechanism for a proton transfer that we discussed in Chapter 7:

$$AH \quad + \quad B \quad \underset{k_{-1}}{\overset{k_1}{\rightleftharpoons}} \quad [AH \cdots B]$$

$$[AH \cdots B] \quad \underset{k_{-2}}{\overset{k_2}{\rightleftharpoons}} \quad [A \cdots HB]$$

$$[A \cdots HB] \quad \underset{k_{-3}}{\overset{k_3}{\rightleftharpoons}} \quad A \quad + \quad HB$$

with $k_{-1} >> k_2$ and only the second step having an appreciable isotope effect of $k_H/k_D = 6$. What will be the isotope effect on the observed forward rate constant, k_f, if $k_3 = k_{-2}$? If $k_3 << k_{-2}$?
Hint: Use Eq. (7.21) for both the normal and isotopically substituted reactions.

Problem 9.6

It is generally believed that hydride ion transfers involve transition states in which the electrophilic hydride acceptor, the hydrogen, and the group originally bonded to the hydrogen form a three-center, two-electron bond. How would you expect isotope effects on these reactions to compare with those on proton or hydrogen atom transfer reactions?
Hint: What motion is likely to be the motion along the reaction coordinate?

Problem 9.7

For the system discussed in connection with Figures 9.5 and 9.6, how often, on the average, would a particle of unit mass, described by the function Φ_1 with $\langle E \rangle = 0.35$ *units*, strike the barrier? How do you think that the 'average transmission coefficient', mentioned in footnote 6 on page 319, was calculated?

Hint: What is the expectation value for the root-mean-square momentum of the particle? What is the average velocity of the particle? We can write the average time required for the system to pass through the barrier as $\int_1^0 t dP(t)$, *where* $P(t)$ *is the probability of the particle being in the left-hand well.*

Problem 9.8

Since the time evolution operator gives a simple result when it operates on an eigenfunction of the Hamiltonian, ψ_k^o:

$$e^{-\frac{i}{\hbar}\mathcal{H}t}\psi_k^o = e^{-i\epsilon_k t/\hbar}\psi_k^o,$$

it is very useful to be able to expand any function as a linear combination of eigenfunctions.

Suppose you had a system in a state described by the function ϕ which is not an eigenfunction of the Hamiltonian. How would you determine the coefficients for the linear combination of eigenfunctions?

Hint: Write the expression for the linear combination, then left-multiply by some one particular eigenfunction and integrate over all space; $\langle \psi_\ell^o | \phi \rangle = ?$

Problem 9.9

Suppose you somehow know that there are three vibrational modes in a particular molecule that are sensitive to a particular isotopic substitution of deuterium for hydrogen, such that the ratio of frequencies is equal to $\sqrt{2}$ for all three modes. Can you be sure that there are no other vibrational modes that are changed by the isotopic substitution?

Hint: Think about the Redlich-Teller Product Rule discussed in Chapter 8. What is the ratio of the product of the squares of frequencies for the normal and isotopically substituted molecule?

Problem 9.10

In the discussion of Section 9.2, one factor other than tunneling is neglected in the calculation of the rate of conversion of reactant to product, this is the so-called *transmission coefficient* which allows for the fact that some systems crossing the barrier could be reflected back across the barrier without reaching product. What types of potential energy surfaces might lead to transmission coefficients of much less than unity?

Hint: Think about things such as very narrow cols, or sharp curvatures of the surfaces in the region of the saddle point. See D. G. Truhlar, **Ann. Revs. Phys. Chem., 35**, *189* **(1984)**, *for a recent review of advances in this area.*

AUTHOR INDEX

Pages in underlined italics are those on which the citations to the original literature are found. Other listed pages generally contain only the reference numbers to the citations at the ends of chapters.

-A-

-B-

-C-

-D-

-K-

-L-

-M-

-N-

-O-

-P-

-R-

-S-

-T-

-U-

-W-

-X, Y, Z-

SUBJECT INDEX

-I, J-

-K-

-L-

-N-

-S-